Nature Swapped and Nature Lost

"Ever wondered why capitalism, neoliberalism, and nature conservation are at odds? Here is the answer! This book is an exquisite rendition of the irremediable tension between sustaining the environment and mobilising nature for capital accumulation, convincingly demonstrating that market environmentalism cannot deliver a socially sane and ecologically sound world."

—Erik Swyngedouw, *Professor of Geography, University of Manchester, UK*

"This book offers a radical, scholarly and socially relevant critique of biodiversity offsetting. Detailed theoretical analysis and compelling examples convincingly demonstrate the political economy behind what many persist in seeing as a neutral technical instrument for nature governance."

—Bill Adams, *Moran Professor of Conservation and Development, University of Cambridge, UK*

"This book offers an innovative Marxist analysis of biodiversity offsetting. It deserves to be closely read by all those interested in the design and outcomes of market-based mechanisms for environmental protection and social contestation against them."

—Sian Sullivan, *Professor of Environment and Culture, Bath-Spa University, UK*

"It is one thing to critique something like biodiversity offsetting as just another mad neoliberal attempt to make nature visible to the market. It is quite another to go beneath this absurd surface, and explain biodiversity offsets within broader political economic processes of capital accumulation and the frenzy of investment in the urban built environment—particularly in the post-2008 crisis world. In *Nature Swapped and Nature Lost*, Elia Apostolopoulou has done the latter—and significantly deepened our understanding of the relationship between neoliberal natures and the uneven geographies of global capitalism today."

—Matthew Huber, *Syracuse University, USA*

Elia Apostolopoulou

Nature Swapped and Nature Lost

Biodiversity Offsetting, Urbanization and Social Justice

palgrave
macmillan

Elia Apostolopoulou
Department of Geography
University of Cambridge
Cambridge, UK

ISBN 978-3-030-46787-6 ISBN 978-3-030-46788-3 (eBook)
https://doi.org/10.1007/978-3-030-46788-3

This Palgrave Macmillan imprint is published by the registered company Springer Nature Switzerland AG
The registered company address is: Gewerbestrasse 11, 6330 Cham, Switzerland

It is and remains the most revolutionary act to always say loudly what is happening

—Rosa Luxembourg (1906)

To Ariadni
Summer 2019, Cambridge

Acknowledgments

This book has been supported by a Marie Curie IEF grant (PIEF-GA-2013-622631, acronym CESINE) and a Carson Writing Fellowship (Rachel Carson Center for Environment and Society, Ludwig-Maximilians-Universität München) as well as from the Hellenic Foundation for Research and Innovation (HFRI) and the General Secretariat for Research and Technology (GSRT), under the HFRI grant entitled "From the right to the city to the right to nature" (GSRT code 235, KE 275 ELKE).

I would like to thank all the people, scholars, activists, and community groups who participated in my research the last six years. This has been a long journey and without them this book, as well as most of my published work, would not have been possible. My discussions with many colleagues and good friends were very often on my mind while writing this book; I won't mention their names here because I am sure they know who they are and they may also find parts of this book that are responses to our common concerns. Along colleagues and friends I have to mention my mentor and also my friend Bill Adams who prompted me to see things differently and start realizing what I am capable of. I left for the end my family, my partner, and my daughter, for being always here, supporting me in these challenging and precarious times for academia.

This book is dedicated to all the people who give everyday struggles for their quality of life and their livelihoods and of course to my little Ariadni who is my inspiration for everything.

Contents

1 Introduction 1

1.1 This Is Where This Book Begins 3

1.2 The Brave New World of Neoliberal Conservation 8

1.3 The Unevenness, Inequality and Injustice of the Reordering of Places and Socionatures: A Historical-Geographical Analysis of Offsetting and Social Resistance Against It 16

References 21

2 Neoliberal Natures and Biodiversity Offsetting 27

2.1 Neoliberalism and Non-human Nature 29

2.2 Neoliberal Conservation 32

2.3 The Dialectics of Green and Un-green Grabbing After the 2008 Financial Crash 37

2.4 Understanding the Neoliberal Hegemony: The Dialectics of Coercion and Consent Under Capitalism in Crisis 41

2.5 A Brief History of the Emergence, Evolution and Neoliberal Origins of Biodiversity Offsetting 46
2.6 Key Definitions 57
2.7 Current Distribution of Biodiversity Offsetting and Compensation Mechanisms Across the Globe 60
References 65

3 Equivalent Natures and Non-places 75
3.1 Biodiversity Offsetting and the Construction of Equivalence: Insights from the Ecological Literature 77
3.2 Offsetting and the Construction of Equivalence: Insights from Critical Social Sciences 86
3.3 Offsetting's Non-places: The Tensions Between Differentiation, Interchangeability, Homogeneity, and Unevenness in the Capitalist Production of Geographical Space 95
References 101

4 Value or Rent? A Marxist Analysis of Offsetting 109
4.1 Nature, Labor and Value 111
4.2 The Implications of Capitalist Commodity Production for Socionatures: Unraveling the Environmental Contradictions of Capitalism 119
4.3 Value or Rent? The Political Economy of Biodiversity Offsetting 124
4.4 Planetary Urbanization, Socio-Spatial Transformations and the Remaking of Places, Livelihoods and Socionatures 133
4.5 A Double Land Grabbing: Insights from Across the Global South and North About the Links Between Offsetting, Social Inequality, and Justice 138

References 148

5 **Biodiversity Offsetting in England: Deepening the Neoliberal Production of Socionatures** 163
5.1 Introduction 165
5.2 Nature Conservation in the UK After the 2008 Financial Crash: Consolidating the Hegemony of Market Environmentalism 167
5.3 Biodiversity Offsetting in the UK and Neoliberal Conservation 173
5.4 The Defra Offset Metric: The Triumph of Reductionism 179
5.5 Offsetting, Depoliticization and the Neoliberal Rescaling of Governance: It Ain't Technical 186
5.6 Offsetting and Habitat Banking: Buying Biodiversity "Off-the-Shelf"? 192
5.6.1 Can Money Capture Nature's Destruction? A Brief Note on the Pervasive Influence of the Logic of Cost–Benefit Analysis 200
5.7 Framing the Social as Irrelevant: Obscuring the Unevenness and Inequality Behind Offsetting's Equivalence 202
5.8 Interregnum: A Discussion with a Designer of Offset Metrics on Ecosystem Services, Offsetting and the Social Implications of the Economic Valuation of Nature 208
References 211

6 **Offsetting, Urbanization and the Neoliberalization of Space in Post-crisis England** 215
6.1 The Neoliberal Restructuring of Planning and the War on Red Tape 217
6.2 The Convergence of Offsetting and Urbanization: Fetishizing Economic Growth 222

6.3 The Selection of Offset Pilots: Deepening the Urban/Rural and Nature/Society Divides 227
6.4 Austerity Localism, Housing and the Presumption in Favor of "Sustainable" Development 229
6.5 Offsetting, Urbanization, and the Right to Nature as a Social Right: Community Struggles Against Large-Scale Infrastructure and Speculative Housing Development in Austerity England 234
6.5.1 Contesting Executive Housing and the Loss of Open Space in North Tyneside 235
6.5.2 The "Feathered" Obstacles to Urban Redevelopment in Lodge Hill 241
6.5.3 Green-Washing Urban Development and Large-Scale Infrastructure? The Case of HS2 Railway 246
6.6 When the Win-Win Rhetoric Meets TINA: The Tyranny of Pragmatism 254
References 258

7 Discussing with the Supporters and the Opponents of Biodiversity Offsetting 265
7.1 Introduction 267
7.2 Discussing with the Supporters of Offsetting 267
7.2.1 Interview with a Conservation Broker 267
7.2.2 Interview with a Consultant Working for the Housing Industry 284
7.3 Discussing with the Opponents of Biodiversity Offsetting 295
7.3.1 Interview with an Environmentalist 295
7.3.2 Interview with an Activist, Member of a Local Committee Opposing a Large-Scale Infrastructure Project 302

8 Afterword 317

8.1 Toward a Radical Retheorization of Offsetting: Uneven Geographies, Inequality, and Social-Environmental Justice 319

8.2 The Contradictions of the Capitalist Production of Nature and the Centrality of Class 326

8.3 The Dystopia of Offsetting's Ahistorical and Asocial Non-places: It's not a Flat World, After All 333

References 342

References 349

Index 387

List of Figures

Fig. 2.1	The evolution of offsetting and compensation policies across the globe (*Source* Design by the author based on data available at: https://portals.iucn.org/offsetpolicy/)	63
Fig. 2.2	Compliance offsets and compensation: Number of active programs by mitigation type, 2007–2016 (*Source* Design by the author based on data available at Bennett et al. [2017b])	63

List of Tables

Table 2.1 BBOP latest advisory group 52
Table 5.1 The Defra biodiversity offset metric 180

1
Introduction

E. Apostolopoulou, *Nature Swapped and Nature Lost*,
https://doi.org/10.1007/978-3-030-46788-3_1

All progress in capitalistic agriculture is a progress in the art, not only of robbing the labourer, but of robbing the soil; all progress in increasing the fertility of the soil for a given time, is a progress towards ruining the lasting sources of that fertility.
[…]
Capitalist production, therefore, develops technology,
and the combining together of various processes into a social whole,
only by sapping the original sources of all wealth: the soil and the laborer

—Karl Marx (1887)

1.1 This Is Where This Book Begins

"The door refused to open. It said 'five cents, please'. He searched his pockets. No more coins; nothing. 'I'll pay you tomorrow', he told the door. Again, it remained locked tight. 'What I pay you', he informed it, 'is in the nature of a gratuity; I don't have to pay you'. 'I think otherwise', the door said. 'Look in the purchase contract you signed when you bought this conapt'. He found the contract. Sure enough; payment to his door for opening and shutting constituted a mandatory fee. Not a tip. 'You discover I'm right', the door said. It sounded smug." In one of my favorite science fiction novels, the Ubik, Philipp Dick (1969) describes a futuristic world where everything, even the door of our house or our fridge, operates with money. This world of the total supreme of money is also a world where science and technology, at their most extreme version, ruin: even death seems to have been defeated since people can come back after they died. This is, nonetheless, only for a limited time period during which the nearly dead are kept in a state of "semi-life" in a "cold-pac moratorium" where those who are still alive can still visit them and talk to them preferably about their unfinished businesses. Ubik, from the Latin word "ubique," appears most often in the book in the form of an aerosol spray and seems to counter time-regression saving the lives of those to whom it is applied. In a very interesting article in Guardian, Sam Jordison[1] argues that Ubik, as reality, does not make much coherent sense: "the unease, the difficulty, the contradictions are partly the point. It's all about the realization that things aren't as they seem – that everything you thought you knew is wrong." As one of the contributors in the discussion below the Guardian article (with the nickname "Mexican2") points out, after the first hint in the book that something is definitely wrong, people are trying to construct "some harmonious whole to cover up the chaos" without ever actually reaching any negotiated relationship with what the world is because something always disrupts them. In this weakness to make sense of a world of constant contradictions and see the

[1] See https://www.theguardian.com/books/booksblog/2014/mar/18/philip-k-dick-ubik-reading-group.

world "upside down," as Eduardo Galeano[2] urged us several years ago, is where this book begins.

In another world, more real than Ubik's world, in our own Earth, some people are more concerned about defeating life rather than death. This is driven by a noble vision of "smaller families" and thus "less carbon" which can be achieved through "rational" family planning. Using fundraising models, like crowd-funding, what these folks are advocating for is better contraception and support services to communities "in need" with the goal to promote a more "sustainable" future for both people and the planet.[3] According to this "new, cost-efficient" fundraising model any of us can donate as little as 10 GBP to save the planet by preventing a child from being born. The goal is of course romantic and benign: to achieve a decent quality of life on a healthy planet, on our one and only Earth. So instead of linking poverty and environmental degradation to capitalism and imperialism to make sense of our world, we can instead put forward a simple solution to what seems to be after all "competition between humans" for the limited resources that Earth has to offer us: what if we could be *less*? In this over-simplistic resurgent neo-Malthusian pragmatism is where this book begins.

And then there is "Eden",[4] a £74 million project in Cornwall (UK) that has been constructed as part of Britain's millennium celebrations. Eden's project directors, Tim Smit and Jonathan Ball, characterized it as a practical application of the primary motto of Rio Agenda 21: "think globally, act locally" (Bartram and Shobrook 2000). The Eden Project could perhaps be part of Ubik's story—it can definitely compete with sci-fi movies. As Jonathan Glancey writes in Guardian[5]: "nothing quite like the Eden Project has been seen this side of Quatermass and It Came from Outer Space. Here is the biggest, oddest greenhouse ever -a mass of steel-framed bubbles moulded into the hollow of a former china-clay pit." The Eden Project, as a scientific and technological "simulation," to borrow

[2]See https://archive.nytimes.com/www.nytimes.com/books/00/11/12/reviews/001112.12fonsect.html?scp=18&sq=fruit%2520company&st=cse.

[3]See https://www.populationmatters.org/the-issue/overview/.

[4]See https://www.edenproject.com/.

[5]See https://www.theguardian.com/culture/2001/mar/12/artsfeatures.

a term from Baudrillard, can be seen as a kind of eco-utopia, a duplication of nature that implies that environmental catastrophe has already occurred and what we are left with is the illusion of "forestalling the end" (Bartram and Shobrook 2000, p. 373). Interestingly, the future plans of Eden Project seem much more pragmatic, including, inter alia, the construction of an energy-efficient, 120-bedroom hotel which will offer "dispersed accommodation with direct access to the Cornish landscape" and "will sit gently within the surrounding countryside".[6] In this understanding of nature conservation as the search of a lost "Eden" through the production of socially constructed, virtual natures, is where this book begins.

And now it's time for a more cynical and realistic story. It goes like that: you are a landowner and you've just inherited land from your parents that has been in your family for generations. Even though there is no doubt that you love the land, you must also make a living and all the property taxes and property maintenance costs are a drain on your resources. You do not want to sell your land so you must think how you can earn money from it. Which are your options? Probably you must see how you can develop the land. You realize that a good idea is to sell part of your land to a real estate developer who will turn it into suburban housing at a good profit. But. There is a huge but here. During the conversations with the real estate developer you discover that there are various endangered species within this land, and that it is illegal to harm them. Which can be possibly the solution to this conundrum? As Bayon et al. (2008) explain the magic word is species banking. But let's allow the authors to describe the situation in their own words: "Nobody except you and your consultants knows that these species are on your land, and you know that if the government finds out, your development options will be severely limited (perhaps even shattered). Clearly, in this situation having the endangered species is not so much an asset as it is a very clear and present liability. If you lacked scruples –or maybe just because your financial need was great– you would be tempted to adopt […] the 'Three S's Approach to Species Management': Shoot, Shovel and Shut

[6] See https://www.white-design.com/architecture/all-projects/eden-hotel-cornwall/.

Up. We're not saying that people around the US are busy killing endangered species as a result of the ESA. We just don't know. What we do know is that the incentive structure unwittingly set up by the ESA could push people to do just that. Fortunately, these perverse incentives are being reversed as a result of species banking. In the situation above, once species banking is in place, the landowner has a third choice: he can turn the land into a conservation bank, generate species credits, and sell those to create an income. In other words, the species that was once a liability has now become an asset" (Bayon et al. 2008, p. 5). In this cynical, but nonetheless, honest admittance that in the era of neoliberal capitalism nature will not be "saved" until capitalists and landlords can gain more from its conservation than from its destruction is where this book begins.

Let's now go back to my favorite sci-fi and to the Mad Max films largely inspired by the 1970s oil crisis. The 1979 original Mad Max movie takes place in an Australian wasteland. As McCausland,[7] the screenwriter of the original film explained in 2006, the Mad Max script was written "based on the thesis that people would do almost anything to keep vehicles moving and the assumption that nations would not consider the huge costs of providing infrastructure for alternative energy until it was too late." In a key scene during the third film, 1985s "Mad Max: Beyond Thunderdome," Max is told the last part of the story. In a desperate grab to secure the final reserves of oil, the world was drawn into a nuclear war that decimated the remaining natural resources. In the latest "Mad Max: Fury Road," released in 2015, when the movie starts the world has already died. The Green Place, the once fertile area of the Wasteland, is now a fetid bog filled with mud and dead trees due to soil contamination, quite like the Colonies, the toxic waste dumps to which "unwomen" are exiled in the recent adaptation of Margaret Atwood's *Handmaid's Tale* on the screen. At some point Max is being asked why he is hurting people. His response is cynical and clear: "It's the oil, stupid. Oil wars. We are killing for gasoline. The world is running out of water. Now there's the water wars. Once, I was a cop. A road warrior searching for a righteous cause, to the terminal freak-out point. Mankind has gone rogue, terrorizing itself [...] Further on was the oily bog. Once there had

[7]See http://uk.businessinsider.com/mad-max-what-happened-to-world-2016-2?r=US&IR=T.

been life and color here. Once it has been home. Now it was a poisoned swamp of death. Only the sky fishermen remained, casting their lines to lure the rare morsel of flying meat." It is in these, futuristic or perhaps not so futuristic in the era of the Capitalocene[8] (Moore 2017) or Chthulucene[9] (Haraway 2016), apocalyptic and dystopian allegories, produced not only by fiction but also by the contradictions of our own (capitalist) world that this book begins.

For the end, I kept a so-called "innovative" market-based conservation tool that, at least as I see it, draws inspiration—unwittingly or not I do not know to be honest- from all the above stories: biodiversity offsetting. As a brilliant online video[10] shows, offsetting is not just another ordinary conservation tool but an "incredibly exciting" policy that can "make dreams come true" according to the CEO of an imaginary organization. This CEO goes on and uses the example of Regent's Park in London to show that offsetting has the potential to transform the historic park to a giant gas plant while achieving a No Net Loss (or even a Net Gain!) of biodiversity. How is this possible? By buying the necessary amount of biodiversity "off-the-shelf" along with other ecosystem "services" to mitigate for the loss of nature in Regent's Park. Offsetting's revolutionary potential lies on the possibility to buy and exchange biodiversity, in terms of offset credits, to mitigate for the ecological losses that development activities are causing. This essentially means that as soon as someone needs more biodiversity the only thing, she/he has to do is to go to the marketplace and buy it as any other product. The same thing can of course happen with carbon, water, wetlands, and the list goes on and on. In this comfortable, and deeply alienating, scenario we should not worry too much about the destruction or degradation of non-human nature

[8]As Moore (2017) argues by drawing on McBrien (2016) and Dawson (2016) the Capitalocene is also a Necrocene—a system that not only accumulates capital but drives extinction. He further writes (Moore 2017, p. 597): "At stake is how we think through the relations of Capitalocene and Necrocene – between the creativity of capitalist development and its deep exterminism. That exterminism is not anthropogenic but *capitalogenic*".

[9]As Donna Haraway (2016) explains the word Chthulucene is a compound of two Greek roots (khthon and kainos) that together name a kind of timeplace for learning to stay with the trouble of living and dying in response-ability on a damaged earth. Kainos means now, a time of beginnings, a time for ongoing, for freshness.

[10]See the video here: https://vimeo.com/99079535.

since, in the era of the *Anthropocene* (a term that, inter alia, naturalizes climate change as Malm and Hornborg 2014 explain; see also Swyngedouw and Ernstson 2018), we (humans) are in a position to remake and reproduce nature lost or even create better nature, for the one we have just destroyed. In this extreme technomanagerial optimism and its naïve faith to capitalism's ability to infinitely reproduce itself is where this book begins.

1.2 The Brave New World of Neoliberal Conservation

Various scholars and activists across the globe, starting from different philosophical positions, political ideologies and conceptual frameworks, are finding common ground on the fact that nature conservation, at least in its mainstream version, is undergoing a profound change in its goals, donors, mottos, communication strategies, and tactical alliances, in order to seek a closer alignment with capital. This has been, inter alia, expressed in the emergence of various controversial partnerships between big industries, including the housing, mining, infrastructure, construction, and oil and gas industry, governments, environmental brokers, investors, and NGOs, particularly BINGOs. These partnerships imply, or sometimes openly admit, that the development/growth and the environmental agendas are not as antithetical as the traditional environmental movement may have thought and that capitalists and environmentalists must work together if any kind of sustainability is to be achieved any time soon. As Adams (2017, pp. 243, 245) aptly argues "in this paradoxical engagement, conservationists are exhibiting an extreme form of pragmatism - a willingness to sleep with the enemy," essentially "turning a blind eye to their own past and to the working of neoliberal capitalism" while "showing a remarkable willingness to entertain future risks to biodiversity from the outworking of neoliberalism."

While for some this shows an increasing attempt to fit conservation to the neoliberal world order leading to the emergence of a new form of *neoliberal conservation* (Apostolopoulou and Adams 2015; Arsel and Büscher 2012; Brockington and Duffy 2010; Büscher et al. 2012,

2014; Corson 2010; Igoe and Brockington 2007; Sullivan 2006), for others it has generated a growing feeling of "renaissance" in the conservation community manifesting hopes for a more direct alignment with economic forces (Daily and Matson 2008). The emergence of various market-based conservation instruments, including carbon and biodiversity offsets, species banking and payments for ecosystem services, under the rubric of the green economy/growth (Corson et al. 2013; Wanner 2015) and green capitalism (Prudham 2009), has been central to the above transformation and emblematic of a widespread broader turn to market-based solutions to environmental problems. As Wanner (2015) argues the green economy/green growth discourse can be actually seen as a passive revolution in Gramscian terms in the sense that the dominant sustainable development discourse, subsumed by capitalist hegemony, is protected in the context of a combined global economic, ecological and development crisis. Nonetheless, this new stage in the capitalist production of nature is for many quite alluring as it is rather emphatically expressed in Madsen's et al. (2010, p. 59) review of biodiversity markets worldwide, where the authors note with overt amazement: "*There was a day when a farmer sitting in his kitchen selling corn futures on an electronic trading platform would have sounded as futuristic as Buck Rogers, but as we all know, that scene is relatively commonplace these days. Biodiversity markets are on that same trajectory from futuristic to unremarkable.*"

The inspiration for this book largely stems from the above observations and the many paradoxes that I have encountered in my engagement with various aspects of nature conservation since my bachelor dissertation and then my PhD thesis back on the Olympic games of 2004 in Athens (Greece). As I realized throughout the years by exploring a variety of intense conflicts between economic development and environmental protection in different European countries within and beyond cities, and as I hope to show in this book, there is now a consolidated hegemonic discourse, at least in mainstream policy and academic cycles, that economic growth and environmental sustainability are largely compatible. The latter, particularly the last two decades and even more forcefully since the 2008 financial crash, has been combined with a shift toward the measurement of a putatively quantified economic value of nature as the most efficient way to consider the "true value" of nature. This has of course its roots in the neoliberal idea that as long as nature remains without price is escaping capitalist rationality making it hard to prevent its exploitation. These ideological and policy shifts signal a wider transformation not only of environmental policy but also more broadly of nature–society relationships whose origins can be traced in the late twentieth century. Important moments include the 1992 Earth Summit, chaired by the millionaire Canadian businessman Maurice Strong, that brought about the institutionalization of a neoliberal approach to sustainable development (Böhm et al. 2012) and the 1997 Kyoto Protocol that introduced carbon markets. In the early twenty-first century, the 2002 World Summit on Sustainable Development reiterated the triumph of neoliberalism by being emblematic of a market-based approach to environmental protection and poverty alleviation whereas at the UNCED Rio+20 conference in 2012, support for market-based approaches to global environmental problems was reaffirmed and environmental markets were seen as key components of a promising "green economy" (Foster 2012). Indeed, the last two decades there has been a consistent and growing interest in natural capital accounting and efforts to determine the "value" of the so-called ecosystem "services." One of the most ambitious endeavors to value (in economic terms) the natural world has been The Economics of Ecosystems and Biodiversity (TEEB) study, firstly published in 2010, and called for governments to include natural capital values in national accounts. In the same year, the World Bank launched the Wealth Accounting and the Valuation of Ecosystem Services (WAVES) project, to help partner countries implement natural capital accounting, based on the UN Statistical Commission of the System for Environmental and Economic Accounts (SEEA[11]) approach. SEEA Central Framework is the first international statistical standard for environmental-economic accounting adopted by

[11]The System of Environmental-Economic Accounting (SEEA Central Framework) is a statistical framework consisting of a set of tables and accounts, which guides the compilation of comparable statistics and indicators for policymaking, analysis and research. It has been produced and is released under the auspices of the United Nations, the European Commission, the Food and Agriculture Organization of the United Nations, the Organization for Economic Co-operation and Development, the International Monetary Fund and the World Bank Group. The SEEA Central Framework is a framework for understanding the interactions between the economy and the environment and for describing changes in stocks of environmental assets. It

the United Nations Statistical Commission in March 2012. The role of the private sector has been pivotal in these developments: The Natural Capital Coalition—formerly the TEEB for Business—has been closely working with companies to develop methods for natural capital accounting, valuation and reporting in business. Similarly, at the European level, the EU Biodiversity Strategy for 2020 (particularly Target 2, Action 5) requires Member States to assess the economic value of ecosystem services and integrate them into accounting systems at the national level.

What we currently witness represents only the latest stage in a long history of contradictions between capitalism and non-human nature. Indeed, as I have discussed elsewhere (Apostolopoulou 2009, 2010; Apostolopoulou and Adams 2015), capitalist expansion always had detrimental effects for the two sources of wealth: nature and labor power (Marx 1887). However, the consolidation of the neoliberal environmental agenda signaled some profound quantitative and qualitative differences in nature–society relationships with indicative example the shift toward the economic valuation of nature. Part of the latter is the now commonsensical assertion that environmental protection and nature conservation must make economic sense in capitalist terms, namely should support or at least not hinder capital accumulation. Crucially, this assertion runs through everything: from the conservation of a remote protected area to the protection from privatization or complete destruction of an urban public green space. Even though there are several important manifestations of this shift, in this book I focus on the case of environmental offsetting and, more specifically, on biodiversity offsetting, a dystopian policy that offers an extreme indication of the ongoing attempt to reframe non-human nature in terms of isolated, asocial biodiversity units and priced credits that can be exchanged across space and time.

The choice of the word offsetting is of course not coincidental. It reflects the wider transformation in the hegemonic narrative of conservation and, more broadly, in nature–society relationships under neoliberal

---

puts statistics on the environment and its relationship to the economy at the core of official statistics. For more information see: https://seea.un.org/content/about-seea.

capitalism. Offsetting is an economic term that means firstly, liquidating a futures position by entering an equivalent, but opposite, transaction that eliminates the delivery obligation, and, secondly, reducing an investor's net position in an investment to zero, so that no further gains or losses will be experienced from that position.[12] Borrowing terms and concepts from neoclassical economics (and not from political economy) to describe and understand non-human nature has been a common practice in the neoliberal era, with most characteristic examples the widespread adoption of the terms natural capital and ecosystem services. These terms have been increasingly and powerfully dominating mainstream environmental policy and governance and have now become the norm in environmental and conservation politics across the globe indicating an increasing consensus around the need to form partnerships with market forces, namely with specific sections of capital, to address the overlapping crises of energy, food, climate, and biodiversity. This is also obvious in everyday conservation practice: any initiative that does not adhere to the market logic or any partnership and governance arrangement that does not involve business partners and the private sector seems out-dated, unrealistic, and condemned to fail.

In environmental policy, the term offsetting emerged from debates of the Kyoto Protocol of the United Nations Framework Convention on Climate Change in 1997, under which industrialized nations could reach emissions reduction targets by purchasing credits created by projects in the developing countries (Bumpus and Liverman 2008). Pollution trading has been first proposed in the 1960s and has been developed by US economists and derivatives traders in the 1970s and 1980s. After a series of failed policy experiments, trading has eventually become the centerpiece of the US Acid Rain Programme in the 1990s at a time of "deregulatory fervour" (Lohmann 2010, p. 78). As Lohmann (2010) points out, the Bill Clinton regime played a key role in the evolution of pollution trading and successfully pressed for the Kyoto Protocol to become a set of carbon trading instruments. It is interesting to note that Al Gore, who carried the US ultimatum to Kyoto, later became a

[12] See http://www.investopedia.com/terms/o/offset.asp.

carbon market actor himself.[13] Importantly, during the 2000s, Europe became the host of the EU Emissions Trading Scheme, one of the world's largest carbon markets (ibid.; see also Newell et al. 2013; Muûls et al. 2016 on EU's trading scheme). Today, even though carbon offset credits can be bought voluntarily, most of them are bought by businesses and governments legally bound to reduce their emissions, or by governments seeking to strengthen the carbon trading market.[14] Offset projects, both under the Kyoto Protocol Clean Development Mechanism and voluntary carbon markets, have, thus, become major areas of policy interest and financial speculation and have received important criticism from critical scholars (see, e.g., Bachram 2004; Brown and Corbera 2003; Bryant 2016; Bumpus and Liverman 2008; Bumpus 2011; Gunderson et al. 2017; Felli 2015; Foster et al. 2009; Klein 2014; Knox-Hayes 2010; Lohmann 2012; Lohmann et al. 2006; McAfee 2016; Spash 2010; Stuart et al. 2019). The central position of carbon markets as a climate policy mechanism has been manifested in the Paris Climate Agreement of the 21st Conference of the Parties of the United Nations Framework Convention on Climate Change (UNFCCC 2015; Stuart et al. 2019). Article 6 of the Agreement, calls parties to pursue co-operative approaches and voluntarily use international transferred mitigation outcomes to help meet their reduction targets while establishing a new mechanism to contribute to the mitigation of greenhouse gas emissions and support sustainable development that allows for the participation of both the public and private sectors (UNFCCC 2015). Interestingly, setting a global carbon price was a key theme in the meeting, with business making calls for countries to introduce a price for carbon and World Bank group president declaring the importance of getting momentum on carbon pricing.[15]

---

[13]Al Gore has invested in a broad array of environmentally friendly energy and technology business ventures, like carbon trading markets, solar cells and waterless urinals (http://www.nytimes.com/2009/11/03/business/energy-environment/03gore.html). Gore co-founded the Generation Investment Management in 2004, with the aim of pushing forward low-carbon finance and promoting alternatives to the fossil fuel economy (https://www.theguardian.com/environment/2013/nov/01/gore-warns-carbon-bubble).

[14]For more information see http://www.fern.org/campaign/carbon-trading/what-are-offsets.

[15]See https://theconversation.com/how-will-carbon-markets-help-the-paris-climate-agreement-52211.

Biodiversity offsetting reflects similar ideas to carbon offsetting but with a focus on biodiversity. It is based on the simple and at the same time quixotic idea that losses to biodiversity in one place, and at one time, can be compensated by creating (or securing) *equivalent* biodiversity gains *elsewhere* (Apostolopoulou et al. 2018). In the offsetting logic, this ecological equivalence can offer the base for the subsequent representation of potential ecological gains as credits that can be sold and bought across space and time to offset equivalent ecological losses and achieve a No Net Loss (NNL) or even a Net Gain of biodiversity. Biodiversity offsetting is often characterized as a groundbreaking approach for its potential to turn environmental destruction to net environmental improvement. The concept of *equivalence* lies at the core of its logic since it not only enables the exchange of credits, but it also offers a positive sign to the otherwise peculiar idea of *offsetting nature* by portraying the whole process as capable of keeping an overall balance between ecological losses and gains (as the NNL goal implies). Biodiversity offsetting has been particularly attractive to both developers and governments who saw in it, especially after the 2008 financial crash, a way to facilitate economic development, often in the form of urban development as I will explain later, while taking into consideration its environmental impacts. Importantly, offsetting has also received significant support from environmental NGOs who saw in it an opportunity to work closely with the private sector and gain more funding in an era of prolonged austerity (Apostolopoulou and Adams 2015; Gray and Barford 2018).

The increasing popularity of offsetting reflects the wider ideological hegemony of neoliberal logic that aims to discipline environmental and conservation politics on focusing on practical and pragmatic solutions sidelining fundamental critiques (Adams 2017; Brockington 2014; Büscher 2010; Fletcher 2014). This hegemony is, inter alia, a product of the inability, and in most cases also unwillingness, of the mainstream environmental movement to highlight the political dimensions of the unresolved contradiction between economic growth and environmental protection, place social needs at the core of social–environmental conflicts, and envisage a production of socionatures that would support social and environmental justice. Biodiversity offsetting is thus a product and a trigger of the current convergence of (market) environmentalism and capitalism. As mentioned above this includes the increasing alignment of NGOs, particularly big ones (BINGOs), with corporations, donors, and major industries, including some of the biggest polluters whose operations are both environmentally and socially destructive (Apostolopoulou and Cortes-Vazquez 2018; Igoe et al. 2010; MacDonald 2008; Re:Common 2019). This alignment, often explained as a pragmatic choice under the hegemony of a depoliticized framing of biodiversity loss as primarily a problem of lacked capital (Apostolopoulou 2019; Dempsey and Suarez 2016), has led to the increasing disassociation of mainstream conservation from social–environmental struggles, ultimately rendering conservation an issue with no social relevance that can be addressed through technocratic or technomanagerial solutions. This has been expressed in many anti-offsetting struggles, including several cases across England that I will discuss in the following chapters, in the distance of a significant part of conservationists from local people's demands and sometimes even the adoption of the rhetoric of offsetting's supporters that the policy concerns only biodiversity, and not social, impacts. Or to put it differently—with the aim to expose the extreme reductionism and nature/society dualism of the policy—that offsetting affects only nature and not society.

A final note is important here. Where I talk about "nature conservation" in this book, I have in mind a broad definition that even though includes the relevant environmental and conservation legislation and activities designed to prevent the degradation of nature, it primarily conceptualizes conservation as a social practice, encompassing all community struggles and social–environmental movements fighting against the loss, destruction, and degradation of natures and green spaces within and beyond cities.

1.3 The Unevenness, Inequality and Injustice of the Reordering of Places and Socionatures: A Historical-Geographical Analysis of Offsetting and Social Resistance Against It

There are two major angles through which offsets have been explored so far. On the one hand, the ecological literature has mainly focused on key aspects related to their design and implementation, including offset metrics and multipliers, ways to measure equivalence and No Net Loss, definitions of offsetability and additionality, and various governance issues (see, e.g., Bull et al. 2013, 2017; Gardner et al. 2013; Quétier and Lavorel 2011; Pilgrim et al. 2013). A review of this literature reveals that debates over the selection of measurement units, exchange rules and offsetting's wider efficacy still persist. This lack of consensus is not unique to biodiversity, but it also characterizes carbon offsetting, ecosystem services and natural capital accounting (see, e.g., Bumpus 2011; Dempsey and Robertson 2012; Gómez-Baggethun et al. 2010; MacKenzie 2009; Muradian et al. 2013). As we will see in the following chapters, this major difficulty in reaching consensus reflects the inherent complexity of finding metrics that can simultaneously offer accurate measurements of ecological relations and satisfy demands for mainstreaming environmental concerns into policymaking. On the other hand, in critical social sciences literature, offsets are widely considered a manifestation of neoliberal conservation. Here, offset metrics are primarily analyzed as valuation tools that play a key role in the production of new value(s) from nature (see, e.g., Bracking et al. 2014; Dauguet 2015; Sullivan 2013; Robertson 2000, 2012; Lansing 2012). Coming from a wide array of theoretical positions, what critiques of the construction of equivalence, in both carbon and biodiversity offsetting, have in common is a focus on the politics of measurement, valuation, and calculation (Apostolopoulou et al. 2018). This approach has opened an important critique of dominant technical analyses by shifting the focus on how offset metrics enable the production of interchangeable natures, transforming nature into an asset whose control can be given to anyone with access to land and credit (Pawliczek and Sullivan 2011). Importantly, a significant part of this literature has strongly criticized offsetting's underestimation for the social and cultural aspects of biodiversity loss (see, e.g., Ruhl and Salzman 2006; Seagle 2012; Bidaud et al. 2017; Holmes and Cavanagh 2016) whereas recently the ecological literature has also raised more openly the need to consider the profound impacts of offsetting on people (Ives and Bekessy 2015; Griffiths et al. 2018).

Despite the considerable contribution of the above literatures to the analysis of offsetting, there are at least four key aspects that remain fundamentally underexplored. Firstly, there are very limited studies on how offsetting may reinforce a socially and geographically uneven production of nature, transforming places and livelihoods. Biodiversity offsetting's profound social and class implications are so far sidelined by the proliferating technical literature obscuring that under the surface of an apparently technical process to calculate ecological equivalence between offset and development sites, offsetting brings unevenness and injustice (Apostolopoulou and Adams 2017, 2019). Secondly, the increasing convergence of offsetting and urbanization has so far escaped analytical attention preventing the understanding of the way offsetting may enable or/and facilitate the reordering of landscapes according to the patterns of urban development and growth. Thirdly, by almost taking for granted that offset are a new type of "ecological commodities," a theoretical discussion on offsetting's role for capital accumulation through the lenses of critical political economy, and particularly through the labor theory of value and the theory of rent, is generally lacking (Apostolopoulou et al. 2018; Greco and Apostolopoulou 2019). Fourthly, despite the emergence of various anti-offsetting struggles across the globe, critical scholarship has so far paid very limited attention to social contestation against offsetting and even less to its class aspects.

In this book, by combining a theoretical analysis of offsetting with extensive empirical work in the UK, I aim to highlight the historically specific interactions and socio-economic and political contexts that explain why and when biodiversity offsetting is selected as the appropriate policy to address the environmental impacts of economic development. My goal is to offer a Marxist historical-geographical analysis of the emergence of offsetting in the context of the global economic crisis, the politics of prolonged austerity, and governmental aspirations for large-scale urban development projects. I hope that this analysis will advance a deeper understanding of the effects of the 2008 financial crash, and the global economic crisis, to the emergence and evolution of neoliberal conservation policies and the gradual emergence of a form of nature conservation that is largely driven by development interests and is being strongly influenced by urbanization patterns. By focusing on the emergence, evolution, and resistance to offsetting in the UK, an advanced economy of the Global North which often escapes an in-depth analysis of its conservation policies from a radical perspective, I also aim to contribute to work on actually existing neoliberal conservation in the Global North, which still remains extremely peripheral in comparison to the Global South (Apostolopoulou and Cortes-Vazquez 2018). It is important to mention here that the choice of the UK is not coincidental. The UK has been, and still is, a pioneer in market environmentalist both in Europe and globally. Particularly during the last decade, UK governments have strongly engaged with the Convention on Biological Diversity's (2010) ecosystem approach and the need to restructure nature conservation around the concept of ecosystem services and natural capital. This approach has been placed at the core of government environmental policymaking and delivery with indicative example the publication of the UK National Ecosystem Assessment (NEA) in 2011 (UK NEA 2011). The UK was the first European country that published a National Ecosystem Assessment offering an analysis of the UK's natural environment in terms of the benefits it provides to society and economic prosperity. The UK is also one of the few European countries that has relatively recently started experimenting with offsetting and it, thus, offers an excellent case for exploring its evolution in Europe in the post-2008 era in the context of the wider entrenchment of neoliberal policies (see Chapter 5 for more details).

Some may argue that biodiversity offsetting is only a minor part of the profound transformations in nature–society metabolism or even that the policy is being blamed for much more than it actually aims to bring about. This is only partly true as it would have been for any other policy. Offsetting has been defended using all sorts of arguments: from the very technical approaches that promote the policy as a more accurate method of calculating biodiversity loss to the more grandiose discourses which see it as part of a wider revolution in the way non-human nature is measured and valued and as an "innovative" mechanism capable to not only support and initiate novel partnerships and alliances between the public and the private sector and between conservationists and developers but also to initiate a large-scale restoration and re-creation of nature. Despite the various and often conflicting discourses that surround the policy, I aim to show in this book that offsetting is not a neutral policy instrument that can be used to serve either benign or malign causes but a reactionary neoliberal policy with highly controversial consequences for the implicated socionatures. Biodiversity offsetting draws on an impressive combination of a wide array of reactionary ideas. It reframes non-human nature as a set of tradable, priced units and treats place as extraneous turning conservation into a system of exchange of ecological losses and gains across space and time in line with contemporary patterns of capitalist urban growth (Apostolopoulou and Adams 2019). It favors technocratic solutions to streamline policy debate about increasing biodiversity loss in a simultaneous effort to facilitate and green wash controversial development projects and claim that it can keep the stock of natural capital constant. Importantly, evidence shows, as we will see in the following chapters, that despite proclaimed intentions, offsetting policies have a very low record in succeeding to fulfill even their strictly reductionist goals. There is also ample evidence that the conservation goals of offsetting policies resemble more to rhetorical devices and do not initiate actual attempts to achieve No Net Loss on the ground. Moreover, and relatedly, biodiversity offsetting is a deeply opportunistic policy: as evidence from around the globe (Chapter 2) but also from England (Chapters 5 and 6) shows, the translation of its conservation goals, including the No Net Loss goal, in practical, on the ground interventions, in the majority of cases, varies considerably according to the objectives of specific projects and programs and the interests of developers. The above do not imply that nature conservation policies were progressive before the emergence and proliferation of offsetting. Biodiversity offsetting did not emerge in a historical or political vacuum but, as mentioned earlier, it built on the long and contradictory history of nature conservation under capitalism. However, and this is of major importance, it has also clearly further deepened these contradictions while creating new ones.

Before closing this introduction, I would like to emphasize that perhaps the most important outcome of this book would be to achieve a clear contribution in advancing a radical analysis of the social, spatial, ecological, and cultural impacts of neoliberal conservation policies from the perspective of social, environmental and spatial justice. However, to combat the logic of contradictory, reactionary and ultimately dystopian policies, like offsetting, and prevent their implementation on the ground, academic research no matter how radical would not suffice. It would rather require a sustained dialogue between radical scholars, activists and social movements about alternatives to the increasing neoliberalization of nature and space which will be directly linked to praxis and actual, lively and long-term collaborations in everyday battles (Apostolopoulou and Cortes-Vazquez 2018). I hope that this book will offer such a theoretically informed and empirically grounded critique of biodiversity offsetting, and more broadly of the contradictions of the capitalist production of space and nature, that will be relevant to radical praxis and, thus, manage to surpass the confines of academia. This is not an easy endeavor but it is based on my longstanding belief that conflicts around the production of space and nature in both urban and rural contexts involve fundamentally political questions, and must be addressed in political terms, by identifying the strategies through which a more egalitarian mode of democratically producing socially and environmentally just socionatures can be achieved. This perceives profound importance today, a period that in retrospect may be considered as a critical transformative point in nature-society relationships. To put it differently, now is probably the time to decide "where must we go, we who wander this wasteland, in search of our better selves" (Mad Max: Fury Road).

References

Adams, W. M. (2017). Sleeping with the enemy? Biodiversity conservation, corporations and the green economy. *Journal of Political Ecology, 24,* 243–257.

Apostolopoulou, E. (2009). *The social conflicts during the implementation of conservation policy in protected areas: Analysis and appraisal of conservation policies in Greece* (PhD thesis). Aristotle University of Thessaloniki, Greece. Available at: https://www.didaktorika.gr/eadd/handle/10442/22497.

Apostolopoulou, E. (2010). A critique of the dominant development ideology for nature-society relationship. *Utopia (Ουτοπία), 91,* 87–106. (in Greek).

Apostolopoulou, E. (2019). Beyond post-politics: Offsetting, depoliticization and contestation in a community struggle against executive housing. *Transactions of the Institute of British Geographers.* https://doi.org/10.1111/tran.12354.

Apostolopoulou, E., & Adams, W. M. (2015). Neoliberal capitalism and conservation in the post-crisis era: The dialectics of 'green' and 'un-green' grabbing in Greece and the UK. *Antipode, 47,* 15–35.

Apostolopoulou, E., & Adams, W. M. (2017). Biodiversity offsetting and conservation: Reframing nature to save it. *Oryx, 51,* 23–31.

Apostolopoulou, E., & Adams, W. M. (2019). Cutting nature to fit: Urbanization, neoliberalism and biodiversity offsetting in England. *Geoforum, 98,* 214–225.

Apostolopoulou, E., Adams, W. M., & Greco, E. (2018). Biodiversity offsetting and the production of "equivalent natures": A Marxist critique. *ACME: An International Journal for Critical Geographies, 17*(3), 861–892.

Apostolopoulou, E., & Cortes-Vazquez, J. (Eds.). (2018). *The right to nature: An academic-activist dialogue on environmental justice, social movements and neoliberal natures.* Abingdon: Routledge-Earthscan.

Arsel, M., & Büscher, B. (2012). Nature™ Inc.: Changes and continuities in neoliberal conservation and market-based environmental policy. *Development and Change, 43,* 53–78.

Bachram, H. (2004). Climate fraud and carbon colonialism: The new trade in greenhouse gases. *Capitalism Nature Socialism, 15,* 5–20.

Bartram, R., & Shobrook, S. (2000). Endless/end-less natures: Environmental futures at the Fin de Millennium. *Annals of the Association of American Geographers, 90,* 370–380.

Bayon, R., Fox, J., & Carroll, N. (2008). *Conservation and biodiversity banking: A guide to setting up and running biodiversity credit trading systems*. London: Earthscan.

Bidaud, C., Schreckenberg, K., Rabeharison, M., et al. (2017). The sweet and the bitter: Intertwined positive and negative social impacts of a biodiversity offset. *Conservation and Society, 15,* 1–13.

Böhm, S., Misoczky, M. C., & Moog, S. (2012). Greening capitalism? A Marxist critique of carbon markets. *Organization Studies, 33,* 1617–1638.

Bracking, S., Bond, P., Brockington, D., Büscher, B., Igoe, J. J., Sullivan, S., & Woodhouse, P. (2014). *Human, non-human and environmental value systems: An impossible frontier?* (LSCV Working Paper Series No. 1). Leverhulme Centre for the Study of Value, School of Environment, Education and Development, The University of Manchester.

Brockington, D. (2014). Celebrity spectacle, post-democratic politics, and Nature™ Inc. In B. Büscher, W. Dressler, & R. Fletcher (Eds.), *Nature Inc.: Environmental conservation in the neoliberal age* (pp. 108–126). Tucson: University of Arizona Press.

Brockington, D., & Duffy, R. (2010). Capitalism and conservation: The production and reproduction of biodiversity conservation. *Antipode, 42,* 469–484.

Brown, K., & Corbera, E. (2003). Exploring equity and sustainable development in the new carbon economy. *Climate Policy, 3,* S41–S56.

Bryant, G. (2016). The politics of carbon market design: Rethinking the techno-politics and post-politics of climate change. *Antipode, 48,* 877–898.

Bull, J. W., Lloyd, S. P., & Strange, N. (2017). Implementation gap between the theory and practice of biodiversity offset multipliers. *Conservation Letters, 10,* 656–669.

Bull, J. W., Suttle, K. B., Gordon, A., Singh, N. J., & Milner-Gulland, E. J. (2013). Biodiversity offsets in theory and practice. *Oryx, 47,* 369–380.

Bumpus, A. G. (2011). The matter of carbon: Understanding the materiality of tCO_2e in carbon offsets. *Antipode, 43,* 612–638.

Bumpus, A. G., & Liverman, D. M. (2008). Accumulation by decarbonization and the governance of carbon offsets. *Economic Geography, 84,* 127–155.

Büscher, B. (2010). Anti-politics as political strategy: Neoliberalism and transfrontier conservation in Southern Africa. *Development and Change, 41,* 29–51.

Büscher, B., Dressler, W., & Fletcher, R. (Eds.). (2014). *Nature Inc.: Environmental conservation in the neoliberal age*. Tucson: University of Arizona Press.

Büscher, B., Sullivan, S., Neves, K., Igoe, J., & Brockington, D. (2012). Towards a synthesized critique of neoliberal biodiversity conservation. *Capitalism Nature Socialism, 23,* 4–29.

Corson, C. (2010). Shifting environmental governance in a neoliberal world: US aid for conservation. *Antipode, 42,* 576–602.

Corson, C., MacDonald, K. I., & Neimark, B. (2013). Grabbing green: Markets, environmental governance and the materialization of natural capital. *Human Geography, 6,* 1–15.

Daily, G. C., & Matson, P. A. (2008). Ecosystem services: From theory to implementation. *PNAS, 105,* 9455–9456.

Dauguet, B. (2015). Biodiversity offsetting as a commodification process: A French case study as a concrete example. *Biological Conservation, 192,* 533–540.

Dawson, A. (2016). *Extinction.* New York: OR Books.

Dempsey, J., & Robertson, M. M. (2012). Ecosystem services tensions, impurities, and points of engagement within neoliberalism. *Progress in Human Geography, 36,* 758–779.

Dempsey, J., & Suarez, D. C. (2016). Arrested development? The promises and paradoxes of "selling nature to save it". *Annals of the American Association of Geographers, 106* (3), 653–671.

Dick, P. (1969). *Ubik.* US: Doubleday.

Felli, R. (2015). Environment, not planning: The neoliberal depoliticisation of environmental policy by means of emissions trading. *Environmental Politics, 24,* 641–660.

Fletcher, R. (2014). Orchestrating consent: Post-politics and intensification of Nature TM Inc. at the 2012 World Conservation Congress. *Conservation & Society, 12,* 329–342.

Foster, J. B. (2012). *Rio + 20: Here we go again, Amandla!* Available at: http://www.amandlapublishers.co.za/special-features/the-green-economy/1317-rio–20-here-we-go-again–by-john-bellamy-foster.

Foster, J. B., Clark, B., & York, R. (2009). The midas effect: A critique of climate change economics. *Development and Change, 40,* 1085–1097.

Gardner, T. A., Von Hase, A., Brownlie, S., Ekstrom, J. M. M., Pilgrim, J. D., Savy, C. E., et al. (2013). Biodiversity offsets and the challenge of achieving no net loss. *Conservation Biology, 27,* 1254–1264.

Gómez-Baggethun, E., De Groot, R., Lomas, P. L., et al. (2010). The history of ecosystem services in economic theory and practice: From early notions to markets and payment schemes. *Ecological Economics, 69,* 1209–1218.

Gray, M., & Barford, A. (2018). The depths of the cuts: The uneven geography of local government austerity. *Cambridge Journal of Regions, Economy and Society, 11,* 541–563.

Greco, E., & Apostolopoulou, E. (2019). Value, rent and nature: The centrality of class. *Dialogues in Human Geography, 10*(1).

Griffiths, V. F., Bull, J. W., Baker, J., et al. (2018). No net loss for people and biodiversity. *Conservation Biology*. https://doi.org/10.1111/cobi.13184.

Gunderson, R., Stuart, D., & Petersen, B. (2017). Ideological obstacles to effective climate policy: The greening of markets, technology, and growth. *Capital and Class, 42,* 133–160.

Haraway, D. J. (2016). *Staying with the trouble: Making kin in the Chthulucene*. Durham: Duke University Press.

Holmes, G., & Cavanagh, C. J. (2016). A review of the social impacts of neoliberal conservation: Formations, inequalities, contestations. *Geoforum, 75,* 199–209.

Igoe, J., & Brockington, D. (2007). Neoliberal conservation: A brief introduction. *Conservation and society, 5,* 432–449.

Igoe, J., Neves, K., & Brockington, D. (2010). A spectacular eco-tour around the historic bloc: Theorising the convergence of biodiversity conservation and capitalist expansion. *Antipode, 42,* 486–512.

Ives, C. D., & Bekessy, S. A. (2015). The ethics of offsetting nature. *Frontiers in Ecology and the Environment, 13,* 568–573.

Klein, N. (2014). *This changes everything: Capitalism vs. the climate*. New York: Simon and Schuster.

Knox-Hayes, J. (2010). Constructing carbon market spacetime: Climate change and the onset of neo-modernity. *Annals of the Association of American Geographers, 100,* 953–962.

Lansing, D. M. (2012). Performing carbon's materiality: The production of carbon offsets and the framing of exchange. *Environment and Planning A, 44,* 204–220.

Lohmann, L. (2010). Neoliberalism and the calculable world: The rise of carbon trading. In K. Birch & V. Mykhnenko (Eds.), *The rise and fall of neoliberalism* (pp. 77–93). Zed Books: London and New York.

Lohmann, L. (2012). Financialization, commodification and carbon: The contradictions of neoliberal climate policy. *Socialist Register, 48,* 85–107.

Lohmann, L., Hällström, N., Österbergh, R., & Nordberg, O. (2006). *Carbon trading: A critical conversation on climate change, privatisation and power* (p. 362). Uppsala: Dag Hammarskjöld Foundation.

MacDonald, C. (2008). *Green Inc.: An environmental insider reveals how a good cause has gone bad*. New York: Lyons Press.

MacKenzie, D. (2009). Making things the same: Gases, emission rights and the politics of carbon markets. *Accounting, Organizations and Society, 34,* 440–455.

Madsen, B., Carroll, N., Kandy, D., & Bennett, G. (2011). *State of biodiversity markets report: Offset and compensation programs worldwide*. Washington, DC: Forest Trends.

Malm, A., & Hornborg, A. (2014). The geology of mankind? A critique of the Anthropocene narrative. *The Anthropocene Review, 1*(1), 62–69.

Marx, K. (1887). *Capital: A critique of political economy* (Vol. I). Moscow: Progress Publishers. Available at: https://www.marxists.org/archive/marx/works/1867-c1/.

McAfee, K. (2016). Green economy and carbon markets for conservation and development: A critical view. *International Environmental Agreements: Politics, Law and Economics, 16,* 333–353.

McBrien, J. (2016). Accumulating extinction. In J. W. Moore (Ed.), *Anthropocene or capitalocene?* (pp. 116–137). Oakland: PM Press.

Moore, J. W. (2017). The Capitalocene, Part I: On the nature and origins of our ecological crisis. *The Journal of Peasant Studies, 44,* 594–630.

Muradian, R., Arsel, M., Pellegrini, L., et al. (2013). Payments for ecosystem services and the fatal attraction of win-win solutions. *Conservation letters, 6,* 274–279.

Muûls, M., et al. (2016). *Evaluating the EU emissions trading system: Take it or leave it? An assessment of the data after ten years* (Grantham Institute Briefing Paper 21). Imperial College London.

Newell, R., Pizer, W., & Raimi, D. (2013). Carbon markets 15 years after Kyoto: Lessons learned, new challenges. *The Journal of Economic Perspectives, 27,* 123–146.

Pawliczek, J., & Sullivan, S. (2011). Conservation and concealment in SpeciesBanking.com, US: An analysis of neoliberal performance in the species offsetting industry. *Environmental Conservation, 38,* 435–444.

Pilgrim, J. D., Brownlie, S., Ekstrom, J. M. M., Gardner, T. A., von Hase, A., ten Kate, K., et al. (2013). A process for assessing offsetability of biodiversity impacts. *Conservation Letters, 6,* 376–384.

Prudham, S. (2009). Pimping climate change: Richard Branson, global warming, and the performance of green capitalism. *Environment and planning A, 41,* 1594–1613.

Quétier, F., & Lavorel, S. (2011). Assessing ecological equivalence in biodiversity offset schemes: Key issues and solutions. *Biological Conservation, 144*, 2991–2999.

Re:Common. (2019). *Turning forests into hotels: The true cost of biodiversity offsetting in Uganda*. Available at: https://www.recommon.org/eng/turning-forests-into-hotels-the-true-cost-of-biodiversity-offsetting-in-uganda/.

Robertson, M. (2000). No net loss: Wetland restoration and the incomplete capitalization of nature. *Antipode, 32*, 463–493.

Robertson, M. (2012). Measurement and alienation: Making a world of ecosystem services. *Transactions of the Institute of British Geographers, 37*, 386–401.

Ruhl, J. B., & Salzman, J. E. (2006). *The effects of wetland mitigation banking on people* (FSU College of Law, Public Law Research Paper No. 179). Available at SSRN: https://ssrn.com/abstract=878331.

Seagle, C. (2012). Inverting the impacts: Mining, conservation and sustainability claims near the Rio Tinto/QMM ilmenite mine in Southeast Madagascar. *Journal of Peasant Studies, 39*, 447–477.

Spash, C. L. (2010). The brave new world of carbon trading. *New Political Economy, 15*, 169–195.

Stuart, D., Gunderson, R., & Petersen, B. (2019). Climate change and the Polanyian counter-movement: Carbon markets or degrowth? *New political economy, 24*, 89–102.

Sullivan, S. (2006). Elephant in the room? Problematising 'new'(neoliberal) biodiversity conservation. *Forum for Development Studies, 33*, 105–135.

Sullivan, S. (2013). After the green rush? Biodiversity offsets, uranium power, and the 'calculus of casualties' in greening growth. *Human Geography, 6*, 80–101.

Swyngedouw, E., & Ernstson, H. (2018). Interrupting the anthropo-obscene: Immuno-biopolitics and depoliticizing ontologies in the anthropocene. *Theory, Culture & Society, 35*, 3–30.

UK NEA. (2011). *The UK national ecosystem assessment: Technical report*. Cambridge: UNEP-WCMC.

UNFCCC. (2015). *Adoption of the Paris agreement*. Geneva: United Nations Office.

Wanner, T. (2015). The new 'passive revolution' of the green economy and growth discourse: Maintaining the 'sustainable development' of neoliberal capitalism. *New Political Economy, 20*(1), 21–41.

2

Neoliberal Natures and Biodiversity Offsetting

E. Apostolopoulou, *Nature Swapped and Nature Lost*,
https://doi.org/10.1007/978-3-030-46788-3_2

> The development of culture and of industry in general has ever evinced itself in such energetic destruction of forests that everything done by it conversely for their preservation and restoration appears infinitesimal.
>
> —Karl Marx (1910)

2.1 Neoliberalism and Non-human Nature

Neoliberalism is in the first instance a theory of political economic practices that proposes that human well-being can be best advanced by liberating individual entrepreneurial freedoms and skills within an institutional framework characterized by strong private property rights, free markets, and free trade (Harvey 2005). The role of the state in neoliberalism is to create and preserve an institutional framework appropriate to such practices and set up the necessary structures and functions that are required to secure private property rights and to guarantee, if needed by force, the proper functioning of markets (ibid.). Neoliberalism is the outcome of a set of converging historical determinants and even though it is not easy to precisely determine its beginnings there are specific events that clearly mark its emergence. Earliest expressions of neoliberal ideology have been evident from the end of World War II but, as David Harvey (2005, p. 19) explains, actually existing neoliberalism can be better understood as a *political-economic project* that emerged during the 1970s initially in the United States and the UK, later in continental Europe and gradually across the globe, with the key aim to re-establish, renew, and expand the conditions for capital accumulation. Various developments are linked with the rise of neoliberalism including events surrounding the crisis of the dollar and policies enacted during the dictatorships in Latin America in the 1970s (notably the 1973 coup in Chile), as well as the 1973 oil crisis during which the members of the Organization of Arab Petroleum Exporting Countries proclaimed an oil embargo (by the end of which the price of oil had risen from US$3 per barrel to approximately $12 globally). For many, neoliberalism has been identified with the policies of Margaret Thatcher, who has been elected prime minister of Britain in 1979, and Ronald Reagan, who has been elected prime minister of the United States in 1980.

The period when neoliberalism emerged is not coincidental. The corporate capitalist class felt intensely threatened both politically and economically toward the end of the 1960s and into the 1970s and made the strategic choice to launch a political project that would curb the increasing power of labor (Harvey 2016). Despite the aspirations of the ruling class neoliberalism has not been as effective as expected by its supporters in revitalizing global capital accumulation but it has nonetheless succeeded in restoring class power and (re)creating the power of economic elites (Duménil and Lévy 2004). As Gérard Duménil and Dominique Lévy (Duménil and Lévy 2001) have extensively shown, neoliberalization has been from the very beginning a remarkable illustration of the class contradictions underlying capitalism: after the implementation of neoliberal policies in the late 1970s, the share of national income of the top 1% of income earners in the United States soared to reach 15% (very close to its share before World War II) by the end of the century (see also Harvey 2005). Neoliberalism expressed the strategy of the capitalist classes in alliance with upper management intending to strengthen and expand their hegemony (Duménil and Lévy 2013). Consequently, the overall dynamics of capitalism under neoliberalism, both nationally and internationally, have been determined by new class objectives that worked to the benefit of the highest income brackets, capitalist owners, and the upper fractions of management (ibid.).

If we consider that in neoliberalism, economic elites seek to increase their wealth, income, and political and economic freedom and flexibility primarily by rolling back the redistributive reforms of the mid-twentieth century, we can then see that neoliberalism constitutes a counter-revolutionary project that intensifies social inequality (Harvey 2016; Heynen et al. 2007). Key expressions of the latter include the new configurations of income distribution that neoliberalism brought about, the direct political assault on organized labor and anti-corporate reforms with the goal to disempower labor, the creation of various bubbles in the asset market, the strong pressure on workers to restore profit rates, the attempt to cover the gap between declining wages and increasing effective demand by pushing the debt economy to its limits (Harvey 2012), as well as the extensive deregulations, market-friendly reregulations and privatizations. It is important to emphasize here that the aggregate purpose and cumulative effect of neoliberal reforms has not been to roll back the state in general but to roll back and restructure a particular kind of state that, in most of the advanced capitalist countries, has been translated to attacks against the Keynesian welfare state. This has been followed by the roll-out of new state forms, modes of regulation, and regimes of governance, attempting to consolidate and manage the increasing expansion of markets (ibid.). In the case of the UK, as Peck and Tickell (2007) explain, this included the restoration of a unilateral "right to manage" that presupposed the decapitation of the labor movement, the shift to market-oriented economic policies and the abolition of corporatist institutions, the privatization of nationalized industries together with the deregulation of key sectors, the extension of financial markets which entailed the transformation and internationalization of the City of London, and the intensification of competitive relations in the labor market which were predicated on the erosion of social entitlements and workplace protections. As Peck and Tickell (2007, p. 28) put it "in Thatcherite discourse, these efforts were part of the wider program of 'rolling back the frontiers of the state' in order to enlarge the space for private enterprise, competition, and individual liberty."

As we can rather easily assume from the above, nature–society relationships have not remained immune from neoliberalism. Rather the opposite. As Heynen et al. (2007, p. 10) explain, the relationship between neoliberal reform and environmental politics, governance, and change, has been inherent in the consistent imperative that run through the history of neoliberalization, namely "to expand opportunities for capital investment and accumulation by re-working state-market-civil society relations to allow for the stretching and deepening of commodity production, circulation and exchange." The latter has been combined with an emphasis on individual rights and freedoms, and particularly with a clear support for private property rights (ibid.). Importantly, even in cases where neoliberal measures were not primarily aiming at expanding opportunities for capital accumulation, or failed to do so, they still brought about new ways of "locking up surplus capital (often highly sheltered from taxation) in the form of socio-natural fixes" (Heynen et al. 2007, pp. 11–12). The pervasive influence of neoliberalism to non-human nature has been expressed well in the new terms that have emerged to describe neoliberal-driven transformations in nature–society relationships: these include the "neoliberalization of nature," the concept of "neoliberal" or "capitalist natures" and, more recently, the term "neoliberal conservation." These terms have been increasingly used to describe nature–society relationships under neoliberal capitalism and led to the formation of an expanding critical literature which, inter alia, explores what exactly constitutes the neoliberalization of nature (see e.g., Apostolopoulou et al. 2014; Apostolopoulou and Adams 2015; Bakker 2010; Birch et al. 2010; Heynen and Robbins 2005; Heynen et al. 2007; McCarthy and Prudham 2004). Even though a review of the critical scholarship on neoliberalism and nature is beyond the scope of this book, I would like to refer here to a series of influential papers published by Noel Castree (2008a, b, 2010) on neoliberalism's environmental logic. Castree, by following a Polanyian-Marxian theoretical perspective, offers six generic elements of neoliberal thought and practice that largely characterize the neoliberalization of nature. The first element is privatization, defined as the assignment of clear private property rights to social or environmental phenomena that were previously state-owned, unowned, or communally owned. The second element is marketization, defined as the assignment of prices to phenomena that were not included in market exchange or that were unpriced. The third element is deregulation, namely the rollback of state interference in several areas of social and environmental life with the goal to reduce state regulation and to allow more actors to become self-governing within centrally prescribed frameworks and rules. The fourth element is market-friendly reregulation, namely the deployment of state policies to facilitate privatization and marketization of wider spheres of social and environmental life. The fifth element is the adoption of market proxies in the residual public sector that refers to a state-led attempt to make public services to function in line with the private sector, namely as efficient and competitive businesses. The sixth element is the construction of flanking mechanisms in civil society, described as the state-led encouragement of civil society groups (including various charities, NGOs, etc.) to provide services that states used to, or could potentially, provide for their citizens.

2.2 Neoliberal Conservation

Critics of conservation have noted, especially the last two decades, an increasing engagement between nature conservation and neoliberalism (see e.g., Apostolopoulou and Adams 2015; Apostolopoulou et al. 2018; Arsel and Büscher 2012; Brockington and Duffy 2010; Büscher et al.

2012, 2014; Corson 2010; Dressler and Roth 2011; Fletcher 2012; Igoe and Brockington 2007; Ojeda 2012; Sullivan 2006). Neoliberal conservation can be considered the latest stage in the long existent history of a contradictory relationship between capitalism and environmental protection, in that it reflects an increased intensity and variety of forms of capitalist invasion into conservation (Brockington and Duffy 2010). Its core axiom, as Büscher et al. (2012) argue, is that for natures to be saved nature conservation should be brought to the markets and conserved or protected natures should attract private investment. As Corson (2010) points out, in the neoliberal era, the international conservation agenda has created new symbolic and material spaces for global capital expansion by, inter alia, carving out new physical territories for capitalist accumulation. The latter can occur either through the demarcation and enclosure of common lands as protected areas or through the so-called "green" capitalist enterprise. The enclosure, commodification, and privatization of non-human nature can be seen in the increasing emphasis on the profits of the ecotourism industry, the formation of a variety of public–private conservation partnerships, and the inclusion of private entities, including major industries, in the governance of protected areas (Apostolopoulou et al. 2012a, b, 2014; Apostolopoulou and Pantis 2010; Hutton et al. 2005; Igoe and Brockington 2007; Maestre-Andrés et al. 2018; Paloniemi et al. 2015). Neoliberal conservation reflects the reconstitution of nature–society relationships according to market rules and is accompanied by a shift to devolved governance systems. The latter is expected to facilitate self-regulation along neoliberal lines and ensure the reproduction of TINA[1] (There Is No Alternative) dogma by subjecting local social, political, and environmental conditions to market dynamics (Büscher and Dressler 2012).

A key feature of the advancement of the neoliberal mode of conservation globally has been the rescaling of biodiversity governance (see Apostolopoulou et al. 2014 for a detailed discussion on the interplay between rescaling and neoliberalization) expressed in the construction of new scales, the revamping of scales and the shifting relationships between

[1] Historically, TINA has been used by the nineteenth-century classical liberal thinker Herbert Spencer and became a slogan of Margaret Thatcher in the 1980s.

scales (McCarthy and Prudham 2004; Smith 2010) as well as in the reduced role of the state in conservation and natural resource management. Indicative examples of these developments include the rearticulation of political scales downwards to the level of cities or regions, where public–private partnerships shape entrepreneurial practice and ideology, upwards to larger social and political levels, and outwards to broader networks of private capital (Swyngedouw 2000, 2005). The rescaling of governance to the subnational level has been accompanied with an increasing influence of supranational policy frameworks, international multilateral agreements and policy initiatives (in the case of conservation see for example the establishment of the Intergovernmental Science-Policy Platform on Biodiversity and Ecosystem Services—IPBES).[2] In all processes of biodiversity governance rescaling, there has been an expanded role for non-state actors and for the voluntary sector, often in partnership with state/quasi-state, private/market, and civil society actors (Apostolopoulou et al. 2014). Governance rescaling also signals the rescaling of power centers. This has been manifested in the alignments of corporate interests, major industries and donor agencies with environmental NGOs, governments, and local civil groups (Apostolopoulou 2016; Corson 2010; Igoe et al. 2010) and in the emergence and consolidation of an "authoritative managerial class" consisting of senior actors in conservation organizations. As MacDonald and Corson (2012, p. 170) explain this class is "distinct from, albeit intertwined with, the transnational capitalist class of corporate executives, bureaucrats and politicians, professionals, merchants, and the mass media referred to in the critical neoliberal conservation literature." Not surprisingly, in neoliberal conservation, the state-led acquisition and management of protected areas become more directly linked to the private sector and conservation NGOs become much more corporate in style, structure, and partnerships (Adams 2017).

The rescaling of biodiversity governance has been interlinked with the spread of market-based instruments and market proxies by aligning national-level conservation policies toward the idea that the market is the only realistic and viable solution to biodiversity loss and climate change.

[2]IPBES is an independent intergovernmental body that has been established by EU member states in 2012. For more information see https://www.ipbes.net/about.

This has been, inter alia, expressed in the reconstruction of conservation and environmental policies across the globe around the economic valuation of ecosystem services and natural capital accounting. Importantly, in the aftermath of the 2008 financial crash, within the wider context of prolonged austerity, namely of sustained and widespread cuts to governments budgets (Gray and Barford 2018), the devolution of state powers to regions and municipalities has been also accompanied by a further state withdrawal from key areas of social provision, and an increasing delegation of state responsibilities to parastatals, NGOs and consultants. Severe cuts in state funding and conservation personnel went hand in hand with an increase in public–private partnerships, the adoption of market-based instruments, like biodiversity and carbon offsets, and the further marketization and privatization of protected (or ecologically important) natures through tourism or real estate businesses since privileged access to rural land can be lucrative for investors (see Harvey 1989; Jonas et al. 2011). The above-described combinations of public, private, and hybrid governance (Hodge and Adams 2012) reflect a dynamic deconstruction and reconstruction of scales that is far from political neutral. By embodying and expressing power relationships (Apostolopoulou and Paloniemi 2012), processes of biodiversity governance rescaling have established mechanisms of inclusion and exclusion (Swyngedouw 2000) with most notable example the consolidation of the power of private interests which entailed the loss of rights for most people.

A final point is important here. Even though almost no place can claim total immunity from neoliberalism (Harvey 2007), the variegated character of neoliberalization across time and space and its uneven geographical development are endemic features of the basic operational logic of neoliberalism (Brenner et al. 2010; Büscher and Arsel 2012; Neves and Igoe 2012). The neoliberalization of nature has, thus, been a dynamic and contingent process that has occurred in highly varied contexts (Castree 2008a; Hodge and Adams 2012; McCarthy 2006). This variation has been very vividly expressed in actually existing neoliberal conservation. However, acknowledging the multiplicity, complexity, and variegation of neoliberalization does not preclude its theorization as an actually existing circumstance (Brenner and Theodore 2002). It rather reiterates the need to explore the context of context, namely the patterned and patterning macrospatial landscapes within which programs of neoliberalization are articulated and the conditions of production of such landscapes (Brenner et al. 2010), as well as the historically geographically specific factors and socioeconomic, environmental, and political conditions that shape place-specific outcomes. Thus, even though neoliberal programs have often been highly opportunistic, involved a lot of experimentation, and translated in contextually specific ways, they clearly share key common elements. It should be noted that this issue has been the focus of persistent debates regarding whether we can talk about a wider neoliberal environmental project or we should instead focus on specific neoliberalizations and their variegations in different contexts (see e.g., the discussion between Bakker 2009 and Castree 2009). The emphasis on the particularities of neoliberalization processes on the ground when it is accompanied by an underestimation of the common aspects of different neoliberal projects has often been counterproductive since it has prevented the integration of research on neoliberal natures. This does not mean that it is possible to extract a composite picture of a "typical" neoliberal conservation policy. This would rather be a "fool's errand" as Harvey (2005, p. 70) argues. It does mean, however, that it is both analytically and politically important to try to unravel the resemblances and discrepancies between different policies, places and countries in order to know "our enemy better, in general, and in its many guises" (Peet 2013, p. 3) and, thus, offer a more effective critique. It is also necessary for unraveling the specific class interests that neoliberal policies serve and their social impacts. This is evident in the case of offsetting especially given that the hegemonic rhetoric tries to portray it as a rational, pragmatic solution to better address the environmental impacts of economic development obscuring its political and ideological role and the interests it serves.

2.3 The Dialectics of Green and Un-green Grabbing After the 2008 Financial Crash

Since the 2008 global financial crisis started to unfold, the state of neoliberalism has been at the epicenter of significant debates.[3,4] In 2008, Neil Smith argued that neoliberalism is "dead but dominant," in a "state of atrophy," but with its economic and military power enduring and continuing to produce human destruction, whether upon the environment or among people.[5] In 2009,[6] Eric Hobsbawm argued that even though it was yet unknown how grave and lasting the consequences of the economic crisis would be, "they certainly mark the end of the sort of free-market capitalism that captured the world and its governments in the years since Margaret Thatcher and Ronald Reagan." During the same period, David Harvey asked whether the crisis signals "the end of neoliberalism" to argue that the answer is not straightforward.[7]

Today, in many places across the globe, neoliberalism appears very much alive even though it is also increasingly losing its legitimacy (Apostolopoulou and Cortes-Vazquez 2018).[8] But how do we end up where we are today? At the aftermath of the 2008 financial crash and once the immediate need to save the banks was over, the crisis shifted from a financial crisis in the banking sector to a fiscal crisis of the state with the aim to protect the interests of banks, bondholders and industrial capital (Lapavitsas et al. 2011). This meant a sudden deterioration in public finances, and fundamental and rapid changes at all levels of government and all policy sectors in many countries, most notably in the embattled EU economies of southern Europe. As Lapavitsas et al. (2010) explain rescue or bailout packages for the economies of the EU periphery (from

---

[3] See http://www.guardian.co.uk/commentisfree/2009/apr/10/financial-crisis-capitalism-socialism-alternatives.

[4] See http://www.counterpunch.org/2009/03/13/is-this-really-the-end-of-neoliberalism/.

[5] See https://www.berghahnjournals.com/view/journals/focaal/2008/51/focaal510113.xml.

[6] See http://www.guardian.co.uk/commentisfree/2009/apr/10/financial-crisis-capitalism-socialism-alternatives.

[7] See http://www.counterpunch.org/2009/03/13/is-this-really-the-end-of-neoliberalism/.

[8] See also David Harvey's recent interview on the topic: https://mronline.org/2019/02/16/the-neoliberal-project-is-alive-but-has-lost-its-legitimacy-david-harvey/.

the EU and the International Monetary Fund-IMF) have been driven by neoliberal ideology convinced of the virtues of fiscal responsibility as both cure and preventative of crises. In return, in many countries, including the UK, strict austerity programs have been widespread (see Apostolopoulou and Adams 2015; Cahill 2011; Calvário et al. 2017). Overall, and despite initial debates and some stimulus measures, EU governments, in classic neoliberal fashion, imposed draconian austerity (Harvey 2011, p. 17) and retrenchment (cutting public expenditure, raising indirect taxes, reducing wages), coupled with liberalization in the form of extensive deregulations and privatizations of public property (Lapavitsas et al. 2011).

The economic crisis that has spread unevenly across the globe has signaled a clear intensification in the exploitation of non-human nature. As Harvey (2010, p.11) argues, financial crises lead to reconfigurations, new models of development, new spheres of investment and new forms of class power: that is also true for environmental and conservation policies. Indeed, the intensification of the neoliberalization of nature, itself a crisis-induced, crisis-inducing process (Peck et al. 2012), has been a fundamental element of the post-crisis era (Apostolopoulou et al. 2014; Apostolopoulou and Adams 2015) testifying both the key role of nature to capital accumulation and neoliberalism's opportunism. The latter has been expressed in the sharp policy shift to environmental policies that questioned the premises of the "green economy" that had been prevalent in the agenda of neoliberal policymakers worldwide during at least the last two decades (Castree 2010). Key examples include the rolling back of environmental and planning regulatory frameworks, the extensive privatization of land and natural resources, as well as the springing of the debt trap as a primary means of accumulation by dispossession confirming that crisis creation, management, and manipulation on the world stage is related to the uneven development of capitalism with profound consequences for both people and nature (Harvey 2007). These changes have been accompanied by a further rescaling of environmental governance manifested in the upgraded role of international organizations, such as the IMF, which have evolved since the 1980s into prime agents of neoliberalization (ibid.; see also Apostolopoulou and Adams 2015; Swyngedouw 2004).

As I argue elsewhere (Apostolopoulou and Adams 2015), a key element in nature–society relationships in the post-2008 era is *the coexistence, and sometimes even symbiotic relationship, between processes of green and un-green grabbing*. The term green grabbing is used to describe the ways through which land and/or resources are appropriated for environmental purposes and can be understood as part of the wider debate on land grabbing (Fairhead et al. 2012; Borras et al. 2012). As Fairhead et al. (2012) explain, in green grabbing, environmental green agendas are the core drivers and goals of grabs and are linked to biodiversity conservation, biocarbon sequestration, biofuels, ecosystem services, ecotourism or offsetting. Green grabbing builds on the history of colonial and neocolonial resource alienation in the name of the environment while also bringing new dimensions into play and involves the alienation of land and the restructuring of rules and authority in the access, use and management of natural resources (ibid.). The intensification of processes of green grabbing and the emergence of new green enclosures have been obvious in the aftermath of the 2008 financial crash and have been linked to the shift toward the economic valuation of nature and the widespread application of market-based policies (see also the Introduction). The latter has been, inter alia, expressed in the Rio + 20 summit (or Earth Summit 2012) with its focus on natural capital accounting initiatives, such as the Natural Capital Declaration, in the key outcomes of the 11th Conference of the Parties to the United Nations Convention on Biological Diversity which included calls to further integrate the economic value of ecosystems into national biodiversity plans (UNEP/CBD/COP 2012) and in the key focus of EU policies in the post-2008 period in market-based instruments (EC 2011). These approaches by promoting the accounting of the value of environmental goods and services as the key solution to biodiversity decline and ecosystem degradation not only overlook capitalism's environmental contradictions but also portray it as key to future ecological sustainability (Apostolopoulou and Adams 2015; see also Igoe et al. 2010).

Importantly, green grabbing processes and capitalism's strategic interest in promoting a neoliberal version of conservation, via its further inclusion into market function, goes hand in glove with parallel processes whereby capitalism seeks to obstruct the conservation of species and ecosystems, and to grab and exploit hitherto protected natures without any green or eco-friendly argumentation. We have termed this latter process "un-green" grabbing (Apostolopoulou and Adams 2015), and suggested that attention needs to be paid to its importance in the engagement between capitalism and nature under neoliberalism, particularly, in the post-2008 era. "Un-green" grabbing is a form of land or resource grabbing that involves the exploitation and appropriation of protected natures and which, unlike green grabbing, is not done in the name of the environment (Fairhead et al. 2012). Un-green grabbing is a distinct subset of general capitalist land or resources appropriation, and its dialectics with green grabbing highlight the dual outcome of a single exploitative engagement of capitalism with nature and its profound challenges for the feasibility of a radical environmentalist position (Apostolopoulou and Adams 2015). Even though the appropriation and exploitation of nature has been at the core of the capitalist enterprise, the current intensification of un-green grabbing, especially in countries with strong environmental legislation (like for example Greece), marks a new era where governments are forcefully taking back environmental regulations that have been shaped by the competition among capitals but also won through environmental and social movements and hard class struggle (Harvey 2005; Vlachou 2005) and entails the loss of rights.

Un-green grabbing processes are evident in the intensification of the classical, longstanding conflict between economic development and environmental protection in many places across the globe and relate to the ability of certain capitalist interests to advance their position under conditions of economic crisis by securing public assets, including land and natural resources, as well as by carving out new areas for capital accumulation. Their combination with processes of green grabbing have led to a situation where strategies to create markets in biodiversity and carbon coexist with the rolling back of environmental regulatory frameworks in the face of increasing debts and decreasing profits. Thus, along with neoliberalism's optimism on green capitalism's ability to "save nature," we also witness the unwillingness of powerful sections of capital to commit to a more sustainable path of development as manifested in several international meetings and UN summits. It is worth mentioning here that in the sidelines of the recent UN Climate Summit in New York (22

September 2019) oil and gas executives (from BP, Shell Chevron) held an exclusive invitation-only forum with environmentalists and government representatives in an attempt to influence negotiations in favor of fossil fuel companies.[9] As Extinction Rebellion[10] put it: "this is how lobbying and greenwash works: private meetings, on the sidelines, back door stuff, handpicked attendees."[6] In parallel, at the national level, responses to the economic crisis have mainly focused on restoring dominant political economic arrangements to where they were several years ago, rather than taking the opportunity to "green" them in any substantive sense (McCarthy 2012). These developments have been obvious in the UK, as we will see in Chapter 6, but also in many other countries across the globe where governments have used the crisis as a "Trojan horse" to enable the further neoliberalization of nature (Apostolopoulou and Cortes-Vazquez 2018; Cortes-Vazquez and Apostolopoulou 2019; Apostolopoulou et al. 2014). The intensification and coexistence of green and un-green grabbing, that is evident in offsetting policies as we will see below, captures well the contradictory ways in which capitalism under crisis engages with non-human natures (Apostolopoulou and Adams 2015) and reflects its inherent weaknesses in formulating a positive vision for nature–society relationships.

2.4 Understanding the Neoliberal Hegemony: The Dialectics of Coercion and Consent Under Capitalism in Crisis

The second element, along with the intensification of green and un-green grabbing, that is crucial for understanding the effects of the prolonged economic crisis to nature–society relationships is the changing dialectics of domination and hegemony or of coercion and consent. To understand these we have to first acknowledge the consolidation of the, rather paradoxical, approach of market-based environmentalism. I am

---

[9]See https://www.theguardian.com/environment/2019/sep/18/fossil-fuel-invite-only-forum-un-climate-summit.

[10]For more information see: https://rebellion.earth/.

saying paradoxical in the sense that environmentalism, as emerged in the radical movements of the 1960s after the classical publication of Rachel Carson's Silent Spring and within the context of the war in Vietnam, and as still expressed in many environmental movements across the globe (Apostolopoulou and Cortes-Vazquez 2018), has been largely opposing the capitalist exploitation of non-human nature and has been directly questioning capitalism's obsession with economic growth. The increasing popularity of market-based environmentalism relates to the shift to consensual politics, marked by the mainstreaming of the sustainable development discourse in the late 1980s (Adams 2009). The sustainable development discourse despite its transformations during the last three decades remains undoubtedly hegemonic: according to mainstream conservation politics not only there is no contradiction between "sustainable" development (perceived in mainstream discourse as sustainable economic growth) and conservation but also conservation has no chance of surviving without being in alliance with capital and rendering itself a profitable business.

To understand how this belief in the market became so commonsensical in nature conservation it is helpful to draw on Gramsci's analysis on the *historic bloc*. As Gramsci argued in the Prison Notebooks (1971, p. 366), structures and superstructures form a historic bloc, a complex, contradictory, and discordant ensemble of superstructures that reflects the ensemble of the social relations of production. The concept of the historical bloc shows the dialectical unity of content and form, of an economic structure and its ratifying superstructure and ideologies (Thomas 2006). As Marx (1859) writes in the *Contribution to the Critique of Political Economy*, "each particular mode of production, and the relations of production corresponding to it at each given moment, in short 'the economic structure of society', is the real foundation, on which arises a legal and political superstructure and to which correspond definite forms of social consciousness, and that the mode of production of material life conditions the general process of social, political and intellectual life." The ideology of market-environmentalism, therefore, and the framing of the economic valuation of nature as the only viable solution to ecosystem degradation cannot be adequately understood without considering the characteristics of the capitalist mode of production. Indeed, the ideology of market-based environmentalism reflects capitalism's exploitative use of nature and natural resources as mere material conditions of capital accumulation (Burkett 1997) as well as the subjugation of use value to exchange value in capitalist commodity production (Harvey 2014). These are not of course discussed, acknowledged, or presented as such in mainstream environmental politics that tend to interpret adherence to the principles of the market as almost natural. As Igoe et al. (2010) argue two aspects of the Gramscian notion of the historic bloc can help us to further understand how this hegemony is actually achieved in practice. Firstly, the fact that the ideologies accompanying a particular historic bloc conceal the nature of existing relations of production by presenting a naturalized view of class hierarchies that comprise these relations. Secondly, that this concealment has the effect of smoothing over the contradictions, paradoxes, and irreconcilable differences that exist within these relations. The emphasis on the economic valuation of nature even though emerges from the historically specific relations created by capitalism as a value system perform an ideological function, which is that of naturalizing the process of concrete abstraction (Sayer 1987) on which value in capitalism is based. Its underlying ideology is neoliberalism and its ultimate goal is to change society's shared, common sense of socionatural interactions and nature–society interrelationships.

However, a historic bloc is characterized by internal contradictions and involves a combination of coercive and noncoercive authority that is configured in space and time (Davies 2013). As Thomas (2006, p. 68) explains, "an historical block is not given as a permanent element, achieved once and for all. Rather, it is actively and continually forged by the 'hegemonic apparatus' of a class – the various institutions and practices by means of which it concretizes its hegemonic project and continues to secure both social and political leadership, that is leadership both in civil society and at the level of the state." The dialectical integration of consent with coercion or of hegemony with domination, united in their distinction (Thomas 2009), is fundamental to Gramsci's conception of the *integral state* as "the entire complex of practical and theoretical activities with which the ruling class not only justifies and maintains its dominance, but manages to win the active consent of those over whom it rules" (Thomas 2009, p. 138). In the neoliberal era, and particularly after the 2008 financial crash, displays of coercive power play a key role in reinforcing and creating the conditions for hegemony (Davies 2013). Indeed, in the post-2008 era, the limits of market-based environmentalism and the contradictions that the hegemonic discourse of consensus aims to hide have become more visible and a widespread shift toward coercion has been evident. As Swyngedouw argues (2005) at moments of increasing socioeconomic tension, the delegation of power and the inclusion of new strata of civil society in governance can go hand in hand with a shift of the state to more authoritarian or more autocratic ways of governing. Indeed, in many countries, in the aftermath of the 2008 financial crash, the strengthening of powerful, in most cases unauthorized actors (with prominent examples conservation brokers, consulting companies, industries, and big NGOs), and their privileged access in shaping conservation and development agendas, has been accompanied by restricting access and control for the majority of local people or even their total exclusion from decisions affecting their livelihoods. The shifting dialectics between hegemony and domination, or consent and coercion, have been also visible in neoliberalism's tendency to favor governance by experts and elites as well as to insulate key institutions from democratic pressures (see also Harvey 2005) making conservation primarily a matter for experts and technocrats. The latter has been manifested in the emphasis on expert knowledge and experts' committees (see Sect. 6.2), and the increasing role of nonelected and unaccountable supranational institutions. Greece offers here an emblematic case: in the post-2008 period, the so-called Troika (European Commission, European Central Bank and International Monetary Fund) and sociétés anonymes have played a key role in imposing an unprecedented deregulation of environmental and planning policy across the country and a massive land grabbing (Apostolopoulou and Adams 2015).

Relying on undemocratic and unaccountable institutions to enforce socially environmentally unequal policies underlines the changing geometry of social power: strengthening the power of the banking, industry (including extractive, infrastructure and housing industries) and tourist sectors goes hand in glove with disempowering the working class. Hegemony and domination are strategically differentiated forms of a unitary political power (Thomas 2009), analytically distinct but functionally intertwined (Davies 2013), as it has been evident in the fact that several places across the globe have become the locus of resistance against the unequal distribution of the costs and benefits of neoliberal policies, and their increasing social-environmental and spatial injustices (Apostolopoulou and Cortes-Vazquez 2018). The Environmental Justice Atlas,[11] a collaborative map of social conflicts around environmental issues from across the globe, has so far (November 2019) documented 2.956 of such cases, including struggles to defend land, air, water, forests, and livelihoods from damaging projects and extractive activities with several social-environmental impacts, including mining, fracking, dams, pipelines, gas flaring, waste, and nuclear energy among others (see Temper et al. 2018). These conflicts are not distributed evenly across the world corroborating that uneven development "is the concrete pattern and process of the production of nature under capitalism" (Smith 2010, p. 8; Harvey 2010). Importantly, uneven geographical development contributes to social and spatial injustices (Soja 2010) and acting against such injustices requires building networks of solidarity (Hadjimichalis 2011). The intensification of these conflicts has been accompanied by the emergence of new movements that have tried to articulate a critical narrative to neoliberalism and produce and defend alternative futures, sometimes at the price of risking the lives of activists[12] (Apostolopoulou and Cortes-Vazquez 2018). Governmental reactions to strident protests reinforce Polanyi's fear that the liberal, and by extension the neoliberal, project could only ultimately be sustained by resort to authoritarianism (Harvey 2005).[13] Social-environmental movements across the globe along with recent climate strikes expose that the contested character of neoliberal hegemony making more and more evident the contradictions and the limits of capitalism.

---

[11] See https://ejatlas.org/.

[12] See www.globalwitness.org/en/blog/new-data-reveals-197-land-and-environmentaldefenders-murdered-2017/.

[13] See also the recent interview of David Harvey in Monthly Review: https://mronline.org/2019/02/16/the-neoliberal-project-is-alive-but-has-lost-its-legitimacy-david-harvey/.

2.5 A Brief History of the Emergence, Evolution and Neoliberal Origins of Biodiversity Offsetting

> If they [EPA regulators] had their way,
> you and I would have to live in rabbit's holes or bird's nests.
>
> —Reagan (1980)[14]

As already explained in the previous sections, in the era of neoliberal conservation, markets have an increasingly central role in mainstream approaches to "save" nature by selling it (McAfee 1999): profit-oriented practices and techniques, such as ecotourism, have been expanded, and new market-based instruments, such as payments for ecosystem services or biodiversity and carbon offsets, have been created. But let's see what exactly biodiversity offsets are.

Biodiversity offsets are defined as measurable conservation outcomes designed to compensate for significant adverse biodiversity impacts arising from project development after prevention and mitigation measures have been taken (BBOP 2012a). Biodiversity offsetting and compensation mechanisms constitute market-based policy instruments whose popularity has increased significantly since the global mainstreaming of the economic valuation of ecosystem services and natural capital (TEEB 2010; Badura et al. 2017). Their diversity is reflected in the terms used to describe them, including mitigation, habitat and conservation banking, offsetting, habitat credit trading, BioBanking, and compensatory mitigation (Maron et al. 2012). Some mechanisms are nascent and others more mature, some are based on compliance with regulation and others on voluntary arrangements (Bennett et al. 2017b). Their key aim, in line with the core tenets of market environmentalism, is to reconcile economic development (and growth) with environmental protection by internalizing ecological externalities (ecological losses due to economic development) into decision-making (Bayon et al. 2008).

[14]Cited in Layzer (2012).

Biodiversity offsetting is a policy that has been increasingly informed by neoliberal logic and defended using neoliberal rhetoric (Apostolopoulou et al. 2018). It typifies the neoliberal turn in environmental policy and the wider shift toward neoliberal conservation that has been described in the previous sections. Offsetting emerged in the context of the increasing deregulation of environmental and planning legislation, the reduction of state's role in the direction, legitimization and exercise of control over the use of nature and its further withdrawal from many areas of social provision (Castree 2008a, b; Harvey 2005). It involved the further deregulation and market-friendly reregulation of legislation to facilitate economic development, the extension of the monetary valuation of nature, an increased involvement of the private sector (Quétier and Lavorel 2011) and market-based approaches to conservation (ten Kate et al. 2004), and a broader reliance on private means of sustaining social reproduction.

As Bonneuil (2015) shows in a very interesting genealogy, the emergence of biodiversity offsetting mechanisms has been inextricably linked to the 1973–1990 rollback of environmental regulations and coincided with a broader shift toward market-based solutions to environmental problems. At that time, as several historical facts corroborate, the neoliberal counter-revolution involved, inter alia, massive deregulations of the environmental and planning legislation and extensive privatizations, with the key objective to dismantle the regulations and Keynesian measures adopted after World War II. This prepared the ground for the emergence of air pollution marketable permit systems that, in turn, paved the way for the introduction of biodiversity offsetting mechanisms.

Even though biodiversity offsetting, in the form of habitat and species banking, was formally put in place in the 1990s, its origins can be traced to the US Clean Water Act (1972) and its Section 404 which required mitigation for development projects with significant impacts on wetlands. Offsetting has been used during the 1970s in the United States under the Carter administration, reflecting the compromise that environmental agencies, such as the Fish and Wildlife Service and the National Marine Fisheries Service, were willing to make given the wider political climate. These agencies lacked a veto authority over the COE's (US Army Corps of Engineers) wetland fill permits, and, as Robertson (2000,

p. 470) explains, they knew that if they denied economic development on wetlands probably the COE would override them. They were, therefore, ready to propose mitigation instead of avoidance or minimization, if there was, at least on paper, a promise that there will be No Net Loss of wetlands (Kruczynski 1990, p. 551). That period has been also marked by the growing support for the introduction of tradeable rights. In a relevant report, published in 1981, in line with the neoliberal logic that was gaining ground in US environmental politics, such instruments were characterized as more compatible with the market forces that govern business decisions (ibid.).

In the 1980s, under the Reagan administration,[15] offsetting mechanisms continue to gain increasing popularity. In 1982, the new administration encouraged the launch of a pilot project for the development of the first mitigation bank in Louisiana in partnership with the Tenneco Oil Company under the supervision of the Fish and Wildlife Service (Bonneuil 2015). This was called the "LaTerre Bank" and its primary goal was to satisfy internal demand for compensatory mitigation, namely to create offset credits for the Tenneco Oil Company (Hough and Robertson 2009). In 1986, the first commercial sale of banked section 404 compensation credits occurred at the LaTerre Bank (ibid.). The same year, New Jersey became the first state that passed a wetland offsetting legislation (Robertson 2000).

Even though the principle of No Net Loss was one of the key recommendations of the National Wetlands Policy Forum which took place in 1987 (Bonneuil 2015), the first announcement of a policy of No Net Loss has been attributed to President George Bush in 1988–1989 (Stokstad 2008) who used it to argue that environmental protection and free market could be successfully combined. In November 1990, the Water Resources Development Act announced the creation of an offset market for wetlands and rivers that has been officially launched in 1991.

[15] As Layzer (2012) points out, Lou Cannon, the biographer of Reagan, argued that although Reagan was an outdoorsman and nature lover, he was contemptuous of environmentalists. Indeed, the Republican Party platform was clearly hostile toward both government and environmental regulation (Bonneuil 2015).

The implementation of conservation banking formally started in California on 7 April 1995 when the Resources Agency and the Environmental Protection Agency jointly issued their official policy on conservation banks (Wheeler and Strock 1995). This was followed by a supplemental policy, issued in January 1996, which provided more specific guidance about the establishment of banks while also openly supporting a free-market approach to mitigation.[16] Between 1996 and 2016, the number of banks based on natural assets grew from 50 to over 1.000 (Bennett et al. 2017b).

It is important to point out here that even though more sophisticated mitigation credit banking systems emerged in the 1980s and 1990s (Madsen et al. 2010), US wetland compensatory mitigation of the 1970s can be seen as the formalization of the concept of biodiversity offsets (Burgin 2008). Importantly, the emergence of offsetting has signaled a clear shift toward the use of market-based mechanisms as the desirable approach to the implementation of environmental regulation: commercial wetland banking as a way of implementing the US Clean Water Act, and species banking as a way of implementing the US Endangered Species Act (Bayon et al. 2008). The increasing emphasis on the establishment of conservation banks, consisting of parcels of land approved by regulators to sell mitigation credits for endangered, threatened or other imperiled species or habitats (Madsen et al. 2010), manifested the consolidation of the neoliberal turn in US environmental policy (Robertson 2004). Interestingly, there was remarkable sustained support for offsetting policies despite governmental changes: the rise of neoliberal environmental regulation began during the presidency of Carter and gained major traction during the Reagan and the Clinton presidencies, both of which laid the foundation for the type of free-market regulation that persists until today (Czarnezki and Fiedler 2016).

It is worth mentioning that the United States still has a leading role in biodiversity offsetting and environmental markets more broadly as exemplified in the fact it currently hosts the largest banking market in the world, namely the Compensatory Mitigation for Losses of Aquatic

[16]As we read in the relevant report "the number of conservation banks that are established will be regulated by the free market and willingness of landowners to participate, not by the wildlife agencies" (cited in Mead 2008, p. 11).

Resources program that focuses on wetland and stream offset credits.[17] However, there are also other countries with a long history in similar policies. These include Canada that has been one of the first countries that published, in 1991, a national wetland conservation policy with an explicit reference to No Net Loss followed by an implementation guide that introduced the mitigation hierarchy (Lynch-Stewart et al. 1996). Despite its early adoption, wetland mitigation banking has not been a very common practice in Canada until the 2000s partly because of its controversial results in the US (see also Rubec and Hanson 2009). However, the early 2000s have seen the emergence of various voluntary offset projects, including cases like the Shell Canada-True North Forest, the Oil Sands Leadership Initiative, the Kinder Morgan Canada, and the Hydro-One-Bruce to Milton Biodiversity Initiative (Poulton 2014). Moreover, particularly since 2010, Canadian regulators have been increasingly imposing biodiversity offsets as a condition of development approvals (ibid.).

Another country to which we should make a reference here is Australia; a country that is hosting some of the largest mitigation markets globally. The majority of Australian states and territories have at least one biodiversity offsets policy (Maron et al. 2016) including two important compliance-driven programs, the New South Wales Biodiversity Banking (BioBanking) and Offsets Scheme and Victoria Native Vegetation Offsets. In 2014, two new credit types have been introduced, the General Biodiversity Equivalence Unit and the Specific Biodiversity Equivalence Unit. Even though the new credit types have seen initial growth in the first three years of their introduction in the market, Victoria's Native Vegetation Offsets market saw an overall decline between 2010 and 2016 (Bennett et al. 2017b).

At the European level, the country with the longest history in offsetting is Germany where a national mandatory requirement for offsetting exists since 1976, when the Impact Mitigation Regulation (IMR) was adopted as part of the Federal Nature Conservation Act and the Federal Building Code (IEEP 2014). IMR expands beyond protected areas

[17] For more information see: https://www.epa.gov/sites/production/files/2015-03/documents/2008_04_10_wetlands_wetlands_mitigation_final_rule_4_10_08.pdf.

and species and aims to compensate for impacts in entire ecosystems and landscapes (Gonçalves et al. 2015). However, it does not include legal provisions specifying how to assess the initial state of the areas that are affected by development, the impacts resulting from the intervention or the appropriate methodology to determine the required compensation. An obligation to offset unavoidable impacts on the environment has been also included in French law since 1976, but offsetting has been largely ignored until relatively recently (Tucker et al. 2014).

A key moment in the more recent history of biodiversity offsetting which contributed to its increasing popularity has been the establishment of the "Business and Biodiversity Offsets Program" (BBOP) in 2004. BBOP has been an international partnership of more than 80 organizations and individuals, including companies, government agencies, financial institutions, service providers, and civil society organizations (Table 2.1). As it was claimed in its website[18] its members were testing and developing "best practice" on biodiversity offsets and conservation banking worldwide. BBOP has been a sister initiative to Ecosystem Marketplace, a web-based information platform established in 2004 (Madsen et al. 2010)[19] and launched by Forest Trends that served as BBOP's Secretariat with the Wildlife Conservation Society.[20] Forest Trends is a Washington D.C.-based international non-profit organization. Its history began in 1996 when "a small group of people from the forest industry, donors, and environmental groups" decided to create this new organization to "expand the work of bridging traditional divides and promoting market-based approaches to forest conservation."[21] In 1998, the original group of participants was expanded to include additional representatives from industry, finance, and community conservation, and this enlarged group became the original Board of Directors. In late 1999, there was a further expansion and representatives from other major

[18]BBOP's website is no longer active as explained in the last paragraph of this section.

[19]Forest Trends' Ecosystem Marketplace publishes breaking news, newsletters, original feature articles and major reports on finance for conservation, such as carbon offsets, forest finance, biodiversity offsets and compensation, watershed investments, and private investment in conservation. For more information see their website: https://www.forest-trends.org/ecosystem-marketplace/about-ecosystem-marketplace/.

[20]See http://www.forest-trends.org/program.php?id=117.

[21]See http://www.forest-trends.org/page.php?id=153.

Table 2.1 BBOP latest advisory group[a]

BBOP advisory group membership (2004–2018)
Companies with biodiversity footprint to offset
Ambatovy Project
Anglo American
AngloGold Ashanti
BP
Cemex
Eiffage
Électricité de France
ERAMET
Inmet Mining
Luossavaara-Kiirunavaara AB (LKAB)
New Britain Palm Oil Ltd.
Newcrest Mining Limited
Newmont Mining Corporation
Rio Tinto
Shell International
Sherritt International
Solid Energy Coals of New Zealand
Sveaskog
Tahi Estate
TOTAL SA
Winstone Aggregates
Service providers
Arup
Biodiversity Works
Biotope
Daemeter Consulting
deVilliers Brownlie Associates
Earthtrade
Ecoacsa
EcoAnalysts
EcoDecisión
Enetjärn Natur
Environ Corporation
Environment Bank
Environmental Banc & Exchange
Environmental Resources Management
Golder Associates

(continued)

Table 2.1 (continued)

BBOP advisory group membership (2004–2018)
Hardner & Gullison Associates
Hogan Lovells
International Conservation Services CC
IPIECA
Markit Environmental Registry
Namibian Uranium Association
Nature Task
New Forests
Nollen Group
Poulton Environmental Strategies
Proforest
Redd Forests
Resource Environmental Solutions, LLC
Response Ability, Inc.
Scientific Certification Systems
SLR Consulting
The Biodiversity Consultancy
Tonkin and Taylor
Treweek Environmental Consultants
Wave Action
Wild Business
Wildlands Inc.
WSP
Financial institutions
CDC Biodiversité
Citi
European Bank for Reconstruction and Development
Global Environment Fund
Inter-American Development Bank
Inter-American Investment Corporation
International Finance Corporation
KfW Bankengruppe
Mizuho Bank, Ltd.
Governments and intergovernmental organizations
City of Bainbridge Island
Department for Environment and Rural Affairs—Defra (UK)
Department of Conservation, New Zealand
Department of Sustainability & Environment, Government of Victoria, Australia

(continued)

Table 2.1 (continued)

BBOP advisory group membership (2004–2018)
Forestry Department, Sabah, Malaysia
Government of Catalonia, Spain
Government of Cross River State, Nigeria
International Union for Conservation of Nature (IUCN)
Ministry of Agricultural and Rural Development (MARD), Vietnam Directorate of Forestry
Ministry of Ecology, Energy, Sustainable Development, and Spatial Planning, France
Ministry of Environment and Tourism, government of Namibia
Ministry of Infrastructure and the Environment, The Netherlands
Ministry of Mines and Energy, Namibia
Ministry of Natural Resources and the Environment (MONRE), Vietnam
Ministry of Nature, Environment and Tourism, government of Mongolia
Ministry of Sustainable Development and Infrastructure, Luxembourg
National Ecology Institute, Mexico
National Environment Management Authority, Uganda
National French Institution for Water Management and Policy
National Office for the Environment, Madagascar (ONE)
Ramsar Convention on Wetlands
South African National Biodiversity Institute
United Nations Development Programme
United Nations Environment Programme—World Conservation Monitoring Centre
United States Agency for International Development
US Fish and Wildlife Service
Wildlife Division, Forestry Commission, Government of Ghana
Conservation and civil society groups
African Wildlife Foundation
Biodiversity Neutral Initiative
BirdLife International
Botanical Society, South Africa
Brazilian Biodiversity Fund (Funbio)
Cambridge Centre for Conservation Policy
Centre for Research-Information-Action for Development in Africa
Conservation International
Earthmind
Earthwatch Institute
Ecoagriculture Partners
EcoTopia Science Institute, Nagoya University

(continued)

Table 2.1 (continued)

BBOP advisory group membership (2004–2018)
Environmental Defense Fund
Fauna & Flora International
Forest Trends
Grupo Ecológico Sierra Gorda, I.A.P., México
Guinée Ecologie
International Institute for Environment and Development
Leibniz Institute of Ecological and Regional Development (IOER)
Nature Conservation Resource Center, Ghana
NEPCon
Pact Inc., Private Agencies Collaborating Together
Rainforest Alliance
Royal Botanic Gardens, Kew
South Rift Association of Land Owners, SORALO
The Nature Conservancy
Tulalip Tribes, US
United Nations Development Programme (Environment and Energy Group)
University of Washington
Wildlife Conservation Society
WWF-UK
Zoological Society of London
Individuals
Steve Botts
Marc Christensen
Toby Gardner
Martin Hollands
Daniela Lerda
Paul Mitchell
Dave Richards
Shelagh Rosenthal
Marian Weber

[a]Available at: https://www.forest-trends.org/wp-content/uploads/2018/11/BBOP-AG-members-life-of-programme.pdf

forest areas besides the United States, including Russia, Brazil, Malaysia, and Canada, joined the board. Forest Trends has three principal roles: convening market players to advance market transformations, generating and disseminating critical information to market players, and facilitating deals between different critical links in the value chains of new forestry.[22]

The BBOP has been broadly acknowledged as the global expert on offsetting and has been playing a key, and quite controversial, role in the widespread adoption of the policy across the globe, very often directly influencing governmental policies, as in the UK case. The BBOP along with the growing power of consultants and conservation brokers exemplifies the heightened role of unelected actors and the emphasis on the authority of a, supposedly neutral, evaluation technoscience (Peck et al. 2012), in global biodiversity governance. Offsetting constitutes an indicative case of this which along with the ignorance for local communities and their exclusion from decision-making manifests the increasingly anti-democratic and depoliticizing nature of neoliberal environmental governance. This is also reflected in the fact that after 15 years of influencing offsetting policies worldwide and promoting a neoliberal agenda disguised as best-practice pragmatism, on 31 December 2018, the BBOP announced that is closing its operations since it achieved its mission, namely the mainstreaming of offsetting and the creation of a "Biodiversity Net Gain movement."[23] This creates major issues of accountability since an organization whose influence remains pervasive until today in national policies, best practice guidance, scientific and policy publications and various case studies, does not exist anymore to respond to any criticisms related to operation. Of course, BBOP is not the only indication of the key role of public–private partnerships and consultancies in offsetting. At the core of the policy lies the promise to bring together a wide array of actors and interests and this promise has played a prominent role in the attention that offsetting has been gaining from environmental organizations, industries, governments and investors, leading to its spreading across the globe (Burgin 2008; ICMM 2005; ten Kate et al. 2004). Indeed, offsets have been particularly attractive to companies with environmentally and socially destructive agendas, including

[22]For more information see: https://www.forest-trends.org/who-we-are/mission-and-history/.

[23]See https://www.forest-trends.org/events/bbop15-bbop-final-conference/.

those involved in mining, oil and gas, industrial agriculture, and large-scale infrastructures, because they help them to access the land and capital they require to carry out their projects and obtain their social license to operate (Re:Common 2019).

2.6 Key Definitions

BBOP's key role in the mainstreaming of biodiversity offsetting worldwide is, inter alia, obvious in the fact that it has offered one of the most widely accepted and frequently used definitions of offsetting. It is indicative that in the No Net Loss initiative that the EU launched in 2014, the definition of offsets has been based on the BBOP definition.[24] According to the latter, the goal of biodiversity offsets is to achieve No Net Loss and preferably a Net Gain of biodiversity with respect to species composition, habitat structure, ecosystem function and, at least in theory, to people's use and cultural values associated with biodiversity (BBOP 2012a, p. 13). The achievement of No Net Loss (NNL) is central to the BBOP's definition and one of the ten BBOP principles for designing "high quality" offsets. It is defined as follows: "a biodiversity offset should be designed and implemented to achieve in situ, measurable conservation outcomes than can reasonably be expected to result in no net loss and preferably a net gain of biodiversity" (BBOP 2012b). NNL has had a pervasive influence on offsetting.[25] Today, even though objectives vary among projects, offsets are being increasingly defined as the set of activities identified to achieve a No Net Loss or a Net Gain of biodiversity in the specific context of the development project concerned (EFTEC and IEEP 2010; Maron et al. 2012).

The NNL objective is very important because it directly links biodiversity offsets to the necessity to achieve ecological equivalence between

[24]The NNL initiative of the EU had been open to consultation on June 2014 and its future state is at the moment uncertain (http://ec.europa.eu/environment/nature/biodiversity/nnl/index_en.htm).

[25]The central position of the NNL objective as an explicit aim of an offset program draws a distinction between offsets and compensation (two terms that in some jurisdictions are considered as synonyms) since the latter does not necessarily involves this aim (http://ec.europa.eu/environment/nature/biodiversity/nnl/pdf/NNL_Glossary.pdf).

what is lost in a development site and what is gained in an offset site. This requires the quantification of biodiversity losses and gains and their representation through numerical scores in order to obtain concrete evidence of measurable on-the-ground conservation outcomes. The creation of ecologically equivalent gains sets the base for the creation of conservation credits contributing to the introduction of habitat or conservation banking. The latter has been defined as a market where the credits from actions with beneficial biodiversity outcomes can be purchased to offset the debit from environmental damage and, thus, "as an extension of biodiversity offsets; turning offsets into assets that can be traded, creating a market system for developers' compensation liabilities"[26] (Eftec and IEEP 2010). In habitat banking, and this is again important, credits can be produced in advance of, and without ex-ante links to, the debits they compensate for, and stored over time (Eftec and IEEP 2010).[27] Not surprisingly, North America where biodiversity offsetting is more developed, particularly the US wetland and species compensation programs, as well as Canada, particularly its fish habitat compensation program, host the largest number of offset credit banks worldwide (Madsen et al. 2010). The interrelationship between offsetting and banking has been also obvious in the Resolution of the European Parliament on the development of a regulatory framework based on the NNL initiative[28] which emphasized the importance of developing innovative financial mechanisms to reach biodiversity targets and, particularly, habitat banking in conjunction with offsetting. Biodiversity offsetting's conjunction with banking makes even more obvious the neoliberal character of the policy. As the Department of Environment and Conservation in New South Wales explains by referring to its BioBanking scheme, the government

[26]The term habitat banking can refer to both species and habitats – therefore it can be used as analogous to conservation banking and biodiversity banking. In the USA mitigation banking refers to wetland restoration and conservation banking is species-specific but, in most jurisdictions, bio/conservation/habitat/mitigation banking are synonyms (Eftec and IEEP 2010). Moreover, in the context of the NNL working group of the European Union conservation credits may include habitats, species and/or ecosystem services (ibid.).

[27]See also http://ec.europa.eu/environment/nature/biodiversity/nnl/pdf/NNL_Glossary.pdf.

[28]Action 7 under target 2 of the EU Biodiversity Strategy to 2020 seeks to "ensure no net loss of biodiversity and ecosystem services". Additionally, the European Commission established a Working Group on No Net Loss of Ecosystems and their Services (NNL WG) in the context of the Common Implementation Framework of the Biodiversity Strategy.

"intends to be involved in establishing BioBank sites only to facilitate the establishment of the scheme and a viable new industry of conservation offset providers" but "ultimately the market will set the price of credits" and determine the function of the scheme (DEC 2006, p. 11).

Another important term in the offsetting literature is the mitigation hierarchy. Biodiversity offsetting is the last resort of the mitigation hierarchy after all relevant measures have been taken to avoid and minimize development impacts and then to restore biodiversity on-site. The position of offsetting as the last stage in this hierarchy has played a crucial role in its popularity among developers and industries because it means that the offsetting option can facilitate the approval of development projects that would otherwise have been refused on the grounds that they do not sufficiently address their residual impacts. This has been evident in my work in England (see Chapters 5 and 6) where offsetting has been widely used to facilitate the approval of controversial planning applications after their initial rejection. Non-surprisingly, protected natures have not been immune from the latter: mitigation and compensatory measures have become exceedingly popular tools for facilitating development in the vicinity of protected sites both in the UK (Apostolopoulou and Adams 2019) and across Europe (Schoukens and Cliquet 2016).

It is useful to point out here that offsetting has some important differences in comparison to previous approaches for calculating ecological compensation. Firstly, the calculation of losses and gains in the development and the offset site, respectively, is based on numerical scores that are the outcome of the application of specific metrics. Secondly, as the NNL goal implies, in offsetting it is necessary to achieve an ecological equivalence between the development and the offset site and to offer—at least in theory—evidence of measurable and comparable on-the-ground conservation outcomes. Thirdly, the numerical representation of gains and losses sets the base for the creation of offset credits enabling the introduction of conservation banking.

2.7 Current Distribution of Biodiversity Offsetting and Compensation Mechanisms Across the Globe

In 2010, Madsen et al.[29] reviewed the status and trends of biodiversity offset and compensatory mitigation programs by geographical region and they found 39 active offset programs worldwide (within each active offset program there are numerous individual offset sites, including over 600 mitigation banks) and another 25 programs in various stages of development or investigation. The authors estimated that the global annual market size was at that time $1.8–2.9 billion at minimum, and that the conservation impact of this market included at least 86,000 hectares of land under some sort of conservation management or permanent legal protection per year (p. iv). According to Madsen et al. (2010) offsetting was more developed in North America and Australasia and was rapidly expanding in Africa, Asia and Europe. One year later, in 2011, Madsen et al. estimated that the global annual market size of biodiversity offsetting and mitigation programs had been raised to $2.4–4.0 billion at minimum, including at least 187,000 hectares of land under conservation management or legal protection per year (p. v).

In a more recent report, Bennett et al. (2017b) identified 99 active offset and compensation policies in 33 policies around the globe as of 2016. These programs claim that they have restored, recreated, and protected habitats on more than 8.3M hectares. Nearly two-thirds of the costs of conservation activities in the context of offsets have been paid by the private sector, primarily the energy, transportation, and mining/minerals sectors that were responsible for more than 97% of offsets and compensation mechanisms measured by cumulative land area under management.

[29]It should be noted that consultants have so far played a leading role in the production of biodiversity offsetting reports and independent research on offsetting, especially from a critical social sciences point of view, remains very limited. It is indicative that the most widely cited reports on the state of offsetting schemes and conservation banks across the globe have been prepared by consultants and multinational organizations that bring together actors from various industries, non-governmental organizations, companies, governments, and financial institutions.

As Bennett et al. (2017b) argue, offsetting and compensation mechanisms channeled $4.8 billion in 2016, representing a doubling of transaction value in only five years with the largest markets found in the US, Australia, Germany, and Canada.

Bennett et al. (2017a) also conducted a research focused at the EU level, in which they identified 65 programs and 180, implemented or in-development, projects that used biodiversity offsets and compensation mechanisms in pursuit of No Net Loss or Net Gain of biodiversity as of 2015. The conservation impact of the identified projects and programs included 73,914 hectares as of the end of 2015 with another 46,903 hectares in the pipeline.[30] The countries where programs and projects were identified included Austria, Belgium, Denmark, Finland, France, Germany, Hungary, Iceland, Italy, The Netherlands, Norway, Spain, Sweden, Switzerland, and the UK. The report distinguished between EU-level compensation frameworks[31] transposed into national laws and national or subnational program frameworks. Importantly, in 93% of the identified projects the primary driver for their implementation was compliance to regulatory obligations.

According to the report, the largest number of implemented projects was found in Italy and Germany. Sweden, France, the UK, and Belgium also had several projects that have been reported as implemented or underway. The 92.6% of hectares that have been reported as restored, enhanced, or conserved were on government-owned and managed lands. Privately owned compensatory mitigation sites and commercial mitigation banks made up almost equal shares of the remainders (Bennett et al. 2017a). The report also provided interesting information regarding offset buyers. For the 36 projects that such information has been collected, buyers most often represented public and private parties involved in major infrastructure projects, such as highway construction, ports

[30] Bennett et al. (2017a) point out that this figure underestimates the actual impact of compensatory mitigation in Europe, since less than 60% of the identified projects reported data about the land area covered by their activities.

[31] There are three major EU-level regulatory frameworks that include compensatory mitigation elements: the Birds and Habitats Directives, the Environmental Liability Directive, and Environmental Impact Assessment frameworks.

and airports expansion, and energy distribution systems. Energy generation/distribution and transportation/shipping projects tended to be very large, averaging 15,014 ha and 7789 ha, respectively (Bennett et al. 2017a). The greatest number of buyers was in Germany, followed by France, the UK, and Switzerland.

In 2017, IUCN (International Union for Conservation of Nature) and The Biodiversity Consultancy (TBC) launched a Global Inventory of Biodiversity Offset Policies (GIBOP)[32] at the Annual General Meeting of the Intergovernmental Forum on Mining, Minerals, Metals and Sustainable Development (IGF) in Geneva. This offers a free online database with information on the status, scope, and implementation of biodiversity offset policies worldwide and contains national environmental laws and legislation on offsets provision from 197 countries. Key findings of the GIBOP include that over 100 countries have or are developing biodiversity compensation or/and offset policies, 37 countries have legal requirements for the compensation and/or offsetting as part of project permitting, and that compensation/offset schemes are primarily embedded in Environmental Impact Assessments[33] (Fig. 2.1). The GIBOP along with the website of Forest Trends which maps environmental markets,[34] market-based mechanisms, and conservation and restoration finance, offer plenty information on the evolution and adoption of environmental markets across the globe. According to the latest available data, as of 2018, there were more than 12,983 offset and compensation projects worldwide extending over 153,679 km^2 (see also Bull and Strange 2018).

At least two key observations can be made from the data provided in Fig. 2.2.

Firstly, we see that the great majority of compensation/offset schemes are driven by compliance to regulatory obligations, are located on government-owned and managed land, and are paid by the private sector, primarily the energy, transportation, and mining/minerals sectors. These

[32]Available at: https://portals.iucn.org/offsetpolicy/policy-reviews.

[33]For more information see: https://portals.iucn.org/offsetpolicy/.

[34]Available at: https://www.forest-trends.org/about-our-project-data.

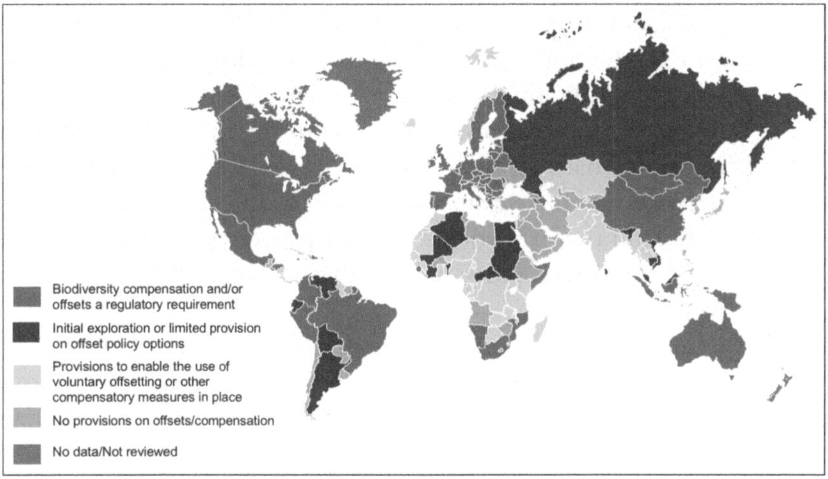

Fig. 2.1 The evolution of offsetting and compensation policies across the globe (*Source* Design by the author based on data available at: https://portals.iucn.org/offsetpolicy/)

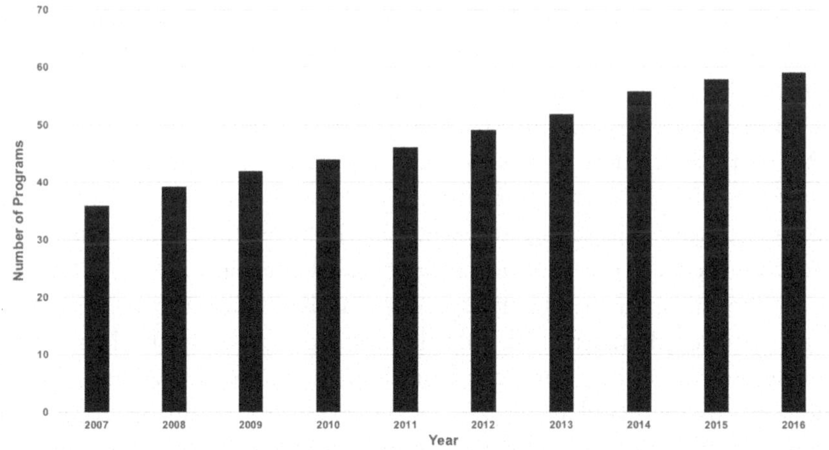

Fig. 2.2 Compliance offsets and compensation: Number of active programs by mitigation type, 2007–2016 (*Source* Design by the author based on data available at Bennett et al. [2017b])

are important data for understanding the role of offsets for current patterns of economic development and growth, including urbanization patterns, and, consequently, their role in processes of rent extraction contra to their popular conceptualization as mechanisms that bring about the commodification of nature, as we will see in more detail in Chapter 3. These characteristics of offsets are also crucial for understanding the double grabbing of land that is involved in these processes, by destroying the development site and restricting public access to the offset site for conservation purposes.

Secondly, the data on the global state of offsetting and compensation mechanisms corroborate a rise in both offsetting projects and national legislation requiring offsets since the financial crash of 2008 (Fig. 2.2; see also The Biodiversity Consultancy 2013). This means that the increasing adoption of compensation/offsetting policies has coincided with the intensification of the neoliberalization of nature in the post-2008 era (Apostolopoulou and Adams 2015), reproducing the environmental backlash that followed the 1973 economic crisis (Bonneuil 2015). However, it is also important to be cautious when interpreting what exactly these developments mean in practice. Bennett et al. (2017a) note in their review of global biodiversity markets that only 36 out of 99 offsets and compensation programs (namely the 36%) posted specific activity in 2016. Even though, therefore, new policies and regulations emerge every year (Fig. 2.2), it is not entirely clear whether these are resulting in compensatory mitigation and whether the NNL/Net Gain objectives are even tested on the ground. Indeed, available data are very limited due to the minimal public transparency under which most offset programs operate (Bennett et al. 2017b) and peer-reviewed multinational assessments concerning the actual implementation of offset policies are generally missing (Bull et al. 2017). This is quite alarming because it shows that offsetting by offering to major industries both planning permission and their social license to operate (Maron et al. 2016) can enable economic development and permanently change land uses in the absence of sufficient data (Mandle et al. 2015; May et al. 2017; Gibbons et al. 2018) to evaluate its ecological and social impacts on the ground.

References

Adams, W. M. (2009). *Green development: Environment and sustainability in a developing world.* London: Routledge.

Adams, W. M. (2017). Sleeping with the enemy? Biodiversity conservation, corporations and the green economy. *Journal of Political Ecology, 24,* 243–257.

Apostolopoulou, E. (2016). Biodiversity offsetting in England: Governance rescaling, socio-spatial injustices, and the neoliberalization of nature. *Web Ecology, 16,* 67–71.

Apostolopoulou, E., & Adams, W. M. (2015). Neoliberal capitalism and conservation in the post-crisis era: The dialectics of 'green' and 'un-green' grabbing in Greece and the UK. *Antipode, 47,* 15–35.

Apostolopoulou, E., & Adams, W. M. (2019). Cutting nature to fit: Urbanization, neoliberalism and biodiversity offsetting in England. *Geoforum, 98,* 214–225.

Apostolopoulou, E., Adams, W. M., & Greco, E. (2018). Biodiversity offsetting and the production of "equivalent natures": A Marxist critique. *ACME: An International Journal for Critical Geographies, 17*(3), 861–892.

Apostolopoulou, E., Bormpoudakis, D., Paloniemi, R., Cent, J., Grodzińska-Jurczak, M., Pietrzyk-Kaszyńska, A., et al. (2014). Governance rescaling and the neoliberalization of nature: The case of biodiversity conservation in four EU countries. *International Journal of Sustainable Development and World Ecology, 21,* 481–494.

Apostolopoulou, E., & Cortes-Vazquez, J. (Eds.). (2018). *The right to nature: An academic-activist dialogue on environmental justice, social movements and neoliberal natures.* Abingdon: Routledge-Earthscan.

Apostolopoulou, E., Drakou, E. G., & Pediaditi, K. (2012a). Participation in the management of Greek Natura 2000 sites: Evidence from a cross-level analysis. *Journal of environmental management, 113,* 308–318.

Apostolopoulou, E., Drakou, E., Santoro, F., & Pantis, J. D. (2012b). Investigating the barriers to adopting a "human-in-nature" view in Greek biodiversity conservation. *International Journal of Sustainable Development and World Ecology, 19,* 515–525.

Apostolopoulou, E., & Paloniemi, R. (2012). Frames of scale challenges in Finnish and Greek biodiversity conservation. *Ecology and Society, 17*(4), 9.

Apostolopoulou, E., & Pantis, J. D. (2010). Development plans versus conservation: Explanation of emergent conflicts and governmental political handling. *Environment and Planning A, 42,* 982–1000.

Arsel, M., & Büscher, B. (2012). Nature™ Inc.: Changes and continuities in neoliberal conservation and market-based environmental policy. *Development and Change, 43,* 53–78.

Badura, T., Ferrini, S., Agarwala, M., et al. (2017). *Valuation for natural capital and ecosystem accounting.* Synthesis Report for the European Commission. Norwich: Centre for Social and Economic Research on the Global Environment, University of East Anglia. Available at: http://ec.europa.eu/environment/nature/capital_accounting/pdf/Valuation_for_natural_capital_and_ecosystem_acounting.pdf.

Bakker, K. (2009). Neoliberal nature, ecological fixes, and the pitfalls of comparative research. *Environment and Planning A, 41,* 1781–1787.

Bakker, K. (2010). The limits of 'neoliberal natures': Debating green neoliberalism. *Progress in Human Geography, 34,* 715–735.

Bayon, R., Fox, J., & Carroll, N. (2008). *Conservation and biodiversity banking: A guide to setting up and running biodiversity credit trading systems.* London: Earthscan.

BBOP (Business and Biodiversity Offsets Programme). (2012a). *Standard on biodiversity offsets.* Washington, DC: BBOP. Available at: http://bbop.forest-trends.org/guidelines/Standard.pdf.

BBOP (Business and Biodiversity Offsets Programme). (2012b). *Resource paper: No net loss and loss-gain calculations in biodiversity offsets.* Washington, DC: BBOP.

Bennett, G., Chavarria, A., Ruef, F., & Leonardi, A. (2017a). *State of European markets 2017: Biodiversity offsets and compensation.* Ecosystem Marketplace, Forest Trends.

Bennett, G., Gallant, M., & ten Kate, K. (2017b). *State of biodiversity mitigation 2017.* Markets and Compensation for Global Infrastructure Development. Ecosystem Marketplace, Forest Trends, Washington DC.

Birch, K., Levidow, L., & Papaioannou, T. (2010). Sustainable capital? The neoliberalization of nature and knowledge in the European "knowledge-based bio-economy". *Sustainability, 2,* 2898–2918.

Bonneuil, C. (2015). Tell me where you come from, I will tell you who you are: A genealogy of biodiversity offsetting mechanisms in historical context. *Biological Conservation, 192,* 485–491.

Borras, S. M., Jr., Franco, J. C., Gómez, S., Kay, C., & Spoor, M. (2012). Land grabbing in Latin America and the Caribbean. *The Journal of Peasant Studies, 39,* 845–872.

Brenner, N., Peck, J., & Theodore, N. (2010). Variegated neoliberalization: Geographies, modalities, pathways. *Global Networks, 10,* 182–222.

Brenner, N., & Theodore, N. (2002). Cities and the geographies of "actually existing neoliberalism". *Antipode, 34,* 349–379.

Brockington, D., & Duffy, R. (2010). Capitalism and conservation: The production and reproduction of biodiversity conservation. *Antipode, 42,* 469–484.

Bull, J. W., Lloyd, S. P., & Strange, N. (2017). Implementation gap between the theory and practice of biodiversity offset multipliers. *Conservation Letters, 10,* 656–669.

Bull, J. W., & Strange, N. (2018). The global extent of biodiversity offset implementation under no net loss policies. *Nature Sustainability, 1,* 790.

Burgin, S. (2008). BioBanking: An environmental scientist's view of the role of biodiversity banking offsets in conservation. *Biodiversity and Conservation, 17,* 807–816.

Burkett, P. (1997). Nature in marx reconsidered a silver anniversary assessment of alfred schmidt's concept of nature in Marx. *Organization & Environment, 10,* 164–183.

Büscher, B., & Arsel, M. (2012). Neoliberal conservation, uneven geographical development and the dynamics of contemporary capitalism. *Tijdschrift voor Economische en Sociale Geografie, 103,* 129–135.

Büscher, B., & Dressler, W. (2012). Commodity conservation: The restructuring of community conservation in South Africa and the Philippines. *Geoforum, 43*(3), 367–376.

Büscher, B., Dressler, W., & Fletcher, R. (Eds.). (2014). *Nature Inc.: Environmental conservation in the neoliberal age.* Tucson: University of Arizona Press.

Büscher, B., Sullivan, S., Neves, K., Igoe, J., & Brockington, D. (2012). Towards a synthesized critique of neoliberal biodiversity conservation. *Capitalism Nature Socialism, 23,* 4–29.

Cahill, D. (2011). Beyond neoliberalism? Crisis and the prospects for progressive alternatives. *New Political Science, 33,* 479–492.

Calvário, R., Velegrakis, G., & Kaika, M. (2017). The political ecology of austerity: An analysis of socio-environmental conflict under crisis in Greece. *Capitalism Nature Socialism, 28*(3), 69–87.

Castree, N. (2008a). Neoliberalising nature: The logics of deregulation and reregulation. *Environment and Planning A, 40,* 131–152.

Castree, N. (2008b). Neoliberalising nature: Processes, effects, and evaluations. *Environment and Planning A, 40,* 153–173.

Castree, N. (2009). Researching neoliberal environmental governance: A reply to Karen Bakker. *Environment and Planning A, 41,* 1788–1794.

Castree, N. (2010). Neoliberalism and the biophysical environment: A synthesis and evaluation of the research. *Environment and Society: Advances in Research, 1,* 5–45.

Corson, C. (2010). Shifting environmental governance in a neoliberal world: US aid for conservation. *Antipode, 42,* 576–602.

Cortes-Vazquez, J. A., & Apostolopoulou, E. (2019). Against neoliberal natures: Environmental movements, radical practice and "the right to nature". *Geoforum, 98,* 202–205.

Czarnezki, J. J., & Fiedler, K. (2016). *The neoliberal turn in environmental regulation* (Utah Law Review No. 1, Article 1). Available at: http://dc.law.utah.edu/ulr/vol2016/iss1/1.

Davies, J. (2013). Rethinking urban power and the local state: Hegemony, domination and resistance in neoliberal cities. *Urban Studies, 51,* 3215–3232.

DEC. (2006). *BioBanking: A biodiversity offsets and banking scheme: Conserving and restoring biodiversity in NSW* (Working Paper). Sydney: Department of Environment and Conservation NSW.

Dressler, W., & Roth, R. (2011). The good, the bad, and the contradictory: Neoliberal conservation governance in rural Southeast Asia. *World Development, 39,* 851–862.

Duménil, G., & Lévy, D. (2001). Costs and benefits of neoliberalism: A class analysis. *Review of International Political Economy, 8,* 578–607.

Duménil, G., & Lévy, D. (2004). *Capital resurgent: Roots of the neoliberal revolution.* Cambridge, MA and London, UK: Harvard University Press.

Duménil, G., & Lévy, D. (2013). *The crisis of neoliberalism.* Cambridge, MA and London, UK: Harvard University Press.

EC (European Commission). (2011). *Our life insurance, our natural capital: An EU biodiversity strategy to 2020.* Communication from the Commission to the European Parliament, the Council, the Economic and Social Committee and the Committee of the Regions, COM (2011) 244 Final, Brussels.

EFTEC and IEEP. (2010). *The use of market-based instruments for biodiversity protection—The case of Habitat Banking.* Technical Report for European Commission DG Environment, EFTEC and IEEP, London.

Fairhead, J., Leach, M., & Scoones, I. (2012). Green Grabbing: A new appropriation of nature? *The Journal of Peasant Studies, 39,* 237–261.

Fletcher, R. (2012). Using the master's tools? Neoliberal conservation and the evasion of inequality. *Development and Change, 43,* 295–317.

Gibbons, P., Macintosh, A., Constable, A. L., & Hayashi, K. (2018). Outcomes from 10 years of biodiversity offsetting. *Global Change Biology, 24,* 643–654.

Gonçalves, B., Marques, A., Soares, A. M. V. D. M., & Pereira, H. M. (2015). Biodiversity offsets: From current challenges to harmonized metrics. *Current Opinion in Environmental Sustainability, 14,* 61–67.

Gramsci, A. (1971). *Selections from the prison notebooks* (Q. Hoare & G. Nowell-Smith, Eds. & Trans.). New York: International Publishers.

Gray, M., & Barford, A. (2018). The depths of the cuts: The uneven geography of local government austerity. *Cambridge Journal of Regions, Economy and Society, 11,* 541–563.

Hadjimichalis, C. (2011). Uneven geographical development and socio-spatial justice and solidarity: European regions after the 2009 financial crisis. *European Urban and Regional Studies, 18,* 254–274.

Harvey, D. (1989). From managerialism to entrepreneurialism: The transformation in urban governance in late capitalism. *Geografiska Annaler: Series B, 71,* 3–17.

Harvey, D. (2005). *A brief history of neoliberalism.* Oxford: Oxford University Press.

Harvey, D. (2007). Neoliberalism as creative destruction. *The Annals of the American Academy of Political and Social Science, 610,* 21–44.

Harvey, D. (2010). *The enigma of capital: And the crises of capitalism.* London: Profile Books.

Harvey, D. (2011). Roepke lecture in economic geography—Crises, geographic disruptions and the uneven development of political responses. *Economic Geography, 87,* 1–22.

Harvey, D. (2012). *Rebel cities: From the right to the city to the urban revolution.* London and New York: Verso.

Harvey, D. (2014). *Seventeen contradictions and the end of capitalism.* London, UK: Profile Books.

Harvey, D. (2016). Neoliberalism is a political project. *Jacobin.* Available at: https://www.jacobinmag.com/2016/07/david-harvey-neoliberalism-capitalism-labor-crisis-resistance/.

Heynen, N., McCarthy, J., Prudham, S., & Robbins, R. (Eds.). (2007). *Neoliberal environments: False promises and unnatural consequences.* London and New York: Routledge.

Heynen, N., & Robbins, P. (2005). The neoliberalization of nature: Governance, privatization, enclosure and valuation. *Capitalism Nature Socialism, 16,* 5–8.

Hodge, I. D., & Adams, W. M. (2012). Neoliberalisation, rural land trusts and institutional blending. *Geoforum, 43,* 472–482.

Hough, P., & Robertson, M. (2009). Mitigation under Section 404 of the Clean Water Act: Where it comes from, what it means. *Wetlands Ecology and Management, 17,* 15–33.

Hutton, J., Adams, W. M., & Murombedzi, J. C. (2005). Back to the barriers? Changing narratives in biodiversity conservation. *Forum for development studies, 32,* 341–370.

ICMM (International Council on Mining and Metals). (2005). *Biodiversity offsets—A briefing paper for the mining industry.* London, UK. Available at: https://www.ecosystemmarketplace.com/wp-content/uploads/archive/documents/Doc_417.pdf.

IEEP. (2014). *Study on specific design elements of biodiversity offsets: Biodiversity metrics and mechanisms for securing long term conservation benefits.* DG ENVIRONMENT: ENV.B.2/ETU/2013/0060r. Final Report. A Report Submitted by ICF Consulting Services in association with IEEP and Associated Experts. London, UK.

Igoe, J., & Brockington, D. (2007). Neoliberal conservation: A brief introduction. *Conservation and society, 5,* 432–449.

Igoe, J., Neves, K., & Brockington, D. (2010). A spectacular eco-tour around the historic bloc: Theorising the convergence of biodiversity conservation and capitalist expansion. *Antipode, 42,* 486–512.

Jonas, A. E. G., Gibbs, D., & While, A. (2011). The new urban politics as a politics of carbon control. *Urban Studies, 48,* 2537–2554.

Kruczynski, W. L. (1990). Mitigation and the sector 404 program: A perspective. In J. A. Kusler & M. E. Kentula (Eds.), *Wetland creation and restoration: The status of the science* (pp. 549–554). Washington, DC: Island Press.

Lapavitsas, C., Kaltenbrunner, A., Labrinidis, G., Lindo, D., Meadway, J., Michell, J., et al. (2010, September). *The Eurozone between austerity and default.* Research on Money and Finance (RMF) Occasional Report. Available at: http://www.researchonmoneyandfinance.org/.

Lapavitsas, C., Kaltenbrunner, A., Lindo, D., Meadway, J., Michell, J., Painceira, J. P., et al. (2011, November). *Breaking up? A route out of the Eurozone crisis.* Research on Money and Finance (RMF) Occasional Report 3. Available at: http://www.researchonmoneyandfinance.org/.

Layzer, J. A. (2012). *Open for business: Conservatives' opposition to environmental regulation.* Cambridge: MIT Press.

Lynch-Stewart, P., Niece, P., Rubec, C., & Kessel-Taylor, I. (1996). *The federal policy on wetland conservation: Implementation guide for federal land managers.* Ottawa: Environment Canada.

MacDonald, K. I., & Corson, C. (2012). 'TEEB begins now': A virtual moment in the production of natural capital. *Development and Change, 43,* 159–184.

Madsen, B., Carroll, N., & Moore Brands, K. (2010). *State of biodiversity markets report: Offset and compensation programs worldwide.* Washington, DC: Forest Trends.

Maestre-Andrés, S., Calvet-Mir, L., & Apostolopoulou, E. (2018). Unravelling stakeholder participation under conditions of neoliberal biodiversity governance in Catalonia, Spain. *Environment and Planning C: Politics and Space.* https://doi.org/10.1177/2399654417753624.

Mandle, L., Tallis, H., Sotomayor, L., et al. (2015). Who loses? Tracking ecosystem service redistribution from road development and mitigation in the Peruvian Amazon. *Frontiers in Ecology and the Environment, 13,* 309–315.

Maron, M., Hobb, R. J., Moilanen, A., Matthews, J. W., Christie, K., Gardner, T. A., et al. (2012). Faustian bargains? Restoration realities in the context of biodiversity offset policies. *Biological Conservation, 155,* 141–148.

Maron, M., Ives, C. D., Kujala, H., Bull, J. W., et al. (2016). Taming a wicked problem: Resolving controversies in biodiversity offsetting. *BioScience, 66,* 489–498.

Marx, K. (1859). *A contribution to the critique of political economy.* Moscow: Progress Publishers.

Marx, K. (1910). *Capital: A critique of political economy* (Vol. II). Chicago: Charles H. Kerr & Co. and London: Swan Sonnenschein & Co.

May, J., Hobbs, R. J., & Valentine, L. E. (2017). Are offsets effective? An evaluation of recent environmental offsets in Western Australia. *Biological Conservation, 206,* 249–257.

McAfee, K. (1999). Selling nature to save it? Biodiversity and green developmentalism. *Environment and Planning D: Society and Space, 17,* 133–154.

McCarthy, J. (2006). Neoliberalism and the politics of alternatives: Community forestry in British Columbia and the United States. *Annals of the Association of American Geographers, 96,* 84–104.

McCarthy, J. (2012). The financial crisis and environmental governance "after" neoliberalism. *Tijdschrift voor economische en sociale geografie, 103,* 180–195.

McCarthy, J., & Prudham, S. (2004). Neoliberal nature and the nature of neoliberalism. *Geoforum, 35,* 275–283.

Mead, D. L. (2008). History and theory: The origin and evolution of conservation banking. In R. Bayon, J. Fox, & N. Carroll (Eds.), *Conservation and biodiversity banking: A guide to setting up and running biodiversity credit trading systems* (pp. 9–31). London, UK and Sterling, VA: Earthscan.

Neves, K., & Igoe, J. (2012). Uneven development and accumulation by dispossession in nature conservation: Comparing recent trends in the Azores and Tanzania. *Tijdschrift voor Economische en Sociale Geografie, 103,* 164–179.

Ojeda, D. (2012). Green pretexts: Ecotourism, neoliberal conservation and land grabbing in Tayrona National Natural Park, Colombia. *Journal of Peasant Studies, 39,* 357–375.

Paloniemi, R., Apostolopoulou, E., Cent, J., Bormpoudakis, D., Scott, A., Grodzińska-Jurczak, M., et al. (2015). Public participation and environmental justice in biodiversity governance in Finland, Greece, Poland and the UK. *Environmental Policy and Governance, 25,* 330–342.

Peck, J., Theodore, N., & Brenner, N. (2012). Neoliberalism resurgent? Market rule after the great recession. *The South Atlantic Quarterly, 111,* 2.

Peck, J., & Tickell, A. (2007). Conceptualizing neoliberalism, thinking Thatcherism. *Contesting Neoliberalism: Urban Frontiers, 26,* 50.

Peet, R. (2013). Comparative policy analysis: Introduction. *Human Geography, 6,* 1–10.

Poulton, D. W. (2014). *Biodiversity offsets, a primer for Canada.* Sustainable Prosperity and Institute of the Environment, Ottawa.

Quétier, F., & Lavorel, S. (2011). Assessing ecological equivalence in biodiversity offset schemes: Key issues and solutions. *Biological Conservation, 144,* 2991–2999.

Re:Common. (2019). *Turning forests into hotels: The true cost of biodiversity offsetting in Uganda.* Available at: https://www.recommon.org/eng/turning-forests-into-hotels-the-true-cost-of-biodiversity-offsetting-in-uganda/.

Robertson, M. (2000). No net loss: Wetland restoration and the incomplete capitalization of nature. *Antipode, 32,* 463–493.

Robertson, M. (2004). The neoliberalization of ecosystem services: Wetland mitigation banking and problems in environmental governance. *Geoforum, 35,* 361–373.

Rubec, C. D., & Hanson, A. R. (2009). Wetland mitigation and compensation: Canadian experience. *Wetlands Ecology and Management, 17,* 3–14.

Sayer, D. (1987). *The violence of abstraction: The analytic foundations of historical materialism.* Basil: Blackwell.

Schoukens, H., & Cliquet, A. (2016). Biodiversity offsetting and restoration under the European Union Habitats Directive: Balancing between no net loss and deathbed conservation? *Ecology and Society, 21*(4), 10.

Smith, N. (2010). *Uneven development* (3rd ed.). New York: Verso.

Soja, E. (2010). *Seeking spatial justice.* Minneapolis: University of Minnesota Press.

Stokstad, E. (2008). New rules on saving wetlands push the limits of the science. *Science, 320*(5873), 162–163.

Sullivan, S. (2006). Elephant in the room? Problematising 'new'(neoliberal) biodiversity conservation. *Forum for Development Studies, 33,* 105–135.

Swyngedouw, E. (2000). Authoritarian governance, power, and the politics of rescaling. *Environment and Planning D: Society and Space, 18,* 63–76.

Swyngedouw, E. (2004). Globalisation or 'Glocalisation'? Networks, territories and rescaling. *Cambridge Review of International Affairs, 17,* 25–48.

Swyngedouw, E. (2005). Governance innovation and the citizen: The Janus face of governance-beyond-the-state. *Urban Studies, 42,* 1991–2006.

TEEB. (2010). *The economics of ecosystems and biodiversity: Mainstreaming the economics of nature: A synthesis of the approach, conclusions and recommendations of TEEB.* Geneva: TEEB.

Temper, L., Demaria, F., Scheidel, A., et al. (2018). The Global Environmental Justice Atlas (EJAtlas): Ecological distribution conflicts as forces for sustainability. *Sustainability Science, 13,* 573–584.

ten Kate, K., Bishop, J., & Bayon, R. (2004). *Biodiversity offsets: Views, experience, and the business case.* Gland, Switzerland and London, UK: IUCN and Insight Investment.

The Biodiversity Consultancy. (2013). *Government policies on biodiversity offsets.* Cambridge: The Biodiversity Consultancy Ltd.

Thomas, P. (2006). Modernity as 'passive revolution': Gramsci and the fundamental concepts of historical materialism. *Journal of the Canadian Historical Association, 17,* 61–78.

Thomas, P. (2009). *The Gramscian moment: Philosophy, hegemony and Marxism.* Leiden: Brill.

Tucker, G. M., Allen, B., Conway, M., Dickie, I., Hart, K., Rayment, M., et al. (2014). *Policy options for an EU no net loss initiative.* Report to the European Commission (with Annexes), Institute for European Environmental Policy (IEEP), London.

UNEP/CBD/COP. (2012). *Report of the eleventh meeting of the conference of the parties to the Convention on Biological Diversity*. Available at: http://www.cbd.int/doc/?meeting=cop-11.

Vlachou, A. (2005). Debating sustainable development. *Rethinking Marxism, 17,* 627.

Wheeler, D. P., & Strock, J. M. (1995). *Official policy on conservation banks.* Issued Jointly by the Resources Agency and California Environmental Protection Agency.

3
Equivalent Natures and Non-places

E. Apostolopoulou, *Nature Swapped and Nature Lost*,
https://doi.org/10.1007/978-3-030-46788-3_3

What speaks to the soul, escapes our measurements.

—Alexander von Humboldt (1808)

3.1 Biodiversity Offsetting and the Construction of Equivalence[1]: Insights from the Ecological Literature

As explained in the previous chapter, achieving equivalence between ecological losses in the development site and ecological gains in the offset site is the cornerstone of the biodiversity offset design process. Even though offsetting programmes differ in terms of their equivalence requirements, ecological equivalence is a fundamental aspect that lies at the core of the definition of an offset and is widely acknowledged as necessary for achieving No Net Loss of biodiversity (BBOP 2012a; Maron et al. 2012).

The establishment of ecological equivalence requires the accurate assessment of losses and gains using common currencies, namely, offset metrics, and units of measurement. The aim of the process is to quantify selected aspects of nature lost allowing its representation through numerical scores, for example a number of biodiversity units in the UK (Defra 2013) or a number of biodiversity credits in Australia BioBanking (DEC 2006). Policymakers and governments interested in implementing biodiversity offsetting insist that the calculation of ecological gains and losses must be as straightforward as possible. However, in practice, selecting biodiversity currencies and offset metrics to render ecological losses and gains equivalent has proven far from simple (Apostolopoulou and Adams 2017a; Bezombes et al. 2017; Bull et al. 2013; Gardner et al. 2013; Gonçalves et al. 2015; Gamarra et al. 2018).

The literature on offset metrics reveals a variety of problematic assumptions and weaknesses starting from the very beginning: what exactly should be measured (e.g., biodiversity and/or ecosystem functions and/or ecosystem services) when calculating ecological gains and losses,

[1]In the context of biodiversity offsetting, the term is synonymous with the concept of "like for like" and refers to areas with highly comparable biodiversity components (Tucker et al. 2014). This similarity can be observed in terms of species diversity, functional diversity and composition, ecological integrity or condition, landscape context (e.g., connectivity, landscape position, adjacent land uses or condition, patch size, etc.), or ecosystem services (including people's use and cultural values) (Conway et al. 2013).

is far from settled. As mentioned in Chapter 2, BBOP's (2012a) definition of NNL refers to species composition, habitat structure, ecosystem function and even people's use and cultural values associated with biodiversity.[2] Given such an ambitious goal the categories of ecological information that can be included are numerous, as are the range of metrics proposed, and disputes over the superiority of different metrics can be endless (see also Robertson 2006). In Germany, for example, there has been a great variety of different offset metrics since the introduction of the policy in 1976 (IEEP 2014) whereas in France, only basic guidance has been available until the time of writing this book on assessing ecological equivalence, and offset requirements have been determined on a case-by-case basis leading to major variations in the methods used (Quétier et al. 2014; Tucker et al. 2014).

Moreover, and relatedly, biodiversity data is rarely comprehensive creating important and complex questions about which biodiversity characteristics should be counted (Walker et al. 2009). To add to this, there is no single metric that could objectively capture the full extent of biodiversity whereas biodiversity itself has no universal, unambiguous definition (Bull et al. 2013). But even if biodiversity could be measured, it is questionable whether it can be used to evaluate ecosystem processes. As Palmer and Filoso (2009) point out "the assumptions that simple proxies, like habitat descriptors, can be used to evaluate restoration success and that single ecological measures, like biodiversity, can be used to evaluate a full suite or 'bundle' of ecosystem processes are not only naïve but have been demonstrated to be false for many ecosystems." Further difficulties arise due to complexities in estimating changes in ecological condition and obtaining direct measures of components for which surrogates cannot be used, or lack of clarity on how to best aggregate biodiversity losses and gains across the landscape (Bull et al. 2013; Gardner et al. 2013).

[2]Even though BBOP's definition of NNL is widely accepted, so far, no metrics have been designed to account for all these aspects. BBOP also argues that other biodiversity surrogates or economic evaluation methods focused on people's cultural and use values can be used to establish "a package of benefits needed to motivate stakeholders to support the offset, compensating them for residual impacts on their livelihoods and amenity and engaging them in offset implementation (e.g. through sustainable livelihood activities from which they benefit)" but again there is no significant progress on this so far.

In practice, ecological relations are oversimplified using surrogates, such as habitat variables, that are typically combined into metrics. Thus, until recently, wetland mitigation offsets in the United States have been based on crude land swaps measured by area, i.e., a given area of a habitat type destroyed equals another area of the same habitat type being either preserved or restored (Hough and Robertson 2009; Wilkinson 2008). This approach ignores variations in the quality or status of the habitat or species, or the balance of habitat types impacted at the development site or conserved in the offset site (Quétier and Lavorel 2011). Simple habitat area ratio-based metrics have been widely used in Germany, although critiqued as highly reductionist and unable to capture ecosystem values reliably (Tucker et al. 2014). In France, the most commonly used offset metrics are based on a simple habitat area with multipliers approach. As Quétier and Lavorel (2011) explain for habitats that are difficult to restore, such as peatlands, area ratios are a very crude way for the determination of the levels of acceptable loss of existing habitat. Similarly, in the case of wetlands, the degree to which they can successfully be recreated or restored varies greatly (IEEP 2014; Jähnig et al. 2011; Moreno-Mateos et al. 2012).

In the case of the BBOP (2009a, p. 49), the proposed method is primarily based on an area × quality approach, using a benchmark approach to calculate habitat hectare,[3] supplemented where necessary

[3]The habitat hectares metric has been initially developed by Parkers et al. (2003) and updated by the Victorian Government Department of Sustainability and Environment. The method offers a way of calculating losses and gains in vegetation, based on units of measurement that consider the area affected and the quality or condition of the vegetation impacted. These are described in a "benchmark" that sets out at least 10 types of habitat attribute, such as: number of large trees, canopy cover, number of understory lifeforms, cover of weeds, recruitment, cover of organic litter, abundance of logs, patch size, proximity of remnant vegetation, and distance to core area. The attributes in the benchmark are weighted according to their significance to the overall condition of the system. Each attribute is measured at the impact site before the impact and the predicted score after the impact, comparing the measurements against the benchmark that represents the pristine condition of the habitat in question. The scores for each attribute are then added to provide an estimate of the site's condition expressed as a percentage pristine condition and the area of the habitat is multiplied by this percentage. The same approach is used to estimate the gains at the potential impact sites, comparing the actual measurements before the offset activities start with predicted realistic outcomes from the offset, again compared with the benchmark levels (see IEEP 2014 for further information).

with a calculation of population/occupancy for certain species. The habitat hectares method has been adapted for use in the pilot offsets in England, calculating "biodiversity units" in terms of the product of a distinctiveness score, a habitat condition score and habitat extent (Defra 2013). The method continues to be used in Victoria, Australia and has been widely adopted and adapted internationally, for example in Western Australia, South Africa and several projects associated with the BBOP (IEEP 2014). Importantly, the use of surrogates means that a combination of different attributes can yield the exact same score as another combination (Gibbons and Lindenmayer 2007). The need for operational expediency can prevent strict adherence to "like-for-like," as for example, in the habitat hectares method (Parkers et al. 2003). This could, for example, mean that a loss in large trees might be compensated by an increase in coarse woody debris (McCarthy et al. 2004) highlighting the possibility to lead to the loss of specific ecosystem attributes. The choice of surrogates, as well as estimations of their magnitude, is, thus, largely arbitrary and subjective whereas debates around "in kind" (for the same habitat type) or "out of kind" (where different habitat types are exchanged) trading are numerous. Crucially, when offsetting is underpinned by metrics that allow variables to be substituted will also inevitably lead to the replacement of biodiversity attributes that are difficult to restore with others that are easier to restore (Gibbons et al. 2018).

The above weaknesses are intensified when surrogates are combined with multipliers (Bull et al. 2017; Laitila et al. 2014). Multipliers are used in addition to offset metrics to increase the size of an offset site (for example, multiply by 3 or 5 the impact area), to consider time lags between ecological losses and the delivery of the offset, or to address the uncertainty regarding the exact ecological condition of the habitats and/or species that are lost due to development as well as other risks of failure in the offset process (BBOP 2012b). In the BBOP reports, multipliers are considered capable to address concerns that the offset may not succeed in delivering an NNL outcome, and as "easy to understand, implement and audit" (BBOP 2012b, p. 21). However, the same reports, also acknowledge that, in practice, multipliers "are difficult to calculate accurately and thus do not meet with broad agreement" and that they are "inappropriate for dealing with many types of risk" (ibid.).

Experience with the implementation of offsetting on the ground shows that such precautions are rarely being considered and that multipliers are very often used to cover a variety of issues that offset metrics are unable to incorporate. An indicative example from the UK is the case of the HS2 new high-speed rail link between London and Birmingham that seeks to achieve NNL at a route-wide level.[4] According to available data at the time of writing, the Phase 1 London and West Midlands route of HS2 directly affects 27 ancient woodlands (under continuous cover since c.1600AD). According to the offset metric used in England (see Chapter 5 on Defra metrics), the value of semi-natural ancient woodlands has been scored as "very high" in terms of habitat distinctiveness which meant that offset must be "like for like" (in the case of habitats of low distinctiveness the offset can involve a "trade up" to medium or high distinctiveness habitat). Due to the ecological importance of ancient woodlands, an amendment to the offset metric has been proposed (HS2 2013) suggesting that an increase in the extent of new woodland (provided as an offset) could compensate for their loss. This implied that creating more new trees could offset the loss of irreplaceable habitat. The proposal was met with major opposition and has been eventually abandoned with ancient woodlands currently excluded from offset calculations. Problems related to the use of multipliers become even more serious if we consider that despite the expanding literature on their design, there is no global assessment regarding their actual use in NNL policies or projects and data concerning how exactly multipliers are implemented in practice are scarce (Bull et al. 2017).

Another highly controversial aspect of offsetting is the issue of additionality, namely, the need to provide evidence that conservation gains resulting from offsetting would not have occurred without it (BBOP 2009a, p. 14). Proving additionality means that conservation gains must be assessed in comparison to a realistic counterfactual scenario (Commonwealth of Australia 2012; Defra 2013; Virah-Sawmy et al. 2014). To understand how claims for achieving additionality work it is useful to consider that offsets in most cases take two forms. Firstly, they can be protection offsets, also known as averted loss offsets. This type

[4] For more information on the HS2 case see Chapter 6.

of offsets involves protecting biodiversity from a particular activity, like grazing, overfishing, deforestation etc., and involves assumptions about future rates of habitat loss with and without the conservation action that offsetting requires. Secondly, they can be restoration offsets, which entail restoring, enhancing, or establishing biodiversity (ICMM and IUCN 2012). In restoration offsets, additionality and the NNL goal assume that ecosystems can be recreated or restored to contain equivalent biodiversity values to those lost due to economic development (Maron et al. 2012). Importantly, even ecologists who support the idea of offsetting, argue that restoration ecology is not sufficiently predictive to support the application of the concept and express concerns that offsetting claims potentially push the limits of both scientific knowledge and practical feasibility (e.g., Hobbs et al. 2011; Gibbons et al. 2018; Maron et al. 2012; Palmer and Filoso 2009; Stokstad 2008). The considerable skepticism on whether restoration ecology is, for most ecosystems, capable of delivering biodiversity gains that are sufficient to achieve NNL is partly based on the still relatively embryonic and patchy evidence base of the discipline (Maron et al. 2012) but also on the inherent complexity and variability of natural systems. Indeed, experience from the implementation of offsets reveals the thin empirical evidence in support of restoration promises (e.g., Bernhardt and Palmer 2011; Curran et al. 2013; Gibbons et al. 2018). Moreno-Mateos et al. (2012) in a global meta-analysis of 621 global wetland sites report that even after a century (sic) of restoration, biological structure, driven mostly by plant communities, and biochemical functioning, driven primarily by the storage of carbon in wetland soils, remained on average 26 and 23%, respectively, lower in restored or created wetlands than in reference wetlands. In a more recent paper, Gibbons et al. (2018) point out that habitats that take a long time to develop, like for example old-growth trees, or habitats for which complete restoration is highly uncertain, like unmodified ecosystems, "are likely to be lost or degraded if managed using biodiversity offsetting" (p. 651). Proving additionality has led to endless disputes not only around the validity of specific restoration claims but more broadly around the accuracy of counterfactual scenarios that describe the potential degradation of ecosystems without the implementation of offsetting.

Not surprisingly given all the above, in a report published by the IEEP in 2013 in the context of the EU's No Net Loss initiative, Tucker et al. (2014) after reviewing all available metrics at the time concluded that even the most sophisticated metrics still result in greatly simplified measurements of biodiversity. Similarly, BBOP's Biodiversity Offset Design Handbook (BBOP 2009a, p. 13) notes that it is impossible to either measure "biodiversity in its entirety" or "count every individual in every population of every species" and that "no two sites are identical in biodiversity terms." In a more recent review of offset metrics published in 2018, Gamarra et al. offer similar conclusions. In particular, the authors identify the following 11 offset metrics: habitat hectares (used by the Department of Natural Resources and Environment in the State of Victoria, Australia), units of global distribution (used by the Rio Tinto mining company in Madagascar), uniform mitigation assessment method (used by the Department of Environmental Protection, in Florida, USA), biodiversity significance index (used by the Department of Natural Resources in the State of New South Wales, in Australia), conservation significance index (developed by researchers), metric for biodiversity offsetting pilots (used by the Department for Environment, Food and Rural Affairs, in England), module assessment method (used by the Federal Office for the Environment, in Switzerland), biotope valuation (used by the Federal Ministry for the Environment, Nature and Conservation in Germany), habitat units (used by the Fish and Wildlife Service Federal Agency in the USA), significant environmental benefit (used by the Department of Water, Land and Biodiversity Conservation in the State of South Australia), and offset ratios (used by the Department of Environmental Affairs and Development Planning in the Province of the Western Cape, in South Africa). As they explain, in five metrics, the formula basis is based on area * quality, where quality is calculated as suitability index or biodiversity significance score, or condition + landscape context + habitat functionality, or condition + landscape context. In two metrics, the formula is based on area * ratio and in each of the others on the following: area * ecological value or area * condition * distinctiveness or % of individuals affected of species × or of habitat loss of species x, or finally, as area * %CSI, where %CSI = of sensitive species/available habitat * 100 (this is the formula basis of the only metric that has been designed by researchers). Importantly, Gamarra et al. (2018) note that all metrics scored substantially below the maximum possible score of 6 according to the criteria that they have defined for assessing the metrics (the highest-rated metric scored 2.1), that none of the metrics enable the calculation of residual losses and gains of the cultural values associated with biodiversity (which is part of the BBOP's definition of NNL as we saw above), and that only a few have the ability to differentiate between natural and anthropogenic-induced cycles or trends. Finally, the authors also concluded that very few metrics include indicators that are specific to the target being assessed, consider only objective numerical values, and benchmark areas, all criteria defined by the stakeholders they involved in their research as essential for accounting biodiversity values and, therefore, for assessing the suitability of offset metrics.

To sum up, the construction of ecological equivalence must deal with at least three different sets of problems that lead to three inevitable compromises. Firstly, decide what exactly to measure to achieve a practically reasonable equivalence of ecological losses and gains while admitting that the achievement of NNL per se is impossible. Secondly, design the proper metrics in order to measure the identified losses and gains while admitting that no metrics can capture ecosystem complexity. Thirdly, find a scientifically robust way to replicate these gains and losses while admitting that the evidence base of restoration ecology is not at a stage that can ensure that non-human nature can be replaceable and re-creatable. To add to the above, the concept of NNL, which lies at the heart of offsetting, still lacks an explicit frame of reference against which this is to be achieved,[5] resembling, as critics argue, more a slogan than a specific and transparent policy goal (Maron et al. 2016). Offsetting is, therefore, a process that is inevitably characterized by a series of compromises, simplifications, negotiations, and subjective judgments regarding both the adequacy of selected currencies and accepted types of exchanges (BBOP 2009a; Bull et al. 2013; Gardner et al. 2013; Pilgrim et al. 2013; Quétier and Lavorel 2011). Ultimately, what is implied is that representing ecological losses and gains in terms of numerical scores through the

[5]As Maron et al. (2016) point out, in practice, regulators rarely interpret No Net Loss to mean no biodiversity loss relative to before the impact; rather, it generally means maintaining some presumed trajectory of "background" decline.

application of offset metrics, can be no more than a convention: as even the BBOP (2009a, p. 13) concedes the genuineness of the representatives "may be difficult to demonstrate."

As Sullivan (2013) points out, in biodiversity offsetting, non-human nature and the entities and relationships of which it is comprised are known only by proxy: as numbers, scores, and interchangeable equivalences. This means that conservation is being gradually transformed into the protection of proxies and offsetting often leads to a battle of the proxies that is inherently political. Indeed, in any negotiation that takes place in the context of any offsetting process there will inevitably be different interests involved and hierarchical power relations will primarily determine the outcome. As empirical evidence shows, in most cases, this means that developers will try to downplay biodiversity values, for example by placing negative impacts into a low significance category, will choose offset metrics that give them the most convenient calculation of credits requirement or methods that are easier and cheaper to implement. This is exactly what happened in most of the areas where I conducted fieldwork in England. This is also the evidence we have from one of the most advanced offsets in the world, the Rio Tinto ilmenite mine[6] in Fort Dauphin (Tolagnaro), in the Anosy region of south-eastern Madagascar, one of the most biologically and culturally diverse islands in the world. In the Rio Tinto case, as Virah-Sawmy et al. (2014, p. 63) demonstrate, "the total habitat hectares impacted could vary from 680 using the Rio Tinto-QMM methodology based on vertical structure of the forest, to 1480 using plant diversity as a measure."

[6]See also https://ejatlas.org/conflict/rio-tinto-qmm-ilmenite-mine-madagascar.

3.2 Offsetting and the Construction of Equivalence: Insights from Critical Social Sciences

> There is no necessary connection between great science and great business opportunities; the general theory of relativity has yet to be turned into a money-spinner.
>
> *The Economist*, February 25, 1995

If we reverse the above quote from the Economist, we could say that great business opportunities can result from bad science. So, what if this bad science is sometimes necessary for doing business? By keeping this question in mind, in what follows I aim to contribute to a politicized analysis of the attempt to construct ecological equivalence as described in the previous section by focusing on the design of offset metrics and the role of science in this process. I pay attention to the use of surrogates that, as already explained, lies at the core of the offsetting process leading to simplified measurements of ecological losses and gains. Ample evidence from offsetting policies and programmes shows that simple currencies tend to omit, obscure or conceal ecosystem features and non-interchangeabilities to serve offsetting's key goal, namely, to facilitate nominal biodiversity accounting (Walker et al. 2009). Indeed, despite ongoing attempts to offer more concrete definitions of offsetability (BBOP 2012c), offsetting's underlying assumption is that biodiversity is largely substitutable in terms of type, space, and time. But what if this is precisely the goal of offsetting? If we accept that viable biodiversity barter and meaningful biodiversity protection are mutually exclusive, as Walker et al. (2009) argue, then we can easily see that offsetting's goal is quite explicitly the former and not the latter. I will try to show that this assumption is quite accurate by discussing four important implications of the use of surrogates that are not addressed, at least explicitly, in most of the existing ecological literature.

Firstly, the assumption that biodiversity and ecosystem processes are substitutable and re-creatable, supported by the endeavor to convert complex ecosystems to simple measurable units, is not trivial, or politically neutral. Critical scholars have demonstrated repeatedly the socially and politically contingent way scientific knowledge is produced (Demeritt 1998). Ecology, as any science, is a social process both causing and caused by social organization (Levins and Lewontin 1985). In the case of biodiversity offsetting, the promise to measure non-human nature quickly and easily draws on the advances of modern ecology only partly. It reflects a political goal, namely, the longstanding demand for a streamlined planning system that would not delay economic development or pose unnecessary barriers to it. Biodiversity offsetting is, thus, precisely designed to serve that goal, namely, to simplify and further deregulate the environmental and planning legislation to facilitate economic development and growth.

Offsetting's reductionism also reflects and, simultaneously, reinforces the neoliberal logic of market environmentalism on which conservation policies are increasingly based (Apostolopoulou and Adams 2015; Brockington and Duffy 2010; McAfee 2012). This argues that the solution to biodiversity loss is proper biodiversity accounting to enable the internalization of environmental externalities into economic systems and decision-making processes (Bayon's et al. 2008 analysis offers an indicative example of this logic). In the case of offsetting, this is expressed by the representation of biodiversity through numerical scores and priced credits that can be bought and sold to offer the necessary environmental compensation, and thus to internalize losses due to economic development (perceived as "externalities"). As we read in the Global review of biodiversity markets for 2017 (Bennett et al. 2017, p. 3): "offsets and compensation mechanisms seek to *internalize* the costs to society of biodiversity loss. For example, if a company is required by regulation to offset or compensate for its residual impacts to species or habitats, it must either bear those costs or choose to develop elsewhere, where impacts will be lower. Similarly, new incentives for conservation can be created. For example, landowners may realize they can profit from conserving biodiversity values by developing a mitigation bank" (added emphasis).

The task of constructing ecological equivalence is, therefore, trapped in an unresolved contradiction between the need to claim that the provision of environmental compensation is adequate, and the requirement to facilitate economic development by simplifying the calculation and delivery of this compensation through the planning system. This contradiction is not, as most supporters of offsetting would claim, an objective problem that primarily stems from existing gaps in scientific knowledge. If offsetting's primary goal was to improve the delivery of compensation for biodiversity loss, then the policy would have a neutral position on whether its outcome would make things easier or harder for developers. Offsetting on the contrary, clearly promises a faster, simpler and, ideally, even a profitable process for developers, as the case of England (see Chapters 5 and 6) illustrates. Moreover, and relatedly, when offsetting's implementation on the ground contradicts these promises then, in most cases, it is abandoned (Apostolopoulou and Adams 2019; Bormpoudakis et al. 2019). One of the most clear indications of the politicized nature of offsetting and the policy goals it is expected to support is the contradiction between the expansion of offsetting policies and programs since 2008 and the increasingly disappointing empirical evidence in support of its ecological outcomes, including, inter alia, the absence of evidence that NNL (or net gain) has been actually ever delivered at a programme level (Curran et al. 2013; Gibbons et al. 2018).

To summarize, offsetting policies are based on a series of political and ideological assumptions that are anything but neutral. Offsetting is a product of the broader hegemony of the TINA dogma, namely, that There Is No Alternative to the current state of affairs. This means that economic development is considered not only unlikely to cease anytime soon but also the only feasible solution to the global economic crisis. Following this logic there is no other realistic solution to the continuing loss of biodiversity than to find a pragmatic and practical method to secure compensation for this loss. Offsetting reproduces the assumption of neoliberal economics that the key cause of biodiversity decline is its insufficient quantification that prevents its inclusion into the market and attempts to validate this as the best available approach. It is for these reasons that the well-recognized shortcomings of offset metrics do not invalidate the approach in the eyes of its supporters. Thus, Bayon et al. (2008, pp. 4–5), even though admit that it is "too early to tell" if offsetting and banking work since the reality is that they "haven't really done the necessary studies," they reassure us that these "may be the worst of all possible systems except for all the others" echoing Churchill's well-known quote about democracy.[7]

Secondly, and relatedly, offsetting's neoliberal premises directly influence the choice of metrics and the direction of scientific research. The critical literature on both carbon and biodiversity offsetting (e.g., Bracking et al. 2014; Lohmann 2005; MacKenzie 2009; Sullivan 2013) shows well how the construction of exchangeability and fungibility between different ecosystem attributes dictates the designation of metrics which triumph the primary methodological axioms of positivist science (Smith 2010): mechanistic reductionism at its extreme and a wholly external conception of nature. Offsetting is based precisely on the promise that simplistic metrics can represent nature in numerical scores which adequately describe ecosystems and diversity and do so in spatially explicit ways as functionally abstract units that can be deemed exchangeable (Bumpus 2011; Bumpus and Liverman 2008; Lohmann 2012; Robertson 2000, 2006).

The way the political premises and the metrics of both carbon and biodiversity offsetting are inextricably linked has been obvious in discussions evolving around the principle of additionality (BBOP 2009a). Critics of carbon offsetting have pointed out the contestable character of counterfactual scenarios by showing how politics are directly involved in eliminating possible future scenarios in favor of the scenario that specific industries promote (see Lohmann 2005; Mackenzie 2009). As previously mentioned, this has been also the case in biodiversity offsetting where different counterfactual scenarios directly influence estimations of NNL. Thus, for example, the mining corporation Rio Tinto, tried to underestimate its responsibility for compensation for the forest loss caused by its operations in Madagascar by claiming an inappropriately high rate of biodiversity loss in the counterfactual scenario (Virah-Sawmy et al. 2014). The class character of such scenarios has been further evident in Rio Tinto's attempt to prove the beneficial character of its interventions in Madagascar by highlighting ongoing unsustainable use of natural

[7]In a speech on November 11, 1947, Winston Churchill said in the House of Commons that "democracy is the worst form of government, except for all those others that have been tried".

resources by local Malagasy people. This is a common rhetoric of offsetting's supporters particularly, but not exclusively, in the Global South (Seagle 2012; see also the BBOP pilot Ambatovy project in Madagascar where offsetting is based, inter alia, on the argument that forests have been negatively impacted by local communities, Dickinson and Berner 2010).

Thirdly, the representation of non-human nature in terms of units favors and perpetuates simply technical, as opposed to theoretically and socially informed, critiques of the conventions of dominant methods of measuring and mapping nature (Katz and Kirby 1991; Smith and Katz 1993). This reinforces ecology's lack of engagement with discussions on the social production of nature and is aligned to a moral-liberal and rational-technocratic view of politics and science which discounts the degree to which social commitments are built into the technical details of particular policies (see Demeritt 1998 about how this works in the case of climate change science). In the language of offsetting, biodiversity loss is framed in narrowly scientific terms as a problem of species and habitats decline largely divorced from their social, ecological, evolutionary, and cultural context. Adopting this logic, in designing offset metrics, ecologists exclude from their analyses the political economy responsible for ecosystem degradation and biodiversity loss. Even though this choice is often justified by reference to the technical character of offsetting's calculations, denying politics is itself a political act.

It is important to point out here that in the above process, there is a privileged relationship with science, and scientists, willing to be an "indistinguishable partner of capitalism" by enthusiastically embracing new objects of inquiry and practice (Katz 1998, p. 46), in this case the designation of the most "appropriate" offset metrics. Ecology itself from an inherently political science (Sears 1964)[8] perceives the role of a pragmatic techno-science dedicated to the exercise of measuring and quantifying the non-human world. This marks a transition from the,

[8]Paul Sears (1964, p. 12) in his article *Ecology-a subversive subject* argued that "by its very nature, ecology affords a continuing critique of man's [sic] operations within the ecosystem".

undoubtedly problematic, idealistic investigation of "unique" biodiversity that has dominated conservation policy and science since the nineteenth century (Fabiani 1985), to the search for "biodiversity units" that can be exchanged across space and time and whose ecological meaning is deemed to remain rather unclear. It is crucial here to pay attention to the fact that the harnessing of science and scientists to the task further fortifies it as technical (sensu Jasanoff 2011), offering an apparent guarantee of soundness. This allows offsetting's proponents to present it as part of a rational and pragmatic policymaking process based on the "best available" scientific knowledge, legitimizing potential controversies, gaps, and weaknesses (see also Demeritt 1998). Ecological science is, therefore, welcome if it is convenient to policy, namely, capable of ensuring offsetting's scientific soundness and confirm that existing gaps are inevitable given the current state of knowledge. Thus, for example, BBOP (2009a, p. 42) characterizes its proposed method to offsetting as "experimental," "evolving," "hesitant," and "preliminary" but, at the same time, does not hesitate to claim that the method would be not only "reasonable practical" but also "applicable in all countries and biomes, from desert to tropical and temperate forest to the marine environment."

Fourthly, and again relatedly, shifting the focus from the environmental impacts of economic development to the selection of metrics and exchange rules has been followed by a heightened role for experts and a limited role for local communities contributing to the increasing depoliticization of conservation–development conflicts. Here, it is again relevant to remember Jasanoff's (2011, pp. 633–634) analysis of the ethical turn in the United States during the 1980s. As she explains, during that period, the analysis of scientific and technological advances was split into a domain of facts, largely committed to technical experts in administrative agencies, and a domain of values, increasingly delegated to ethics committees. An effect of this ethical turn has been the privatization of the public sphere by downplaying the role of communal norms and public welfare and by conducting ethical debates in closed forums, out of public view.[9] In the process, research has become subject to ethical

[9]As Jasanoff (2011, pp. 633–634) argues "ethics committees engage in a polite process of opinion formation, oriented toward elite consensus-building, in which the values and sensibilities

reflection instead of political debate and, consequently, ethical expertise became subordinated to technical expertise, so that those with privileged scientific understanding came "to be seen, and also to see themselves, as the best representatives of the public's moral commitments" (Jasanoff 2011, pp. 633–634).

I consider the above analysis not only very insightful but also very relevant to offsetting. Organizations like the BBOP, Forest Trends, or the Ecosystems Task Force in the UK (see Chapter 5) have often supported the implementation of market-based policies, like carbon and biodiversity offsetting, habitat banking and payments for ecosystem services, without any proper public consultation or even under a veil of secrecy. Consequently, local communities have often found themselves completely unable to challenge these policies since they are lacking not only access to information and decision-making processes but also the technical expertise and the resources to hire consultants and solicitors to help them challenge these policies. They have also often found themselves trapped in technical discussions where the loss of green space and ecosystems is framed as an issue that appropriate metrics and calculations can capture and resolve. Not surprisingly, local communities have perceived and experienced offsetting as a crude, dystopian policy that aims to measure their relationship with nature by using reductionist metrics and represent it through units and priced credits, considering their knowledge and experience as irrelevant. At the same time, consultants, experts, brokers, offsetting officials, and often NGOs consider themselves as capable to decide the criteria that will define which places and natures can be rendered exchangeable and equivalent and how this can be achieved without any democratic public debate. As a farmer in North England told me once: "officials think that you need to have the necessary scientific knowledge to judge whether it is possible and ethical to exchange this piece of land here with that piece of land there. Well, we told them that we certainly have the knowledge to distinguish one piece of land from another when these may look the same to them."

---

of a very few, highly educated, articulate individuals stand in for the untrained, and allegedly uninformed, preferences of the multitude." [...] "Indeed, for many scientists, 'understanding the science' and 'understanding how science works' serve almost as threshold tests for the right to speak about the ethical dimensions of emerging science and technology."

Before closing this section a final observation about offset sites is important. Given the shortcomings and weaknesses of offsetting's calculations, how can we characterize offsetting's products? There is a significant part of the critical literature, largely inspired by Latour's writings (see for example Latour 1987) that tends to characterize the results of the application of offset metrics as largely artifactual. Despite the critical insights that these approaches may offer, I strongly believe that they carry the risk of offering largely idealistic explanations of offsetting policies. These analyses remind me Smith's (1998, p. 274) critique of artifactual constructionism and, particularly, his point about the importance to open the contestable dialectical translation between discourse and reality, rather than close it off under the sign of discursive construction. In this sense, I think that approaching offset sites as "artifacts" may easily devolve into neo-Kantian idealism (see Demeritt 1998) for two main reasons. Firstly, because it either implies that non-human nature is not, and will never be, knowable or that there is some ideal "Nature" of our imagination that will always escape scientific calculations. Secondly, because it also implies that the socially constructed knowledge that guides offsetting methodologies is simply false discounting the historically specific conditions that shape it (Fine 2005). A class understanding of knowledge, science and ideology (Levins and Lewontin 1985; Marx and Engels 1970) can reveal that offsetting's calculations and offset sites, are not just "untruth" or "artifacts" but a product of capitalism's historically specific, socially determined relation to nature. To put it differently, the nature that is created based on offset's calculations reflects capitalism itself. Offsetting's claim that it is possible to separate, measure, and exchange natural parts and that these parts can exist outside of the ecosystem, reflects the ontological commitments of reductionism and deepens the ontological separation between humans and non-human nature. Describing nature as a Cartesian machine with replaceable parts, where causes are separated from results and subjects from objects (Levins and Lewontin 1985), makes possible the alienability of units of nature from the ecosystem, with its complex material connectivity and geographical specificities, and their individuation (Castree 2003) from their social and ecological context. Moreover, and relatedly, the conceptualization of the biodiversity units that offset metrics produce as ahistorical natural entities reflects the external nature of post-eighteenth century capitalist ideology (Lohmann 2012; Smith 2006) and capitalism's understanding of non-human nature as a stock of resources that can be measured in quantitative terms and whose value can be captured by price. They result from its exploitative use and development of natural forces as "mere material conditions of capital accumulation" (Burkett 1997, p. 168) and even though they reflect how capitalist commodity production subjugates use value to exchange value, they do not, however prove that ecological "commodities" produce value, as I will show in the next chapter. As Smith (1998, p. 271) argued referring to environment management, such approaches reflect the "technocratic, neoliberal rendition of the construction of nature," evoking a "narrow productionism which keeps nature and culture in separate spheres." Offset sites are, therefore, reflections of the form that nature conservation, or at least its mainstream version, has taken under neoliberal capitalism, namely, of a practice that shows no interest in people's relations to nature and is willing to endorse the destruction of a public green space for the creation of a private conservation bank.

In order to understand the logic of the technical discussions evolving around offsetting, it is necessary to look beneath "the delusive appearance of things" (Marx 1899, p. 54), and accept that, fundamentally, offsetting is not about either literally achieving No Net Loss or ecological equivalence. This means that the core weaknesses of offsetting are not technical issues that better science will resolve. Offsetting in the search of pragmatic win-win solutions purposefully confines NNL to those parts of nature that can be measured, quantified, and replaced most easily. These aspects of nature are reckonable, and can be protected, others cannot. Offsetting is, therefore, trapped in an inescapable ambiguity: it cannot be known whether what is saved is truly equivalent to what has been lost. Ultimately, NNL and ecological equivalence are part of an idealistic attempt to solve capitalism's environmental contradictions. Equivalence is achieved symbolically and only for those "natural parts" selected to be measured by largely arbitrary units (Virah-Sawmy et al. 2014). The role of science in this process is important and it is not neutral: it is science that makes such exchanges possible, more likely and more acceptable, despite their obvious shortcomings. In this way, ecology, and ecologists who despite its apparent contradictions do not reject offsetting but offer recommendations for its improvement, by participating in the exercise of measuring, quantifying, and slicing non-human nature, contribute to its production, materially cum discursively (Smith 1998) and ultimately to the creation of an alienated world (sensu Levins and Lewontin 1985).

3.3 Offsetting's Non-places: The Tensions Between Differentiation, Interchangeability, Homogeneity, and Unevenness in the Capitalist Production of Geographical Space

> The differentiation of geographical space in the last century or so is a direct result of the need, inherent in capital, to immobilize capital in the landscape. [...] In order to produce surplus value, it is necessary that vast quantities of productive capital be spatially immobilized for relatively long periods in the form of factories, machinery, transport routes, warehouses, and a host of other facilities. The spatial immobilization of capital in this way, or as national capitals delimited by the boundaries of the nation state, is simultaneously the production of differentiated geographical space.
>
> Smith (2010, p. 120)

Non-place is a term coined by the French anthropologist Marc Augé in his book "Non-places, introduction to an anthropology of supermodernity" to refer to anthropological spaces of transit, such as motorways, hotel rooms, modern airports, and shopping malls, that are fleeting, disconnected, and historically void, with homogeneous architecture and no local referents (Jameson 1991; Virilio 1997; Sharma 2009). National parks and conservation areas have been also characterized as non-places (Backhaus 2003; Dickinson 2011) since these are often produced in a way that obscures any meaningful sense of place. I draw on these ideas here to argue that offsetting's approach toward ecosystems and places has all the above characteristics: offset sites are understood as the outcome of objective calculations based on the application of modern science (ecology). Following a technocratic and dualistic approach that sees nature both as asocial and as the sum of its parts (excluding humans from the equation), offsetting claims that it is possible to reproduce and recreate nature destroyed by economic development by considering only its biological characteristics while ensuring No Net Loss. Of course, offset sites are often very real places that people were using before or will continue to use after the offsetting regime. In other cases, however, these sites function as habitat or conservation banks with no public access constituting peculiar non-places of conserved or restored nature at least for as long as these remain under offsetting's management rules (because there is no guarantee that they will not be offset again, developed or destroyed). Offsets are also non-places in terms of their relation to time: the offset site is essentially a promise to a *future*, currently non-existent place, that will contain the desirable characteristics to replicate another place that has already gone.

The issue of place remains, therefore, one of the most controversial aspects of offsetting. As already mentioned, most analyses have focused to the ecological and management aspects of offsetting and despite the broad acknowledgment of the need consider cultural values and social and economic aspects related to biodiversity loss and ecosystem degradation (BBOP 2012a), there has been no widely acceptable proposal on how such values can be measured, exchanged, or re-created. This means that the "exchangeability" of the development and the offset site is so far measured by primarily focusing on their ecological characteristics. It is indicative that in BBOP publication (2012c) stating the partnership's position about "offsetable" and "non-offsetable impacts," the main criteria mentioned are the irreplaceability and the vulnerability of biodiversity, two concepts that also dominate the relevant literature, meaning that offsetting is considered straightforward when habitats and species lost are of low or moderate irreplaceability. This reflects the longstanding practice of conservation of equating the values of non-human nature with unique or typical biodiversity, usually in terms of the presence of endangered, rare, or endemic species (Apostolopoulou et al. 2012) and further intensifies the division between what is thought to be common and what is thought to be unique. These mechanistic separations have important implications because they remove the social in two key ways (see also Katz 1998): firstly, by perpetuating and hardening the boundaries between urban, agricultural, and wilderness landscapes, and, secondly, by separating generations of social actors from nature, essentially denying social history to landscapes. In offsetting, the ontological division between non-human nature and society is sharp and purposeful: it is used to portray development and offset sites as places whose value can be represented through numerical scores that are higher for unique or rare biodiversity, and lower for common landscapes. Such landscapes, local green spaces and everyday natural places, considered as non-significant from a biodiversity perspective, in most cases are of great social significance, and as increasing evidence shows (Apostolopoulou and Adams 2019; Apostolopoulou 2019), their low calculated biodiversity value makes them an easy target for economic development.

Offsetting, like classical conservation, involves the conceptual and practical placing of non-human nature within specific spatial bounds, making both places and spaces (Apostolopoulou and Paloniemi 2012; Adams et al. 2014). In both the development and the offset site, nature is demarcated, classified, and subjected to specified regimes of exploitation and management. What happens in these sites reflects capitalism's broader aim to produce space, place, and environments, according to its "own vital needs, desires, and interests" (Harvey 2001, p. 296). What is new is that is taking place under an, increasingly contradictory, ecofriendly disguise: in the offsetting discourse non-human nature is not destroyed due to urban development or extractive activities but it is moved somewhere else. It is exactly this discourse that makes offsetting a deeply contradictory policy that involves, simultaneously, strategies of green grabbing, in the offset site, and un-green grabbing, in the development site. This is also what makes it particularly exposed to the dissonant geographical practices of capitalism: offsetting does not claim that it is possible to halt ecological losses per se or prevent biodiversity loss but that it can achieve an efficient overall quantitative balance between preserved nature and permitted development.

By framing losses and gains as equivalent, offsetting's underlying assumption is that two different places, the development and the offset site, are interchangeable and, thus, that a place can be destroyed *somewhere* if another place is protected in return *elsewhere*. This bizarre exchange of (non-)places is not only feasible in offsetting's logic but also legitimate and, potentially, even desirable if it achieves an improvement in environmental terms (measured in an increased number of credits). This is a crucial point because it lucidly shows offsetting's disinterest for the social that I mentioned above. Livelihoods, local traditions and meanings, the uneven socioeconomic consequences of land-use change, the cultural importance of place, social ties between communities and ecosystems, access to green space and the diversity of both natural and social relations that social space contains (Lefebvre 1991) are irrelevant to offsetting's calculations. This understanding of space and place is not unique to offsetting. It reflects the broader patterns that drive the capitalist production of space and, as Lefebvre (1970, 1991) has extensively shown, shows that capitalism's obsession with economic growth is inextricably linked to the occupation and production of space that ultimately leads to the interchangeability of places. Indeed, the need for comparability has historically led to the triumph of homogeneity: under capitalism places are deprived of their specificity and space is "produced and reproduced as reproducible" (1991, p. 337) whereas spatial interchangeability brings a powerful tendency toward quantification that is "technical in appearance, financial in reality, and moral in essence" (ibid., pp. 338–339). In offsetting, spatial interchangeability is achieved both through economic development and through the associated conservation of offset sites. The latter by being framed as territories that are productive of ecological credits are places that can be exchanged, bought and sold, based on differences that are measured in monetary terms. To borrow Lefebvre's (1991, p. 75) words: "repetition has everywhere defeated uniqueness [...] the artificial and contrived have driven all spontaneity and naturalness from the field [...] Are these places interchangeable because they are homologous? Or are they homogeneous so that they can be exchanged, bought and sold, with the only differences between them being those assessable in money – i.e. quantifiable – terms (as volumes, distances, etc.)? At all events, repetition reigns supreme."

It is interesting to note here that keen supporters of offsetting tend to argue that criticizing offsetting for treating places and natures as equivalent and exchangeable, is based on a misinterpretation of its goals since what is being exchanged is a number of offset credits. However, this is hardly a valid or convincing defense. It rather confirms offsetting's extreme reductionism that has little to do with what is happening on the ground: how is it ever possible to separate a number of credits or units from the ecosystem or from the place where they are located? This is an irrational assumption and the fact that is being discussed in the ecological literature and in mainstream policy discourses as meaningful shows how disconnected from the real world these discussions are and how deeply idealistic extreme scientific positivism can become. The outcome of the offsetting process is that socionatures are being destroyed based on the argument that these can be recreated somewhere else only by measuring selected ecological parameters. In practical terms, this means that a place is destroyed, and compensation is based on calculations of land area, or habitat quality and distinctiveness or some ecosystem aspects. Offsetting offers, therefore, a concrete manifestation of how capital creates its own distinctive ecosystem by treating biodiversity as a mobile resource that can be redistributed across space and time and accounted for by scientists, managers, and technocrats. To paraphrase Lefebvre (1991), *offsetting ultimately produces and reproduces non-human nature as reproducible.*

Nonetheless behind the appearance of equivalence lies the hard materiality of the fact that the land on which biodiversity depends cannot be produced, or reproduced, defining the material limit of any representation of reproducibility (Apostolopoulou et al. 2018). Moreover, and relatedly, offsetting's persistence in imposing a deeply reductionist, often just rhetorically constructed, equivalence between ecological losses and gains brings in practice profound unevenness. Here it is helpful to refer to Smith (2010, p. 50) who explained that when we place the appearance of nature in its specific historical context, then the development of the material landscape presents itself as a process of the production of nature and the differentiated results of this production are the material symptoms of uneven development. Smith, here, offers a Marxist analysis of the impacts of the capitalist production of nature and space pointing to one of its most important and neglected manifestations: the differentiation and unevenness produced across scales as manifested in uneven geographical development. Smith here draws on Lefebvre's (1991) analysis about the contradiction between the increased differentiation of space and the equalizing tendency of capital that drives the capitalist production of space, leading toward the emancipation from space. Both Smith's and Lefebvre's[10] analyses show that capitalism's tendency toward the loss of geographical differentiation is accompanied by the production of new forms of socioeconomic difference (see also Herrera et al. 2007), deepening unevenness and injustices.

Place, therefore, from "the locale of the truth of Being" (Heidegger) becomes largely extraneous in offsetting following the long history of conservation without people. And even though this becomes more obvious in the case of biodiversity, species, habitats, or wetlands banking it also underlines carbon offsetting and emissions trading since there is a common logic behind all offsetting policies, namely, that quantitative calculations of environmental losses and gains can offer the basis for their representation in terms of priced credits which can be exchanged across space and time. Again, this is the rather delusive appearance of things that masks the actual processes at work: offsetting authorizes a radical rescripting of non-human nature as *placeless* only to construct place in accordance with the aims and needs of capital. Offsetting is not, therefore, an *asocial* policy that excludes from its calculations *humans* in general. On the contrary, uneven development, caused, facilitated, or enabled by offsetting, rather than situated in the false ideological dualism of society and nature is the product of the unity of capital with nature (Smith 2010). Its impacts reflect the class character of land-use change and the way landlords and different sections of capital govern the production of nature and space. This is evident in the increasing popularity of compensation and offsetting policies in the aftermath of the 2008 financial crash which has been clearly linked to capital's need to find new

[10]As Smith (2010, p. 126) argues even though this is an original idea of Lefebvre, one can detect its embryos in the work of Marx, Luxembourg and Lenin.

avenues of capital accumulation to combat economic recession. This tendency forces capital to constantly expand, and, thus, to nestle, settle, and establish connections everywhere (Marx and Engels 2002).

A final observation is important here. Although framing the social as irrelevant or ignoring the issue of place have been strongly attacked by critics of offsetting, specific analyses of its class implications and the particular ways through which it reinforces a socially and geographically uneven production of nature are scarce (for exceptions see for example Apostolopoulou and Adams 2017b; Apostolopoulou 2019; Bidaud et al. 2017; Seagle 2012; Ruhl and Salzman 2006). Similarly, there has been equally limited attention to the fact that offsetting, particularly in the case of biodiversity, is tightly interwoven with competition for, and speculation over, land, something partly related to the fact that critical literature has so far almost totally ignored the exploration of offsetting through the lenses of the theory of rent. This is an important omission because it contributes to the inability of existing scholarship to engage in a historical-geographical analysis of offsetting that could consider the spatial organization of capitalism and the various forms of social control it entails (Apostolopoulou et al. 2018). In the next chapter I aim to contribute to such an analysis by showing why analyzing offsetting by drawing on the theory of rent can offer new and important insights of the environmental contradictions of capitalism as expressed in the widespread adoption of neoliberal conservation policies worldwide.

References

Adams, W. M., Hodge, I. D., & Sandbrook, L. (2014). New spaces for nature: The re-territorialisation of biodiversity conservation under neoliberalism in the UK. *Transactions of the Institute of British Geographers, 39,* 574–588.

Apostolopoulou, E. (2019). Beyond post-politics: Offsetting, depoliticization and contestation in a community struggle against executive housing. *Transactions of the Institute of British Geographers.* https://doi.org/10.1111/tran.12354.

Apostolopoulou, E., & Adams, W. M. (2015). Neoliberal capitalism and conservation in the post-crisis era: The dialectics of 'green' and 'un-green' grabbing in Greece and the UK. *Antipode, 47,* 15–35.

Apostolopoulou, E., & Adams, W. M. (2017a). Biodiversity offsetting and conservation: Reframing nature to save it. *Oryx, 51,* 23–31.
Apostolopoulou, E., & Adams, W. M. (2017b). Biodiversity offsetting and the reframing of conservation: A reply to ten Kate & von Hase and Dempsey & Collard. *Oryx, 51,* 40–42.
Apostolopoulou, E., & Adams, W. M. (2019). Cutting nature to fit: Urbanization, neoliberalism and biodiversity offsetting in England. *Geoforum, 98,* 214–225.
Apostolopoulou, E., Adams, W. M., & Greco, E. (2018). Biodiversity offsetting and the production of "equivalent natures": A Marxist critique. *ACME: An International Journal for Critical Geographies, 17*(3), 861–892.
Apostolopoulou, E., Drakou, E., Santoro, F., & Pantis, J. D. (2012). Investigating the barriers to adopting a "human-in-nature" view in Greek biodiversity conservation. *International Journal of Sustainable Development and World Ecology, 19,* 515–525.
Apostolopoulou, E., & Paloniemi, R. (2012). Frames of scale challenges in Finnish and Greek biodiversity conservation. *Ecology and Society, 17*(4), 9.
Backhaus, N. (2003). 'Non-place jungle': The construction of authenticity in National Parks of Malaysia. *Indonesia and the Malay World, 31*(89), 151–160.
Bayon, R., Fox, J., & Carroll, N. (2008). *Conservation and biodiversity banking: A guide to setting up and running biodiversity credit trading systems.* London: Earthscan.
BBOP (Business and Biodiversity Offsets Programme). (2009a). *Biodiversity offset design handbook* (Appendices). Washington, DC: BBOP.
BBOP (Business and Biodiversity Offsets Programme). (2012a). *Standard on biodiversity offsets.* Washington, DC: BBOP. Available at: http://bbop.forest-trends.org/guidelines/Standard.pdf.
BBOP (Business and Biodiversity Offsets Programme). (2012b). *Resource paper: No net loss and loss-gain calculations in biodiversity offsets.* Washington, DC: BBOP.
BBOP (Business and Biodiversity Offsets Programme). (2012c). *Resource paper: Limits to what can be offset.* Washington, DC: BBOP. Available at: http://bbop.forest-trends.org/guidelines/Resource_Paper_Limits.pdf.
Bennett, G., Gallant, M., & ten Kate, K. (2017). *State of biodiversity mitigation 2017.* Markets and Compensation for Global Infrastructure Development. Ecosystem Marketplace, Forest Trends, Washington DC.
Bernhardt, E. S., & Palmer, M. A. (2011). The environmental costs of mountaintop mining valley fill operations for aquatic ecosystems of the Central

Appalachians. In R. S. Ostfeld & W. H. Schlesinger (Eds.), *Year in ecology and conservation biology* (pp. 39–57). New York: New York Academy of Sciences.

Bezombes, L., Gaucherand, S., Kerbiriou, C., Reinert, M. E., & Spiegelberger, T. (2017). Ecological equivalence assessment methods: What trade-offs between operationality, scientific basis and comprehensiveness? *Environmental Management, 60,* 216–230.

Bidaud, C., Schreckenberg, K., Rabeharison, M., et al. (2017). The sweet and the bitter: Intertwined positive and negative social impacts of a biodiversity offset. *Conservation and Society, 15,* 1–13.

Bormpoudakis, D., Tzanopoulos, J., & Apostolopoulou, E. (2019). The rise and fall of biodiversity offsetting in the Lodge Hill large-scale housing development, South East England. *Environment and Planning E: Nature and Space.* https://doi.org/10.1177/2514848619884890.

Bracking, S., Bond, P., Brockington, D., Büscher, B., Igoe, J. J., Sullivan, S., & Woodhouse, P. (2014). *Human, non-human and environmental value systems: An impossible frontier?* (LSCV Working Paper Series No. 1). Leverhulme Centre for the Study of Value, School of Environment, Education and Development, The University of Manchester.

Brockington, D., & Duffy, R. (2010). Capitalism and conservation: The production and reproduction of biodiversity conservation. *Antipode, 42,* 469–484.

Bull, J. W., Lloyd, S. P., & Strange, N. (2017). Implementation gap between the theory and practice of biodiversity offset multipliers. *Conservation Letters, 10,* 656–669.

Bull, J. W., Suttle, K. B., Gordon, A., Singh, N. J., & Milner-Gulland, E. J. (2013). Biodiversity offsets in theory and practice. *Oryx, 47,* 369–380.

Bumpus, A. G. (2011). The matter of carbon: Understanding the materiality of tCO_2e in carbon offsets. *Antipode, 43,* 612–638.

Bumpus, A. G., & Liverman, D. M. (2008). Accumulation by decarbonization and the governance of carbon offsets. *Economic Geography, 84,* 127–155.

Burkett, P. (1997). Nature in marx reconsidered a silver anniversary assessment of alfred schmidt's concept of nature in Marx. *Organization & Environment, 10,* 164–183.

Castree, N. (2003). Commodifying what nature? *Progress in Human Geography, 27,* 273–297.

Commonwealth of Australia. (2012). *Environment protection and Biodiversity Conservation Act 1999 environmental offsets policy.* Department of the Environment, Canberra, ACT, Australia. Available at: http://www.environment.

gov.au/system/files/resources/12630bb4-2c10-4c8e-815f-2d7862bf87e7/files/offsets-policy_2.pdf.

Conway, M., Rayment, M., White, A., & Berman, S. (2013). *Exploring potential demand for and supply of habitat banking in the EU and appropriate design elements for a habitat banking scheme* (Final Report Submitted to DG Environment). UK: ICF GHK and BIO Intelligence Service.

Curran, M., Hellweg, S., & Beck, J. (2013). Is there any empirical support for biodiversity offset policy? *Ecological Applications, 24,* 617–632.

DEC. (2006). *BioBanking: A biodiversity offsets and banking scheme: Conserving and restoring biodiversity in NSW* (Working Paper). Sydney: Department of Environment and Conservation NSW.

Defra (Department for Environment, Food and Rural Affairs). (2013). *Biodiversity offsetting in England green paper.* Department for Environment, Food and Rural Affairs. London, UK. Available at: https://consult.defra.gov.uk/biodiversity/biodiversity_offsetting/supporting_documents/20130903Biodiversity%20offsetting%20green%20paper.pdf.

Demeritt, D. (1998). Science, social constructivism and nature. In B. Braun & N. Castree (Eds.), *Remaking reality: Nature at the millennium.* London and New York: Routledge.

Dickinson, E. (2011). Displaced in nature: The cultural production of (non-) place in place-based forest conservation pedagogy. *Environmental Communication: a Journal of Nature and Culture, 5*(3), 300–319.

Dickinson, S., & Berner, P. O. (2010). Ambatovy project: Mining in a challenging biodiversity setting in Madagascar. In S. M. Goodman & V. Mass (Eds.), *Biodiversity, exploration, and conservation of the natural habitats associated with the Ambatovy project.* Malagasy Nature 3 (pp. 2–13). Antananarivo: Association Vahatra.

Fabiani, J. L. (1985). Sciences des écosystèmes et protection de la nature. In A. Cadoret (Ed.), *protection de la nature. Histoire et idéologie* (pp. 75–93). Paris: L' Harmattan.

Fine, B. (2005). From actor-network theory to political economy. *Capitalism Nature Socialism, 16,* 91–108.

Gamarra, M. J. C., Lassoie, J. P., & Milder, J. (2018). Accounting for no net loss: A critical assessment of biodiversity offsetting metrics and methods. *Journal of Environmental Management, 220,* 36–43.

Gardner, T. A., Von Hase, A., Brownlie, S., Ekstrom, J. M. M., Pilgrim, J. D., Savy, C. E., et al. (2013). Biodiversity offsets and the challenge of achieving no net loss. *Conservation Biology, 27,* 1254–1264.

Gibbons, P., & Lindenmayer, D. B. (2007). Offsets for land clearing: No net loss or the tail wagging the dog? *Ecological Management & Restoration, 8*, 26–31.

Gibbons, P., Macintosh, A., Constable, A. L., & Hayashi, K. (2018). Outcomes from 10 years of biodiversity offsetting. *Global Change Biology, 24*, 643–654.

Gonçalves, B., Marques, A., Soares, A. M. V. D. M., & Pereira, H. M. (2015). Biodiversity offsets: From current challenges to harmonized metrics. *Current Opinion in Environmental Sustainability, 14*, 61–67.

Harvey, D. (2001). Cosmopolitanism and the banality of geographical evils. In J. Comaroff & J. L. Comaroff (Eds.), *Millenial capitalism and the culture of neoliberalism* (pp. 271–309). Durham and London: Duke University Press.

Herrera, L. M. G., Smith, N., & Vera, M. Á. M. (2007). Gentrification, displacement, and tourism in Santa Cruz de Tenerife. *Urban Geography, 28*, 276–298.

Hobbs, R. J., Hallett, L. M., Ehrlich, P. R., & Mooney, H. A. (2011). Intervention ecology: Applying ecological science in the twenty-first century. *BioScience, 61*, 442–450.

Hough, P., & Robertson, M. (2009). Mitigation under Section 404 of the Clean Water Act: Where it comes from, what it means. *Wetlands Ecology and Management, 17*, 15–33.

HS2. (2013). *London-West Midlands environmental statement* (Vol. 5). Technical Appendices. Scope and Methodology Report Addendum (Ct-001-000/2). Available at: https://www.gov.uk/government/uploads/system/uploads/attachment_data/file/260153/Vol5_Scope_and_methodology_report_addendum_CT-001-000.2.pdf.

ICMM (International Council on Mining and Metals) and IUCN (International Union for Conservation of Nature). (2012). *Independent report on biodiversity offsets*. Prepared by the Biodiversity Consultancy. Available at: www.icmm.com/biodiversity-offsets.

IEEP. (2014). *Study on specific design elements of biodiversity offsets: Biodiversity metrics and mechanisms for securing long term conservation benefits*. DG ENVIRONMENT: ENV.B.2/ETU/2013/0060r. Final Report. A Report Submitted by ICF Consulting Services in association with IEEP and Associated Experts. London, UK.

Jähnig, S. C., Lorenz, A. W., Hering, D., Antonis, C., Sundermann, A., Jedicke, E., et al. (2011). River restoration success: A question of perception. *Ecological Applications, 21*, 2007–2015.

Jameson, F. (1991). *Postmodernism, or, the cultural logic of late capitalism*. Durham, NC: Duke University Press.

Jasanoff, S. (2011). Constitutional moments in governing science and technology. *Science and Engineering Ethics, 17,* 621–638.
Katz, C. (1998). Whose nature, whose culture?: Private productions of space and the 'preservation' of nature. In B. Braun & N. Castree (Eds.), *Remaking reality: Nature at the millennium* (pp. 45–62). London and New York: Routledge.
Katz, C., & Kirby, A. (1991). In the nature of things: The environment and everyday life. *Transactions of the Institute of British Geographers, 16,* 259–271.
Laitila, J., Moilanen, A., & Pouzols, F. M. (2014). A method for calculating minimum biodiversity offset multipliers accounting for time discounting, additionality and permanence. *MEE, 5,* 1247–1254.
Latour, B. (1987). *Science in action: How to follow scientists and engineers through society.* Cambridge: Harvard University Press.
Lefebvre, H. (1970). *La révolution urbaine* (Vol. 216). Paris: Gallimard.
Lefebvre, H. (1991). *The production of space.* Oxford: Blackwell.
Levins, R., & Lewontin, R. (1985). *The dialectical biologist.* Cambridge, MA: Harvard University Press.
Lohmann, L. (2005). Marketing and making carbon dumps: Commodification, calculation and counterfactuals in climate change mitigation. *Science as Culture, 14,* 203–235.
Lohmann, L. (2012). Financialization, commodification and carbon: The contradictions of neoliberal climate policy. *Socialist Register, 48,* 85–107.
MacKenzie, D. (2009). Making things the same: Gases, emission rights and the politics of carbon markets. *Accounting, Organizations and Society, 34,* 440–455.
Maron, M., Hobb, R. J., Moilanen, A., Matthews, J. W., Christie, K., Gardner, T. A., et al. (2012). Faustian bargains? Restoration realities in the context of biodiversity offset policies. *Biological Conservation, 155,* 141–148.
Maron, M., Ives, C. D., Kujala, H., Bull, J. W., et al. (2016). Taming a wicked problem: Resolving controversies in biodiversity offsetting. *BioScience, 66,* 489–498.
Marx, K. (1899). *Value, price and profit.* London: Swan, Sonnenschein & Co.
Marx, K., & Engels, F. (1970). *The German ideology* (Vol. 1). New York: International Publishers.
Marx, K., & Engels, F. (2002). *The communist manifesto.* London: Penguin.
McAfee, K. (2012). The contradictory logic of global ecosystem services markets. *Development and Change, 43,* 105–131.

McCarthy, M. A., Parris, K. M., van der Ree, R., et al. (2004). The habitat hectares approach to vegetation assessment: An evaluation and suggestions for improvement. *Ecological Management & Restoration, 5,* 24–27.

Moreno-Mateos, D., Power, M. E., Comín, F. A., & Yockteng, R. (2012). Structural and functional loss in restored wetland ecosystems. *PLoS Biology, 10*(1), e1001247.

Palmer, M. A., & Filoso, S. (2009). Restoration of ecosystem services for environmental markets. *Science, 325*(5940), 575.

Parkers, D., Newell, G., & Cheal, D. (2003). Assessing the quality of native vegetation: The 'habitat hectares' approach. *Ecological Management & Restoration, 4,* 29–38.

Pilgrim, J. D., Brownlie, S., Ekstrom, J. M. M., Gardner, T. A., von Hase, A., ten Kate, K., et al. (2013). A process for assessing offsetability of biodiversity impacts. *Conservation Letters, 6,* 376–384.

Quétier, F., & Lavorel, S. (2011). Assessing ecological equivalence in biodiversity offset schemes: Key issues and solutions. *Biological Conservation, 144,* 2991–2999.

Quétier, F., Regnery, B., & Levrel, H. (2014). No net loss of biodiversity or paper offsets?—A critical review of the French no net loss policy. *Environmental Science & Policy, 38,* 120–131.

Robertson, M. (2000). No net loss: Wetland restoration and the incomplete capitalization of nature. *Antipode, 32,* 463–493.

Robertson, M. (2006). The nature that capital can see: Science, state, and market in the commodification of ecosystem services. *Environment and Planning D: Society and Space, 24,* 367–387.

Ruhl, J. B., & Salzman, J. E. (2006). *The effects of wetland mitigation banking on people* (FSU College of Law, Public Law Research Paper No. 179). Available at SSRN: https://ssrn.com/abstract=878331.

Seagle, C. (2012). Inverting the impacts: Mining, conservation and sustainability claims near the Rio Tinto/QMM ilmenite mine in Southeast Madagascar. *Journal of Peasant Studies, 39,* 447–477.

Sears, P. (1964). Ecology—A subversive subject. *BioScience, 1,* 11–13.

Sharma, S. (2009). Baring life and lifestyle in the non-place. *Cultural Studies, 23,* 129–148.

Smith, N. (1998). Nature at the millenium: Production and re-enchantment. In B. Braun & N. Castree (Eds.), *Remaking reality: Nature at the millennium* (pp. 269–282). London and New York: Routledge.

Smith, N. (2006). Nature as accumulation strategy. In L. Panitch & C. Leys (Eds.), *Socialist register 2007: Coming to terms with nature* (pp. 16–36). London: Merlin.

Smith, N. (2010). *Uneven development* (3rd ed.). New York: Verso.

Smith, N., & Katz, C. (1993). Grounding metaphor: Towards a spatialized politics. In M. Keith & S. Pile (Eds.), *Place and the politics of identity* (pp. 66–81). London: Routledge.

Stokstad, E. (2008). New rules on saving wetlands push the limits of the science. *Science, 320*(5873), 162–163.

Sullivan, S. (2013). After the green rush? Biodiversity offsets, uranium power, and the 'calculus of casualties' in greening growth. *Human Geography, 6,* 80–101.

Tucker, G. M., Allen, B., Conway, M., Dickie, I., Hart, K., Rayment, M., et al. (2014). *Policy options for an EU no net loss initiative.* Report to the European Commission (with Annexes), Institute for European Environmental Policy (IEEP), London.

Virah-Sawmy, M., Ebeling, J., & Taplin, R. (2014). Mining and biodiversity offsets: A transparent and science-based approach to measure "no-net-loss". *Journal of Environmental Management, 143,* 61–70.

Virilio, P. (1997). *Open sky* (J. Rose, Trans.). New York: Verso.

Walker, S., Brower, A. L., Stephens, R. T. T., & Lee, W. G. (2009). Why bartering biodiversity fails. *Conservation Letters, 2,* 149–157.

Wilkinson, J. (2008). In-lieu fee mitigation: Coming into compliance with the new Compensatory Mitigation Rule. *Wetlands Ecology and Management, 17,* 53–70.

4

Value or Rent? A Marxist Analysis of Offsetting

E. Apostolopoulou, *Nature Swapped and Nature Lost*,
https://doi.org/10.1007/978-3-030-46788-3_4

Use-value is, by no means, the thing 'qu'on aime pour lui-même' in the production of commodities. Use-values are only produced by capitalists, because, and in so far as, they are the material substratum, the depositories of exchange-value.

Our capitalist has two objects in view: in the first place, he wants to produce a use-value that has a value in exchange, that is to say, an article destined to be sold, a commodity; and secondly, he desires to produce a commodity whose value shall be greater than the sum of the values of the commodities used in its production, that is, of the means of production and the labor-power, that he purchased with his good money in the open market. His aim is to produce not only a use-value, but a commodity also; not only use-value, but value; not only value, but at the same time surplus-value.

—Karl Marx (1887)

4.1 Nature, Labor and Value

> The notion of metabolism set up the circulation of matter, value and representations as the vortex of social nature. But, as the original German term, 'Stoffwechsel', better suggests, this is not simply a repetitive process of circulation through already established pathways. Habitual circulation there certainly is, but no sense of long-term or even necessarily short-term equilibrium. Rather, 'Stoffwechsel' expresses a sense of creativity… The production of urban nature is deeply political, but it has received far less scrutiny and seems far less visible, precisely because the arrangement of asphalt and concrete, water mains and garbage dumps, cars and subways seem so inimical to our intuitive sense of (external nature).
>
> Smith (2006a)

The relationship between capitalism and nature has been extensively investigated by Marxists and ecosocialists (see e.g., Marx and Engels 1970; Marx 1964; Engels 1940; Foster 1999, 2000, 2002; Foster et al. 2011; Burkett 1999; Harvey 1996, 2014; Clark and York 2005; Smith 2010; O'Connor 1998; Huber 2009, 2017; Malm 2016). In direct opposition to the reductionist hegemonic discourse that keeps humans and non-human nature in separation, the starting point for a Marxist approach to nature–society relationships is the key position of labor. As Marx explains in the first volume of Capital, labor is "an eternal natural necessity which mediates the metabolism between man and nature, and therefore human life itself" (Marx 1990, p. 133). This idea of a "metabolism" between humans and non-human nature with labor as the mediator is central to Marx's historical-materialist argument and transcends his work from his earlier to his more mature writings. Importantly, in Marx's view, there is no clear separation in the labor process between "man and nature," "culture and nature" or "natural and artificial": the labor process is understood as, simultaneously, wholly natural and wholly human and as such "it is construed dialectically as a moment of metabolism in which it is impossible to separate the natural from the human" (Harvey 2010a, p. 111).

Within this unitary conception of the labor process there is, however, a duality that relates to the issue of agency: humans are active agents in relation to non-human nature since they confront "the material of nature as a force of nature," setting in motion their natural forces (namely, their own body) in order to appropriate the materials of nature in a form adapted to human needs. Through this movement humans act upon non-human nature and transform it, and in this way they simultaneously transform their own nature. Marx's approach is thus fundamentally dialectical: humans cannot change nature without changing themselves and cannot change themselves without changing nature. As Harvey (2010a) explains this dialectic, of perpetually transforming oneself by transforming the world and vice versa, is key in understanding the evolution of human societies as well as the evolution of non-human nature. The notion of metabolism, despite being rarely discussed in the field of conservation studies, it is of profound importance for a political ecology critique of conservation under capitalism, most obviously because it directly challenges the mechanistic separation between society and nature. This separation has been prevalent in the history of conservation from the establishment of the first National Parks in the US, including Yellowstone and Yosemite (Kemf 1993; Keller and Turek 1998), until today and has been dramatically expressed in conservation-driven displacements (Agrawal and Redford 2009) and evictions (Brockington and Igoe 2006).

Importantly, labor takes place within the context of a specific mode of production and, therefore, its nature and implications cannot be understood without considering the characteristics and the contradictions of the mode of production. This means that the relation between humans and non-human nature is mediated by the relations of production. This is a crucial point that receives limited attention in mainstream analyzes of environmental degradation restricting their analytical power and their ability to offer accurate and historically specific explanations of environmental problems and crises. The most important of these relations that influence all others are property relations that determine who owns the means of production (e.g., land, forests, instruments of labor). Property relations have a fundamental contribution to the class character of the relations of production which is expressed most vividly in the contradiction between capital, the class who possess the means of production, and labor, the class who possess only their labor power which they must sell to survive (Smith 2010). This contradiction runs through the production process and gives to labor process its specific form under capitalism. In particular, under capitalism, the owner of the means of production exploits labor power with the goal to appropriate the surplus value[1] that is produced during the production process. Crucially, the goal of the latter is not the satisfaction of social needs but the acquisition of profit (Bitsakis 1997). The class structure of capitalism, most fundamentally the contradiction between labor and capital, makes capital accumulation the necessary condition for the reproduction of material life: the demand for profit along with the pursuit of accumulation for accumulation's sake lie at the heart of the capitalist mode of production and distinguishes

---

[1]On surplus value and value we read in Capital (Marx 1887, p. 104): "In the simple circulation of commodities, the two extremes of the circuit have the same economic form. They are both commodities, and commodities of equal value. But they are also use-values differing in their qualities, as, for example, corn and clothes. The exchange of products, of the different materials in which the labor of society is embodied, forms here the basis of the movement. It is otherwise in the circulation M-C-M, which at first sight appears purposeless, because tautological. Both extremes have the same economic form. They are both money, and therefore are not qualitatively different use-values; for money is but the converted form of commodities, in which their particular use-values vanish. To exchange £100 for cotton, and then this same cotton again for £100, is merely a roundabout way of exchanging money for money, the same for the same, and appears to be an operation just as purposeless as it is absurd. One sum of money is distinguishable from another only by its amount. The character and tendency of the process M-C-M is therefore not due to any qualitative difference between its extremes, both being money, but solely to their quantitative difference. More money is withdrawn from circulation at the finish than was thrown into it at the start. The cotton that was bought for £100 is perhaps resold for £100 + £10 or £110. The exact form of this process is therefore M-C-M', where M' = M + D M = the original sum advanced, plus an increment. This increment or excess over the original value I call — **surplus value**. The value originally advanced, therefore, not only remains intact while in circulation, but adds to itself a surplus value or expands itself. It is this movement that converts it into capital". Then in page 121 we also read: "The consumption of labor-power is at one and the same time the production of commodities and of surplus value. The consumption of labor-power is completed, as in the case of every other commodity, outside the limits of the market or of the sphere of circulation". And in page 149: "That portion of the working day, then, during which this reproduction takes place [Marx refers to the reproduction of worker's labor power], I call — necessary labor time, and the labor expended during that time I call — necessary labor [...] During the second period of the labor-process, that in which his labor is no longer necessary labor, the workman, it is true, labors, expends labor-power; but his labor, being no longer necessary labor, he creates no value for himself. He creates surplus value that, for the capitalist, has all the charms of a creation out of nothing. This portion of the working day, I name *surplus labor-time*, and to the labor expended during that time, I give the name of *surplus labor*". It is important, therefore, for a meaningful understanding of **surplus value**, to conceive it as a mere congelation of surplus labor-time, as materialized surplus labor, and, also, for a proper comprehension of **value**, to conceive it as a mere congelation of hours of labor, as materialized labor.

capitalism from its representation in mainstream economics as a simple system for satisfying human needs. It also directly affects nature–society relationships since non-human nature is treated as a means to the ultimate end of profit-making and capital accumulation (Sweezy 2004). It is in this key characteristic of capitalism, namely, the constant trend toward capital accumulation, that we can trace the key difference between the ecological crisis under capitalism and environmental degradation caused by human activities before its emergence (Magdoff 2002).

It is crucial to emphasize here that it is not capital that has invented surplus-labor: wherever a part of society possesses the monopoly of the means of production, the laborer, free or not free, must add to the working-time necessary for his own maintenance an extra working-time in order to produce the means of subsistence for the owners of the means of production (Marx 1990). Surplus-labor, namely, labor beyond the time required for the laborer's own maintenance, and appropriation by others of the product of this surplus-labor, namely, labor's exploitation, is, as Engels aptly explains in anti-Dühring (Engels 2017 [1878]), common to all forms of society that have existed hitherto, in so far these have moved in class antagonisms. However, "it is only when the product of this surplus-labor assumes the form of *surplus-value*, when the owner of the means of production finds the free laborer—free from social fetters and free from possessions of his own—as an object of exploitation, and exploits him for the purpose of the production of *commodities*, it is only then, according to Marx, that the means of production assume the specific character of *capital*" (ibid., added emphasis).

So how is production organized in capitalism? The fact that the main goal is the production of *commodities* is of major importance here. It means that use-value takes a secondary role with profound implications, as we will see later, for the relation to nature. But let's explore in more detail the key characteristics of commodity production. As Marx (1887, p. 199) writes in Capital: "the product appropriated by the capitalist is a use-value, as yarn, for example, or boots. But, although boots are, in one sense, the basis of all social progress, and our capitalist is a decided 'progressist', yet he does not manufacture boots for their own sake. *Use-value is, by no means, the thing 'qu'on aime pour lui-même' in the production of commodities. Use-values are only produced by capitalists, because, and in so far as, they are the material substratum, the depositories of exchange-value*. Our capitalist has two objects in view: in the first place, he wants to produce a use-value that has a value in exchange, that is to say, an article destined to be sold, a commodity; and secondly, he desires to produce a commodity whose value shall be greater than the sum of the values of the commodities used in its production, that is, of the means of production and the labor power, that he purchased with his good money in the open market. *His aim is to produce not only a use-value, but a commodity also; not only use-value, but value; not only value, but at the same time surplus-value*" (added emphasis).

In order to understand the role of non-human nature in the production of surplus value we need to understand what value is. This necessitates paying attention to the crucial distinction between *use value*, *exchange value*, and *value* (Burkett 2006; Marx 1887) which is, in most cases, ignored in both mainstream and critical social sciences literature on the economic valuation of nature. Use value can be defined as "the usefulness of a thing" (Harvey 2010a, p. 16) and Marx understands it as a combination of two elements: the material provided by nature and labor. Even though humans can change the form of materials and modify them, in this process they are constantly helped by non-human forces. Labor is, therefore, according to Marx, "not the only source of material wealth, i.e., of the use-values it produces. As William Petty says, labor is the father of material wealth, the earth is its mother" (Marx 1990, pp. 133–134). Marx introduces here a crucial distinction between *wealth*, the total use values at one's command, and *value*, the socially necessary labor time these use values represent (Harvey 2010a). Exchange value is, on the other hand, the "mode of expression, or form of appearance, of value" (Marx 1990, p. 128) and a representation of the human labor embodied in commodities.[2] Value is then defined as the socially necessary labor time objectified in commodities, namely, as Marx explains (1887, p. 28), the labor time required to produce any use-value "under the normal conditions of production, and with the average degree of skill

[2]As Harvey (2010a, pp. 18–19) explains in one of the examples he often cites also in his talks and lectures, "when you go to the supermarket you can find out the exchange-values, but you can't see or measure the human labor embodied in the commodities directly. It is that embodiment of human labor that has a phantom-like presence on the supermarket shelves".

and intensity of labor prevalent at that time." This means that something has value "only because human labour in the abstract has been embodied or materialized in it" (Marx 1894, p. 28) and what determines value's magnitude is the amount of the socially necessary labour time for its production (Marx 1887). Value, and this is of major theoretical importance, is thus defined as *capitalism's specific form of economic valuation*[3] whose substance can be found in *abstract labor* (Burkett 2006; Fine and Lapavitsas 2000; Greco and Apostolopoulou 2019; Harvey 2010a, b).

It is now starting to become more obvious from the above discussion that there is a constant tension between the true source of wealth (the combination of nature and labor) and capitalism's monetary representation of wealth as abstract labor time (Burkett 1999). The notion of the *free appropriation of nature* by capital is crucial in understanding this tension. The basis for free appropriation lies to the fact that useful properties are often acquired by non-human nature without any human labor involved. As Marx explains in the first paragraphs of Capital, Volume I (Marx 1887, Chapter 1): "one thing can have use value without having value. This is the case when its usefulness to man is not due to labor. Such is air, virgin soil, etc." And then in page 142 of Volume I, we read: "means of production never transfer more value to the production than they themselves lose during the labor process by the destruction of their own use-value. If such an instrument has no value to lose, if, in other words, it is not the product of human labor, it transfers no value to the product. It helps to create use-value without contributing to the formation of exchange value. In this class are included all the means of production supplied by Nature without human assistance, such as land, wind, water, metals in situ, and timber in virgin forests." It is important to highlight here that even though the capitalist production of nature has undoubtedly deepened since Marx was writing Capital and the capitalist penetration into non-human nature has taken novel forms with indicative examples geoengineering and synthetic biology (Adams 2017; Asafu-Adjaye et al. 2015; Smith 2006b), both Marx and Engels were aware of the tendency of capitalism to intensify both the real and formal

[3]It is important to point out here that this definition of value clearly distinguishes Marx from the physiocrats and the contemporary nature valuation debate (see also Burkett 2006).

subsumption of nature. As they wrote in German Ideology: "The nature that preceded human history ... today no longer exists anywhere (except perhaps on a few Australian coral-islands of recent origin)" (Marx and Engels 1845 [1970], p. 63).

Let's go back to the meaning of the free appropriation of nature under capitalism. Even though the value form instituted by capital predicates that the "natural environment has no value in itself - at least not in the calculus of capitalism" (Altvater 1990, p. 24), the production of value is endlessly transforming nature. The originality of Marx's analysis lies in denouncing the mystification operated by capital that appropriates the free gifts of nature within the production process to then make this appropriation invisible through the value form. This offers a neat understanding of economic processes as transformations of values (value formation and revaluation), materials and energy since, as it has been explained above, "labor process is a metabolic interaction between humans and nature" (Altvater 1993, p. 188). So, the key point here is not whether nature contributes to material wealth, because it is established that it does in a quite fundamental way through the free gifts appropriated by capital as invisibilized use values, *but whether nature relates to the value-form under capitalism* (Greco 2015; Greco and Apostolopoulou 2019). As Harvey (2010a) emphasizes, it is not an idea of Marx that the only valid notion of value derives from labor inputs; it is a historical social product specific to capitalism's relations of production. This is a profound contribution of the labor theory of value to current discussions about the economic valuation of nature, including ongoing debates on whether market-based instruments, such as biodiversity or carbon offsets, can be perceived as new ecological "commodities" that producing new "value(s)" from nature (Apostolopoulou et al. 2018). As we explain elsewhere (Greco and Apostolopoulou 2019) use values contribute qualitatively and quantitatively to the creation of exchange value and value but they are not systematically accounted for, and they cannot be, as has been shown by many ecosocialists (see e.g., Burkett 1999). Nature does not have labor power and does not produce value and surplus value and, therefore, when the "gifts" of nature are not produced through human labor, they cannot be unambiguously characterized as commodities even if they take a commodity form (Apostolopoulou et al. 2018; Harvey

2006). But I will come back to the issue of whether offsets can be seen as commodities, offering a more detailed analysis, a bit later.

Before closing the discussion on value and nature it is important to pay attention to one more critical distinction: *it is different to argue that nature has value from arguing that nature serves as a bearer of value and surplus value*. To quote Marx (1887, p. 354) in some length: "apart from the degree of development, greater or less, in the form of social production, the productiveness of labor is fettered by physical conditions. These are all referable to the constitution of man himself (race, etc.,), and to surrounding nature. The external physical conditions fall into two great economic classes: (1) natural wealth in means of subsistence, i.e., a fruitful soil, waters teeming with fish, etc., and (2) natural wealth in the instruments of labor, such as waterfalls, navigable rivers, wood, metal, coal, etc., [...] The fewer the number of natural wants imperatively calling for satisfaction, and the greater the natural fertility of the soil and the favorableness of the climate, so much less is the labor time necessary for the maintenance and reproduction of the producer. So much greater therefore can be the excess of his labors for others over his labor for himself. [...] *Favorable natural conditions alone, give us only the possibility, never the reality, of surplus labor, nor, consequently, of surplus value and a surplus-product.* The result of difference in the natural conditions of labor is this, that the same quantity of labor satisfies, in different countries, a different mass of requirements, consequently, that under circumstances in other respects analogous, the necessary labor time is different."[4] This also means, as Burkett (1999) points out, that if a natural condition of production becomes increasingly scarce, then the average productivity of

[4]See also the following excerpt from Capital, Volume 1 (Marx 1887, p. 28): "The value of a commodity would therefore remain constant, if the labor time required for its production also remained constant. But the latter changes with every variation in the productiveness of labor. This productiveness is determined by various circumstances, among others, by the average amount of skill of the workmen, the state of science, and the degree of its practical application, the social organization of production, the extent and capabilities of the means of production, and by physical conditions. [...] In general, the greater the productiveness of labor, the less is the labor time required for the production of an article, the less is the amount of labor crystallized in that article, and the less is its value; and vice versâ, the less the productiveness of labor, the greater is the labor time required for the production of an article, and the greater is its value. The value of a commodity, therefore, varies directly as the quantity, and inversely as the productiveness, of the labor incorporated in it".

the labor appropriating or utilizing this natural condition will be also reduced in terms of the material use values that are being produced per hour of work. The value of the commodity produced with the increasingly scarce natural condition will, accordingly, be increased due to the greater amount of social labor time that is now required to produce the same use values. This includes any additional necessary labor expended in the appropriation and utilization of substitute natural conditions. These effects are set out not only in Marx's discussions of the *natural basis of surplus value* but also in his analysis of the crises of capitalist reproduction due to crop failures (see Burkett 1999).

4.2 The Implications of Capitalist Commodity Production for Socionatures: Unraveling the Environmental Contradictions of Capitalism

The abstract logic that capitalism attaches to the creation and accumulation of social value also determines the relation with nature (Smith 2010). But what does this mean for socionatures? As Smith (2010) explained in his book *Uneven Development* in the capitalist mode of production nature becomes a universal means of production in the sense that it not only provides the subjects, objects, and instruments of production, but it is also in its totality an appendage to the production process. Non-human nature is used in every possible way to the ends of profit making leading to severe environmental problems, expressed, inter alia, in the depletion of natural resources, the destruction of habitats, the loss of species and climate disruption. Importantly, as we also read in Smith (2010, p. 50), this process does not occur evenly across the globe. As a matter of fact when we place the appearance of nature in its specific historical context, then the development of the material landscape presents itself as a process of the production of nature and the differentiated results of this production are the material symptoms of *uneven (geographical) development*. Smith, here, draws on Lefebvre (1991) and offers a Marxist analysis of the impacts of the capitalist production of nature and space pointing to one of its most important manifestations: the differentiation and unevenness produced across scales.

Importantly, the development of capitalism does not involve only a *quantitative*, in the sense of an expansion of human control over nature, but also a *qualitative* intensification of the appropriation of nature in the sense of the production of nature from within and as part of "second nature"[5] (Smith 2010, p. 65). This lies in the altered relation between use-value and exchange-value which renders the relation with nature primarily an exchange-value relation: in search of profit, capital stalks the earth, and attaches a price tag to parts (or even processes) of non-human nature that largely determines its fate (Smith 2010; Harvey 2014). The, sometimes unintended, consequences of the capitalist production process are thus a product of the way the capitalist mode of production works, namely, the fact that the aim of the labor process is to produce commodities and surplus value. To fulfill this aim, social-ecological considerations are inevitably undermined making capitalism a system based on the constant exploitation of the two sources of wealth: non-human nature and humans. This is aptly described by Engels (1940) in the *Dialectics of Nature*: "individual capitalists, who dominate production and exchange, are able to concern themselves only with the most immediate useful effect of their actions. Indeed, even this useful effect, in as much as it is a question of the usefulness of the commodity that is produced or exchanged, retreats right into the background, and *the sole incentive becomes the profit to be gained on selling*. The social science of the bourgeoisie, classical political economy, is predominantly occupied only with the directly intended social effects of human actions connected with production and exchange. This fully corresponds to the social organization of which it is the theoretical expression. When individual capitalists are engaged in production and exchange for the sake of the immediate profit, only the nearest, most immediate results can be taken into account in the first place. When an individual manufacturer or merchant

[5] Second nature, namely both the material creations of human labor and the societal institutions that facilitate and regulate the exchange of commodities, both directly and indirectly, is produced out of first nature and it is the nature produced by human activity, in opposition to the inherited non-human nature (Smith 2010).

sells a manufactured or purchased commodity with only the usual small profit, he is satisfied, and he is not concerned as to what becomes of the commodity afterwards or who are its purchasers. The same thing applies to the natural effects of the same actions [...] *In relation to nature, as to society, the present mode of production is predominantly concerned only about the first, tangible success; and then surprise is expressed that the more remote effects of actions directed to this end turn out to be of quite a different, mainly even of quite an opposite, character...*" (added emphasis).[6]

This brings us to Harvey's (2014) recent analysis on the contradictions in capital's relation to nature. By adopting a critical stance on whether the environmental contradictions of capitalism are "fatal" or not (p. 246) and by challenging analyses that imply that an environmental catastrophe and a subsequent apocalyptic end of capitalism are in the way, Harvey offers a historical, dialectical, and materialist explanation of the environmental contradictions of capitalism based on four key arguments. Firstly, the long history of capital in successfully resolving its ecological difficulties. Secondly, that fact that non-human nature is internalized within the circulation and accumulation of capital (despite mainstream analyses that portray nature as "external" to capitalist production). As he put it "capital is a working and evolving ecological system within which both nature and capital are constantly being produced and reproduced." Thirdly, the ability to turn environmental issues into big business (here Harvey also refers to Smith's analysis—and I would add also Cindy Katz's analysis—of nature as an accumulation strategy). Fourthly, the possibility that capital continues to circulate and accumulate in the midst of an environmental catastrophe. Here Harvey refers to disaster capitalism a term coined by Naomi Klein in her book *The Shock Doctrine* to refer to the rapid-fire corporate reengineering of societies still reeling from shock.[7]

---

[6]It is important to note here that Engels has also showed a clear awareness of the unequal distribution of environmental harm in capitalism and of the class character of environmental problems particularly within the city. In *The Condition of the Working-Class in England* (Engels 1844) he writes in detail about the miserable environmental conditions that workers have been facing in towns, while arguing that "the members of this money aristocracy can take the shortest road through the middle of all the labouring districts to their places of businesses, without ever seeing that they are in the midst of the grimy misery that lurks to the right and the left" (p. 85).

[7]See http://content.time.com/time/arts/article/0,8599,1666221,00.html.

I think that the crucial point in the above analysis is that, without denying or underestimating the various unintended consequences of the capitalist production of nature, we should approach the direction of the production of nature under capitalism as an open and not as a closed question. The destruction of the means of production can promote ways of securing natural resources (through for example state regulation) and lead to reforms that are compatible with non-structural changes whereas intra-capitalist competition and class struggle have also often been the driver of environmental regulation (see also Vlachou 2005) preventing the degradation of ecosystems. However, at the same time, there is no mechanism in the capitalist production process ensuring that once ecological disaster has been translated into an economic crisis for capitalism, a regulatory mechanism will emerge requiring ecological reorganization, at least for the whole of the capitalist mode of production (Foster 2002). And we cannot also expect, as theories of sustainable development and ecological modernization argue, that as environmental problems intensify some kind of ecological enlightenment will eventually arrive within existent institutions. Capitalism, like any other mode of production, relies on nature, and as Marx points out, the depletion and degradation of non-human nature does not make more sense in the long run than the destruction of the collective powers of labor, since both lie at the root of the production of all wealth (see also Harvey 2010a). Even though ecological decline can threaten capitalist production, individual capitalists, working in their own short-term interests and impelled by competition laws, are perpetually tempted to take the position of "*après moi Ie deluge!*" (Marx 1867, p. 282) with respect to both workers and nature. So even if we accept that a greener capitalism would ever be possible this will be a highly *contradictory, uncertain and geographically uneven* process: any kind of ecological sustainability in capitalism would not be for everyone and will not reverse class inequality.

Nature conservation, as we see in this book through the lenses of biodiversity offsetting, shows vividly the contradictions in the capitalist production of nature particularly since it is a policy that aims to, inter alia, address capitalism's concerns about the depletion of natural resources and often does so without avoiding reproducing or even deepening these contradictions. Natures under conservation status are land-controlling mechanisms (Kelly 2011) for specific purposes. Even though the ostensible purpose is the protection of nature for the public good, a variety of often-contradictory driving forces have shaped the history of protected areas and nature conservation. Their establishment by the state, and in some cases by non-state actors, played an important role in securing ecosystems, and often also public access to land, in the face of degradation caused by capitalism's expansion (Apostolopoulou and Adams 2015). However, the establishment of protected/conservation areas has very often involved the eviction or displacement of local people (Adams and Hutton 2007; Agrawal and Redford 2009; Brockington and Igoe 2006) enacting a separation between humans and non-human nature, which as Marx (1973 [1857]) argued in the Grundrisse, has been essential for the transformation of the natural world into objects of exchange (see also Igoe et al. 2010).

Nature conservation is penetrated by the core environmental contradiction in capitalism: namely, the need to combine the preservation of nature as a resource for future capital accumulation with the exploitation to support current accumulation. As mentioned above, conservation policies, as many other environmental policies, partly reflect pragmatic concerns of capitalists regarding the depletion of resources. These concerns are well described by Harvey when he talks about the importance of the means of production for capitalism (2010a, pp. 320–321): "when capitalists go into the market they need to find the means of production (constant capital)—this is a very broad and complicated category. By oversimplifying the point that we want to raise here is that the lack of availability of these material inputs and conditions constitutes a potentially serious barrier to sustained capital accumulation. In our own times, we see also the obvious impact of energy shortages and rising prices on capitalist dynamics. Barriers of this sort are plainly sources of perpetual concern within a capitalist system, and the equally perpetual need to overcome and circumvent barriers of this sort is often in the forefront of political activity (state subsidies and planning-particularly of physical infrastructures-research and development activity, vertical integration through mergers, etc.,)." Of course, this is only a part of the story, the other being that since the driving force of the capitalist production is profit and surplus value, then, conditions of social-ecological sustainability are never a priority.

At least since the 1970s, capitalism has dealt with the above key contradiction through environmental legislation and regulation to limit environmental despoliation at the hands of capital (Smith 2006b). The state has thus played a dual role as a guardian of both general capitalist class interests and nature, in the sense that it is the state that primarily regulates human interaction with non-human nature. This is not of course a unique characteristic of conservation or environmental policies, but generally characterizes the so-called welfare state (including for example public education and health). The clash between these two roles has often led to contradictory regulations, severe compromises and exceptions in the implementation of regulations (Apostolopoulou and Pantis 2009) leading to the degradation of ecosystems and the emergence of various environmental crises, with major current examples biodiversity loss and climate change. Environmental legislation has effectively created a new scarcity: allowable natural destruction (Smith 2006b) as evidenced in the case of conservation in the various regulations that aim to protect habitats, species, and ecosystems by constraining human activities. The neoliberal shift to market-based environmental and a conservation policy constitutes an attempt to supersede this scarcity and even profit from environmental destruction, as it is evident in the case of offsetting. But whether this means that offsets are commodities or offsetting creates value is a far more complicated issue, as we will see in the next section.

4.3 Value or Rent? The Political Economy of Biodiversity Offsetting

The fact that in offsetting, biodiversity or carbon units, in the form of offset credits, are given a price enabling their exchange across space and time has been a major focus of the critical literature on both biodiversity and carbon offsetting. Here my goal is not to offer a exhaustive review of this literature but to make a theoretical contribution to the debate around the role of offsetting by departing both from neoclassical economic theory that equates the value of something with its price, and the majority of the existing critical literature that tends to theorize offsets as commodities and the process of their creation and exchange as a process that generates value from nature (see e.g., Bracking et al. 2014; Dauguet 2015; Sullivan 2013; Robertson 2012). By drawing on Marxist political economy, and particularly on the labour theory of value and the theory of rent (see also Purcell et al. 2019; Andreucci et al. 2017; Huber 2017; Swyngedouw 2012; Harvey 2006; Smith 2006b; Burkett 2003; Foster 2000), I will rather show that offsetting should be primarily seen as a case where, as quite commonly happens under capitalism, quantitative ratios are established between commodities and the products, consequences, and results of other areas of economic activity. In such cases, social mechanisms for concentrating money can appear as markets even though they are not actually functioning as real markets (Fine and Lapavitsas 2000).

But let's see why this is the case. Following the Marxist understanding of value, described in Sect. 4.1., we can see that something has value only when human, abstract labor has been embodied or materialized in it (Marx 1894). This means that value's magnitude is determined by the amount of the socially necessary labor time for its production (Marx 1887). As mentioned earlier, this definition of value is not a theoretical or mental construction: it derives from an analysis of the capitalist system of production and, therefore, the argument that the substance of value can be found in abstract labour, refers to capitalism's specific form of economic valuation (Burkett 2006; Fine and Lapavitsas 2000). This clarification along with the Marxist distinction between use value, exchange value and value (Burkett 2006; Marx 1887), is crucial if we want to distinguish price from value, value from rent, and commodities, namely, products of human labour, from commodity forms, namely, objects that are not products of labour and which even though may have a price they do not have value (Harvey 2006; Marx 1887).

So far the above analysis can tell us that the construction of an equivalence between offset credits that have been attributed a price is not enough for the creation of value. In order to move further let's see what exactly an offset is. In most cases, an offset site constitutes a restored or conserved habitat (or species) represented as a score of credits that has a price. For example, the unit traded in US mitigation banking, in New South Wales BioBanking or in biodiversity offsetting in England, is a hectare of land under some kind of management or restoration or a habitat acre. Offset credits can be either bought in the context of a specific project or be available for sale "off the shelf" (Bayon et al. 2008; Defra 2013). This means that offset credits can be sold to buyers in advance of the approval of a project, resold later if not used, or acquired to build a credit portfolio to offset future development (see e.g., DEC 2007; Fish and Wildlife Service 2003).

For Marx (1887), commodity production involves producing commodities that both possess use-values and can be sold as use-values for others. Biodiversity (as well as carbon) offsets serve this double role. Buyers can either use them for internal mitigation, as it happens when buyers are purchasing their right to pollute or degrade nature or can sell them to others to use them for providing the required compensation for their development project, or they can do both. Offset credits have, therefore, both a use and an exchange value. The crucial question is whether their exchange value bears any relation to human labor, a question that needs to be explicitly addressed if we want to address the issue of the origin of value and that requires to consider the variations in how offsets are created and managed. To explore, therefore, whether the exchange value of offsets bears any substantive relation to labour in different instances, we need to examine the specific ways through which offsetting is delivered in each case. According to the additionality principle (BBOP 2009a; see also Chapter 2), ecological gains from offsetting could result from management interventions to restore an area, stop its degradation, or avert risks of future degradation and should be assessed in comparison to a counterfactual scenario (see e.g., Defra 2013; Commonwealth of Australia 2012). This means that producing offsets in some cases would require human labor. Indeed, two of the most important costs in the creation and maintenance of offsets are land costs and site work. As Bennett et al. (2017) explain, site costs are especially high for restoration or enhancement credits and can include site preparation and construction contingency funds, planting and irrigation, or invasive species controls.

Even though the above analysis should be the starting point in discussing whether value is created (or not) in offsetting, yet the only reference to this can be found in Smith (2006b), who argued that surplus value could be harvested from habitat restoration or re-creation, an argument that directly derived from the production of nature thesis. The issue of human labor expended on the production of offsets is particularly important in the Global South (see e.g., Bridge 2011; Corbera and Brown 2010) where workers often contribute to restoration or management activities and offsetting is primarily based on exploitative labor relations (see e.g., Dickinson and Berner 2010 on the Ambatovy nickel and cobalt mine project in Madagascar). An exemplary case here is Rio Tinto in Madagascar. The company has largely based its restoration activities on the labor of indigenous people who either remained unpaid for the work they offered or received one euro per day to plan trees at restoration sites (see WRM and Re:Common 2016; Seagle 2012).

The possibility to consider permits produced by human labor expended on planting carbon sinks as a new repository of value has been also acknowledged by Jones (2009), who nonetheless argues that such cases are rare in the emissions trading market. In the biodiversity offsetting literature, this issue has been largely ignored making it impossible, at least until now, to evaluate whether, and at what extent, labor is spent in offset creation. This is a key gap if we aim to understand whether new value is created in offsetting and what social relations of exploitation may be involved in the process and would require empirical work on specific offsetting programs. Two further clarifications are important here. Firstly, an in-depth discussion on the role of labor would also require distinguishing between absolute and differential rents, and between differential rent I and II (see Felli 2014). Secondly, given that not all labor undertaken in a capitalist economy necessarily results in value (Marx 1887, 1894), research is needed to explore whether in offsetting there is the material basis for the commensuration of diverse concrete labors characteristic of commodity production (see Fine and Lapavitsas 2000).

The above questions despite having a major theoretical and political importance remain so far largely unaddressed in the critical literature whose analytical approach is rather different. A shared view in most of the existing critical literature is that the measurement and calculation of biodiversity units is performative in the sense of performative economics (MacKenzie 2009). By often drawing on the work of Callon (Callon 2007; Callon and Muniesa 2003), the tools used to measure offset units are theorized as calculative devices or dispositifs de calcul that, through their performativity, create value (MacKenzie 2009). In this literature, broadly inspired by actor-network theory (Latour 2005), value originates from the performative action of a biodiversity (or carbon) unit, simply because that unit attracts a price. Scholars who explicitly draw on this tradition understand value as something that can be "made" by a socio-technical arrangement, a configuration of people, institutions and technologies that conduct the performation of markets through calculative devices (see e.g., Bracking et al. 2014). The theoretical focus lies, therefore, on the performativity of valuation tools. Without underestimating the contribution of this literature in showing the contested nature of measuring and calculating non-human world in terms of priced credits, it is important to note that this approach implies that capitalism can create value from the act of measuring nature (which means create value from nothing) leaving almost no space for a critical engagement with the historical-material reality behind appearances, namely, the political, economic and social structures that validate and naturalize the value system of capitalism, of which offsetting is just an expression. Indeed, analyses that attribute the creation of value to the application of offset metrics distance themselves from a critical political economic perspective and from the labor theory of value, failing to relate value to capitalist social relations. This ignores the pivotal role of human labor, and consequently of workers, in the creation of value pointing to a very different system instead of the capitalist system of exploitation. It also underestimates capitalism's accumulation crisis tendencies by implying that it is the choice of capital (and capitalists) to invent ways for endlessly producing value. Distinguishing between the form (the exchange value, namely, the quantitative ratio/equivalence of one commodity with another) and the substance of commodity value (abstract labor) is key in understanding that exchange value is frequently adopted by economic processes which may be unrelated to value creation

(Fine and Lapavitsas 2000), and that, therefore, *valuation and calculation devices are not necessarily related to value* (Apostolopoulou et al. 2018). Finally, it is also important to emphasize here that exploring whether offsetting involves the production of value in cases where human labour is involved in the process is fundamentally different from arguing that nature produces value. The importance of this distinction becomes more evident if we consider the key arguments of the literature that instead of focusing on human labour power focuses on the work of nature, a research topic that has recently resurfaced (see for example the dialogue between Kallis and Swyngedouw in Nature, Capitalism, Socialism, 2018). What is crucial to understand here is that work is not the same as labour power. As Harvey (2010a) explains, in Marx's analysis there is a clear distinction between the value of labor and the value of labor power that is of key importance. As Marx (1990, p. 677) writes: "in the expression 'value of labor', the concept of value is not only completely extinguished, but inverted, so that it becomes its contrary. It is an expression as imaginary as the value of the earth. These imaginary expressions arise, nevertheless, from the relations of production themselves. They are categories for the forms of appearance of essential relations." This distinction is crucial in understanding how labor power (and not labor) becomes a commodity under capitalism (Harvey 2010a) and is of profound importance in any discussion about the work of non-human nature and its contribution to the value-form under capitalism. As explained in the previous section, there is no question that the (invisibilized) use values of nature that enter into the production process are a major element in the production of super-profits and in the appropriation and redistribution of value through the mechanisms of rent. However, as already explained, there is no *labour power* involved in this process, and, therefore, no production of *value* and no extraction of *surplus value* as such (see Greco and Apostolopoulou 2019).

Let's now go back to the question of whether value is produced through the creation of biodiversity offsets. Importantly, in many cases, what is more relevant to address this question and understand the role of offsets for capital accumulation is the theory of rent (Andreucci et al. 2017; Swyngedouw 2012; Harvey 2006; Smith 2006b; Marx 1894). This becomes rather evident in the fact that the price that is given to offsets is primarily influenced by the price of land (see Bennett et al. 2017; Conway et al. 2013; DEC 2006) since developers and offset providers often must buy land for offset creation. Obviously, not just any land is suitable for the creation of offsets, and land itself cannot be produced but only commodified and privatized through the creation of a land market and the institution of private property (Harvey 2006). Through this process, land that is not the bearer of socially produced value, but of various use values arising from its innate condition, is turned into the bearer of a price, hence the source of rent. As Marx (1894, p. 461) explains in Volume III of Capital, "landed property is based on the monopoly by certain persons over definite portions of the globe, as exclusive spheres of their private will to the exclusion of all others." Moreover, and relatedly, offset providers often register their land on platforms, such as the Environmental Markets Exchange[8] in England, the BioBanking Public Registers[9] in New South Wales or the global Speciesbanking.com,[10] and assign prices to credits no matter whether any labor has been expended on them or not. Drawing parallels with Marx's theory of rent (1894), we can argue that this creates a basis for the formation of a potentially powerful rentier class, which owns and regulates access to land and nature (see also Harvey 2014). The surplus profit stems from controlling a limited natural resource that is transformed into ground rent, that is, falls into the possession of the owner of offset credits (Smith 2006b; Marx 1894). These credits can thus be considered as a *condition of production* in the sense that developers have to buy them to secure planning permission for their development projects. In that case, rent is a payment for the developer's right to access these necessary conditions in order to use the land of the development site (see also Felli 2014). In cases where landowners only sell their land but are not themselves involved in offsetting, revenue accruing to them is an appropriation of a portion of the total surplus value extracted in the production process, and land rent is a drain on capital accumulation (see

[8]See https://www.environmentbank.com/register/.

[9]See https://www.environment.nsw.gov.au/bimsprapp/biobankingpr.aspx.

[10]This website does not seem to operate anymore despite the enthusiasm with which the supporters of the economic valuation of nature had introduced it, for more information see: https://www.ecosystemmarketplace.com/articles/speciesbanking-com-for-all-your-species-needs/.

also Swyngedouw 2012). This means that even though exchange value is being attached to offsets, it is, nonetheless, in most cases a process unrelated to the creation of value.

By changing land uses and land access, offsetting can also impact the level of rent and the profits generated through landownership, potentially furthering processes of land speculation. So, biodiversity offsetting can have both positive and negative roles to play in relation to capital accumulation and severe socio-ecological conflicts can unfold between capitalists qua capitalists and rentiers (like land owners or ecological/ecosystem services owners), and between rentiers and users of the services these rentiers have exclusive rights to. As Andreucci et al. (2017) argue while the social relations of capital valorization in production develop through the capital-labor relation, the rent-based social relation unfolds through struggles over ownership of assets and the payments for the right and modalities of their use. This has been evident in the pilots in England where some landowners and developers perceived offsets as a new tax and showed unwillingness to adopt the policy while others saw in it an opportunity for speculation (see also Chapters 5 and 6). Importantly, in locations where land values have been increasing due to reasons independent of offsetting, acquiring land for offsets can become prohibitively expensive. This raises serious questions about the socio-spatial unevenness that offsetting creates since it often allows developers to locate offsets where it is cheapest for them. However, there is no univocal trend in this regard, given that the dynamics of rent are place and class specific (Apostolopoulou et al. 2018).

Crucially, whether offsets produce value or not does not change the fact that these policies represent nature in terms of exchange value rendering non-human nature a source of private profit and, in some cases, an instrument of speculation for buyers who aim to gain from boom conditions in biodiversity or carbon markets. The latter means that it is the potential exchange value that matters and, therefore, that conservation credits can be used as a personal ATM machine (as Harvey 2014 argues referring to housing) boosting aggregate demand for offsets. As explained above, credits secure for their owner a claim to receive annual rent and, therefore, any stream of revenue can be considered as the interest on some imaginary, fictitious capital (Marx 1894). This means that, in principle, conservation/habitat banks function as a particular branch of the circulation of interest-bearing capital, with nature treated as a financial asset whose control can be given to anyone who has access to land or credit. Like in all such forms of fictitious capital, what is traded is "a claim upon future revenues, which means a claim upon future profits from the use of the land or, more directly, a claim upon future labor" (Harvey 2006, p. 347). Conservation banks restrict access to and control of non-human nature, both now and in the future, creating a peculiar form of a fetishized nature (Katz 1998) or, as Bayon et al. (2008, p. 38) argue, for US conservation banking "a system of land stewardship that allows for the privatization of providing ecosystem services."

Although conservation banks are commercial ventures and should charge prices in a way that ensures they remain in business (Department of Defense and EPA 2008) and those who are involved in conservation banking mostly aim to gain profit and additional income (Bayon et al. 2008; DEC 2007), a successful outcome cannot be guaranteed (see also Smith 2006b). Conservation banking is a highly speculative and opportunist activity, as shown by the variety of initiatives that are launched with enthusiasm and then suddenly disappear because they fail, and it is particularly exposed to a discounting of values since it gives rise to credits that were not created in response to specific debits, and are, therefore, influenced by past and future conditions, including demand for compensation (EFTEC and IEEP 2010). Moreover, as stated in most banking schemes, the price of credits is determined by bank sponsors and buyers and sellers of credits are free to negotiate their price (see e.g., Department of Defense and EPA 2008; DEC 2007) whereas the same credit system that supposedly protects a particular species or habitat can lead to its destruction when it collapses (Smith 2006b). Literature on carbon credits offers ample evidence on the implications of speculation, discounting of values and financial crashes (Bond 2012; Lohmann 2012; McAfee 2012) that should have at least prevented the supporters of a further market intrusion into nature from creating similar markets for biodiversity.

I would like to close this section by referring to Marx's (1894) analysis in Capital, Volume III where he links the degradation of land with private property and with the emphasis on short-term profitability that prevents any long-term plan that would place at the core issues of social-ecological sustainability: "very conservative agricultural chemists, such as Johnston, admit that a really rational agriculture is confronted everywhere with insurmountable barriers stemming from private property. So do writers who are ex professo advocates of the monopoly of private property in the world…Johnston, Comte, and others, only have in mind the necessity of tilling the land of a certain country as a whole, when they speak of a contradiction between property and a rational system of agronomy. But the dependence of the cultivation of particular agricultural product upon the fluctuations of market-prices, and the continual changes in this cultivation with these price fluctuations—the whole spirit of capitalist production, which is directed toward the immediate gain of money—are in contradiction to agriculture, which has to minister to the entire range of permanent necessities of life required by the chain of successive generations. A striking illustration of this is furnished by the forests, which are only rarely managed in a way more or less corresponding to the interests of society as a whole, i.e., when they are not private property, but subject to the control of the state" (p. 617).

4.4 Planetary Urbanization, Socio-Spatial Transformations and the Remaking of Places, Livelihoods and Socionatures

Whether starting from a theoretical analysis of the role of offsets for capital accumulation as above, or from empirical observations of when and why offset programs and projects are being mostly used and how exactly they operate on the ground, an increasing number of scholars have started acknowledging that the creation of value from biodiversity offsets (as well as from other forms of market-based instruments including carbon offsets), is highly questionable (see e.g., Dempsey and Suarez 2016; Felli 2014; Lapeyre et al. 2015). Drawing on the analysis of the previous sections and my extensive empirical research on biodiversity offsetting, I would like to bring forward here a *retheorization of offsets* not as tools that commodify nature but *as mechanisms that by transforming nature to a set of replaceable and exchangeable parts enable and facilitate the reordering of the landscape according to economic development, and particularly, in the majority of cases, according to urban patterns, enacting and often legitimizing a profound land use change driven by the interests of specific sections of capital*. This retheorization also brings forward an issue of major theoretical and empirical importance that so far has remained largely unaddressed, namely, the increasing convergence of offsetting and urbanization. But to understand this interrelationship it is important to explain first how we conceptualize urbanization in the context of the current unprecedented urban expansion.

The way in which nature is socially mobilized, discursively scripted, economically enrolled, and physically transformed to produce socio-ecological assemblages that support the urbanization process has been the core focus of urban political ecology (Heynen et al. 2006; Kaika 2005, Loftus 2012; Swyngedouw 1996; Keil 2003). The field has made major contributions in analyzing the interrelationships between urbanization processes and social-ecological metabolism (Heynen 2014; Smith 2005) but, despite stressing the need to explore urbanization as a process that transcends city boundaries, it has mainly centered on the city (Ibañez and Katsikis 2014). An emerging strand of urban theory, by reworking the Lefebvrian (1970) process-oriented view of generalized urbanization (Brenner and Schmid 2014, 2015; Merrifield 2013; Angelo and Wachsmuth 2015; Wachsmuth 2014) and by building, inter alia, on the work of David Harvey (1985), Henri Lefebvre (1970), and Doreen Massey (2005), has recently theorized the need to shift the analytical and empirical attention to a set of worldwide socio-spatial transformations that have occurred since the 1980s manifesting the emergence of a planetary form of urbanization. Several critical scholars are increasingly paying attention to the fact that urbanization has been projected into novel geographies that perforate, crosscut, and ultimately explode the line between the urban and the rural profoundly transforming the earth (Brenner and Schmid 2015; UN 2018). The latest commodity boom, one of the most wide-ranging and persistent since the postwar period (IMF 2011; World Bank 2009), accompanied by large investments in infrastructure (IDB 2017) figure among the key determinants of this urban explosion (Arboleda 2016b). The term urban is thus increasingly used to refer to the broad process of the creation of a material physical infrastructure for production, circulation, exchange and consumption (Harvey 2012), and, therefore, to describe the multiscalar production and reproduction of the built environment regardless of population size or density (Arboleda 2016a; Angelo and Wachsmuth 2015).

Brenner and Schmid (2015, 2011) in a series of influential publications have shown that the novel socio-economic conditions, territorial formations and socio-metabolic transformations that serve the growth imperatives of an accelerating, increasingly planetary formation of capitalist urbanization, inter alia, include: (i) the formation of large-scale megacities and polynucleated metropolitan regions (Soja and Kanai 2006), (ii) the unprecedented densification of inter-metropolitan networks, requiring large-scale infrastructural investments across territories and continents as well as oceanic and atmospheric environments, (iii) the restructuring and repositioning of traditional "hinterlands," (iv) the remaking and spatial extension of large-scale land use systems devoted to resource extraction, the production and circulation of energy, and water and waste management, (v) the profound social and environmental transformation of vast, erstwhile "rural" areas through the expansion of large-scale industrial agriculture, the extension of global agribusiness networks, and the imposition of associated forms of land grabbing and territorial enclosure, and (vi) the operationalization of erstwhile "wilderness" spaces, including rainforests, deserts, alpine regions, polar zones, the oceans, and even the atmosphere itself (see Brenner and Schmid 2015). Interestingly, the constant attempt of capitalism to urbanize the countryside has been pointed out by Marx as early as 1857: "the history of classical antiquity is the history of cities, but of cities founded on landed prosperity and agriculture … the Middle Ages (Germanic period) begins with the land as the seat of history, whose further development then moves forward in the contradiction between town and countryside; the modern {age} is the urbanization of the countryside, not ruralization of the city as its antiquity (Marx 1973 [1857], p. 470)." Marx's observation in the Grundrisse even though made more than 150 years ago is timely as ever if we take into consideration the ongoing discussion on planetary urbanization.

The socio-spatial transformations that occur under conditions of planetary urbanization have profound implications for socionatures. Urban expansion is a major driver of global environmental change and biodiversity loss: it accelerates the loss of farmland, alters the climate and ecosystem processes, fragments habitats, and affects food security, access to land and social relations (Hahs et al. 2009; Grimm et al. 2008; Seto et al. 2011; d'Amour et al. 2017). Planetary urbanization signals a quantitative and qualitative intensification of the socionatural and spatial transformation of erstwhile rural places with the highest rates of land conversion projected to occur in remote biodiversity hotspots (Seto et al. 2012; Brenner and Schmid 2014; Merrifield 2013). As urbanization expands beyond geographically distinct urban areas into vast stretches of the Earth (Ibañez and Katsikis 2014) an alteration of the scope and scale of the social-ecological transformation of socionatures is also occurring. Urbanized regions and cities tend to extend their metabolic reach and become more connected to a planetary system of production and exchange and the development of regions of service and supply. These places can be seen as operational landscapes for resource extraction, agroindustrial production, energy circulation, and waste management (Brenner and Katsikis 2014). Importantly, even though several critical scholars have emphasized that these novel urban processes bring forward profound socio-metabolic transformations (Brenner and Schmid 2015; Schmid et al. 2018; Brenner and Katsikis 2014), the specific ways planetary urbanization remakes non-human nature and nature–society relationships remain fundamentally underexplored. The result is that the socionatural dimensions of urbanization that span the city and the countryside are so far largely unknown (Angelo and Wachsmuth 2015). These gaps have important implications for understanding how urbanization impacts socionatures, reflected in recent arguments about the importance of urbanization for wider political ecologies (see Apostolopoulou and Adams 2019; Arboleda 2016a; Angelo and Wachsmuth 2015; Robbins 2012; Kaika and Swyngedouw 2014).

In this book, by focusing on places that are becoming the geographical imprints of an expanded urban reality, I aim to contribute to the above emerging research agenda by exploring the convergence of market-based compensation mechanisms, particularly biodiversity offsets, and urban development. To this end I bring forward a (re)theorization of urbanization and offsetting as processes that, at different magnitudes and scales, remake nature(s) and place(s) by altering material metabolisms, social relations (and everyday lives), and spatial forms. I argue that urbanization and offsetting in their conjunction signal a profound transformation of socionatures through a double process of destruction (through urbanization) and conservation (through offsetting). I also believe that this approach can shift the analytical focus of the existing critical literature from the role of offsets in the production of value and unravel how offsetting may enable or/and facilitate the reordering of landscapes according to the patterns of economic (often urban) development, reinforcing a socially and geographically uneven production of nature.

As I have already mentioned there are still very limited studies on how offsetting transforms places, livelihoods and socionatures. Combining research on offsetting with the very rich critical literature on the socio-spatial dialectic (Soja 1980; Ruddick et al. 2017), place-making (Bebbington and Humphreys-Bebbington 2018; Pierce et al. 2011; Massey 1984), and the production of nature and space (Smith 2010; Ekers and Loftus 2013; Lefebvre 1991; Harvey 1996, 2012; Gandy 2003), can contribute to shifting the focus of research on offsetting to an exploration of how urbanization and offsetting are projected into novel geographies profoundly changing the contours of local lives and natures. Lefebvre's three-dimensional dialectics of space (Lefebvre 1970, 1991) are also of major importance here for understanding the historically geographically specific set of material, spatial, social, cultural, and economic processes that create and recreate places, natures, and livelihoods in each particular case. Moreover, and relatedly, the concepts of dispossession, displacement and emplacement (Çaglar and Schiller 2018; Perreault 2013; Agrawal and Redford 2009), as key characteristics of contemporary neoliberalism, and the right to nature, a term that I have discussed extensively elsewhere (Apostolopoulou 2019; Apostolopoulou and Adams 2019; Apostolopoulou and Cortes-Vazquez 2018; Cortes-Vazquez and Apostolopoulou 2019), can be used as key analytical tools for linking together space, place, nature and power. The combination of the above can lead to a retheorization of offsetting that has the analytical power to offer beyond the-state-of-the-art explanations of how the reordering of the landscape and the re-distribution of ecological losses and gains impacts local inhabitants, unraveling the variations in people's relation to urbanization, extractivism and infrastructure expansion, and bringing forward the centrality of people's struggles (Ruddick et al. 2017; Harvey 2012; Purcell 2014) in capital's endless attempts to reshape socionatures to its image. It can also unravel the role of power asymmetries and broader processes of uneven development bringing urban studies and urban political ecology closer to both classical studies in political ecology (Blaikie and Brookfield 1987; Peet and Watts 2000; Escobar 1999; Martinez-Alier 1995) and more recent research on (neo)extractivism (Burchardt and Dietz 2014; Arsel et al. 2016) and environmental justice (Apostolopoulou and Cortes-Vazquez 2018; Swyngedouw and Heynen 2003; Temper et al. 2018), ultimately shedding light on the novel forms of injustice(s) that are generated by the spatial (re)distribution of environmental resources that occurs under the convergence of offsetting and urban development.

4.5 A Double Land Grabbing: Insights from Across the Global South and North About the Links Between Offsetting, Social Inequality, and Justice

Here, following the argumentation of the previous sections, I offer some exemplary cases of how offsetting has been used to facilitate, enable or greenwash infrastructure expansion, extractivism and planetary urbanization leading to a profound reordering of the landscape with detrimental consequences for the contours of local lives and natures. Before describing these cases, it is important to make two observations. The first one is a theoretical note on land grabbing. Borras et al. (2012, p. 850) define contemporary land grabbing as "the capturing of control of relatively vast tracts of land and other natural resources through a variety of mechanisms and forms that involve large-scale capital that often shifts resource use orientation into extractive character, whether for international or domestic purposes, as capital's response to the convergence of food, energy and financial crises, climate change mitigation imperatives, and demands for resources from newer hubs of global capital." This means that land grabbing involves forms of green grabbing (Fairhead et al. 2012), namely, grabbing of land in the name of the environment as in the case of biodiversity offsetting. But, importantly, it also involves processes of un-green grabbing, which as I have already mentioned is a key aspect of offsetting, namely, processes of grabbing and exploiting land, including hitherto protected areas, without any green or eco-friendly argumentation (Apostolopoulou and Adams 2015). The second observation refers to the fact that the last few years offsets instead of being mainly used for voluntary corporate initiatives (as part of their corporate social responsibility), have been largely institutionalized due to a corporate push materialized by the World Bank Group. In particular, in 2012, the International Finance Corporation, the private sector arm of the World Bank Group, amended its Performance Standard 6, to allow lending to projects that would cause the destruction of what the IFC defines as "critical habitat" as long as the company that would receive the loan had an offsetting plan.[11] Since then, major International Financial Institutions, including the Asian Development Bank, the European Investment Bank, and the African Development Bank, have modified their Standards to incorporate the possibility of offsetting (Re:Common 2019).

The first case I would like to refer to here is a case that I have already mentioned in Chapter 3, namely, the case of Rio Tinto in Madagascar. Rio Tinto is an emblematic example of how offsetting can be used by big industries to obtain not only their license to trash, or address their social corporate responsibilities, but also to gain the support of big conservation NGOs. The story of Rio Tinto starts in the 1990s when the company, in the face of serious difficulties in securing new investment and mining licenses, promised that in its operations at selected mining sites, including the Fort Dauphin ilmenite mine, would not only compensate for the severe environmental impacts but would also deliver a "Net Positive Impact" on biodiversity. Following that logic, despite the

[11]For more information see: https://www.thebiodiversityconsultancy.com/approaches/ifc-ps6/.

fact that mining operations were projected to destroy more than 1650 hectares of a rare and unique coastal forest, Rio Tinto published in 2009 what seemed, at least to those who knew the details of the case, an ironic press release entitled "a mine at the rescue of the unique biodiversity of the littoral zone of Fort Dauphin" (QMM 2009). The rhetoric of environmentally friendly mining operations that would initially destroy the area but would ultimately lead to its improvement was primarily based on offsetting.

Both mining and offsetting had as a result dramatic shifts in the relationship between local communities and their land. QMM (a partnership organization between Rio Tinto and the Government of Madagascar) leased 6000 ha of land for ilmenite extraction: 12% of this went to conservation and an additional 31,275 ha outside of the mine were used for the establishment of biodiversity offsets. As Elnaschie (2013)[12] explains these drastic adjustments in land availability resulted in widespread displacement, threatening traditional livelihoods in the area. It is indicative that one of the offset sites, Bemangidy-Ivohibe, is in the north-eastern area of the Tsitongambarika forest, in Madagascar's Anosy region. Tsitongambarika is not only the largest expanse of lowland humid forest remaining in southern Madagascar, but also a key source of local livelihoods and, therefore, the restrictions that the management rules of the offset site imposed to local forest use posed a severe threat to food security.

QMM clearly used offsetting to facilitate and enable a double land grabbing: both in the mining and the offset site while blaming local communities for unsustainable use of the forest. In QMM's Social and Environmental Impact Assessment, the removal of forest for mining was justified on the grounds that the remaining forest, even in the absence of the mining operations, would have disappeared in 20 years' time due to intensive local use (Seagle 2012). In essence, a double land grabbing took place in Madagascar: the mining corporation took away land from communities both for the mine and for the area they used for locating the

[12]See http://www.ejolt.org/2013/11/rio-tinto-compensation-manipulation-in-southeast-madagascar/.

biodiversity offset raising fundamental issues of social, spatial, and environmental justice. QMM's mining activities caused not only restricted access to land and forest resources, justified by conservationist goals, but were also linked to rigid forest management regimes that criminalized traditional activities and enforced fines that further impoverished the affected communities (Elnaschie 2013). As WRM and Re:Common[13] explain based on their field visits in the area subsistence livelihoods of villagers have been made more precarious so that Rio Tinto could increase its profits. Importantly, despite the major social and economic impacts of the company's interventions, the delivery of restoration activities (in the context of offsetting) has been largely based on the labor of local people as mentioned in Chapter 3.

As the World Rainforest Movement (WRM) and Re:Common (WRM and Re:Common 2016) explain in another very informative joint publication[14] the need to mine in protected areas has increased the need for biodiversity offsetting for the mining industry with conservation NGOs playing a decisive role in securing the credibility and acceptability of offsets. Indeed, Rio Tinto launched its conservation strategy and "Net Positive Impact" goal at the 3rd IUCN World Conservation Congress in 2004, and has established various partnerships with conservation NGOs, including BirdLife International, Conservation International, Earthwatch Institute, Fauna & Flora International, and Royal Botanic Gardens Kew to achieve common conservation goals. Interestingly, in one of the joint reports of IUCN and Rio Tinto a key argument for using offsetting was that Rio Tinto by developing "robust methods of valuing ecosystem services" and "well-functioning markets" could use large non-operational land holdings to create new income streams for conservation.

Another case from Africa that I would like to refer to here is the Kalagala biodiversity offset in Uganda, one of the most important contemporary cases that show not only the double grabbing that offsets

[13] See https://wrm.org.uy/wp-content/uploads/2016/04/RioTintoBiodivOffsetMadagascar_report_EN_web.pdf.

[14] Available at: http://wrm.org.uy/wp-content/uploads/2016/06/Article_Rio_Tinto_in_Madagascar.pdf.

enable and bring about but also the strategic alliance between industries and offsetting by unraveling the clear links between the plans of particular sections of capital and the expansion and institutionalization of offsetting. The Kalagala offset was supposed to address the impacts of a 250-MW dam (the Bujagali dam) constructed by the US-based company AES Nile Power. The dam was completed in 2012, with financing from the World Bank Group and other major International Finance Corporations, and led to the flooding and loss of Bujadali Falls which were not only an important tourist attraction in the country but where also held sacred by the Busoga people living in the area. The dam has negatively affected more than 3000 households whose livelihoods were mostly based on subsistence farming and fishing. The World Bank Group and the Government of Uganda signed an agreement for the establishment of a biodiversity offset in the Kalagala Falls that did not only included conservation activities but also the development of the area into an eco-tourism site where the tourism companies that were based in Bujagali would relocate. As Re:Common (2019) reported after their visit to Kalagala the impacts on community's livelihoods have been detrimental and many families have been deprived of their farming land and fishing ground. This case becomes further controversial and complicated due to the fact that in 2013, the Government of Uganda signed a new agreement with the Chinese government to construct a new dam in the Isimba falls, 15 km north of Kalagala falls. The dam will destroy a large part of the Kalagala offset and after years of discussion and debates it has been recently agreed that the offset would be relocated upstream, meaning that a process of offsetting the offset would take place (ibid.)! Importantly, this case is not isolated but should be seen in the broader context of Uganda's emergence as a new biodiversity offsetting hotpot (Re:Common 2019). A key development in that respect is the launch of the Uganda Biodiversity Fund[15] in 2017 by the Wildlife Conservation Society and USAID with the goal to help extractive companies to save time and reputational risk by providing a cost-effective fund transfer and management mechanism to meet their commitments. The institutionalization of offsetting is expected to benefit enormously large corporations

[15]See https://ugandabiodiversityfund.org/wp-content/uploads/2019/05/UBF-Business-Plan.pdf.

with ongoing, major interests in the country, including the European oil companies that are currently very active in Uganda.

Another project of major importance in Africa that should be mentioned here is the Akyem Mining Project in Ghana (BBOP 2009b) that has affected 242 (1331 persons) households, including a settlement, living within the mining area and an additional 1443 households (7937 persons) with farms within the mining area. Operations in the area had dramatic impacts on local livelihoods including relocation, loss of farmlands, new diseases, and especially an upsurge in malaria cases as a result of the operation of open pits and other stagnant pools of water.[16] Even though according to anecdotal evidence offsetting further worsened the situation, critical research or publicly available data on the current situation are extremely limited.

The double land grab generated by the convergence of offsetting, infrastructure development, and urbanization has been also evident in Latin America. The resource exploration boom in the region has signaled one of the most intensive expansions of the urban fabric globally (Arboleda 2016b) with profound impacts on the involved socionatures. Previously unexplored swaths of land have been granted as concessions to different companies, often overlapping with native communal territories, protected nature reserves and important water resources, leading to a major pressure on the environment and livelihoods (Dourojeanni et al. 2010). This has coincided with the rapid institutionalization of offsetting in many countries across the region (Villarroya et al. 2014). It is worth mentioning here two emblematic cases of the above developments. The first case is the construction of highways in the Peruvian Amazon. The urbanization of the Amazon has been a multi-faceted process comprising the growth of old centers and the formation of new ones (Padoch et al. 2008) leading, already in the 1980s, to its description as an urbanized forest (Becker 1988). Highways represent one of the biggest drivers of deforestation (Vergara et al. 2014) causing major land use changes and driving migration, the creation of new urban settlements and the increase of new urban populations without proper land rights (CAF 2013). The case of the Corredor Vial Interoceánico Sur, a 2600 km

[16] See https://earthworks.org/stories/akyem_mine_ghana/.

long integrated highway connecting three southern ports in the Pacific coast with the three main cities in the South of Peru and the border with Brazil and Bolivia (Crezee 2017), offers an indicative case of how offsetting has intertwined with urban development in the Peruvian Amazon. Corredor's construction started in 2005 and finished in 2011. Several reports refer to the high levels of deforestation and illegal logging caused by improved access to previously isolated natural areas (Crezee 2017), and to the major opposition of indigenous groups (Cruz et al. 2008). Biodiversity offsetting has been the primary tool to address the impacts of the highway. As part of the offsetting strategy, 5.770 ha have been conserved through tourism enterprises, and 7 private protected areas, and a conservation network, have been established in the Madre de Dios region (Escalas et al. 2017). According to a recent assessment, a further investment in the conservation of four sites is required to compensate for the impacts of the project (ibid.).

A second case of major importance is the Cerrejón mine in Colombia, the biggest open-pit coal mine in Latin America located in the Guajira region (Christians et al. 2018). The mine has exploited around 13,000 ha of land, involving train and port infrastructure development, and has caused an important migration of labor and the creation of new suburban areas (Carbones del Cerrejón Limited y Conservación Internacional 2012). Cerrejón mine has been hugely contested locally due to its impacts on more than 100 Wayuu indigenous populations and Afro-Colombian communities (Betancur and Villa 2016; Teherán 2014). These included forced evictions, the destruction of sacred sites, the impoverishment of soils, and the contamination of water, and had devastating impacts on food sovereignty and local economies, previously based on farming and livestock keeping (Solly 2016; Vicente et al. 2011). Cerrejón operates since 1986 but in 2012 it developed a biodiversity offsetting strategy with a big international NGO, including reforestation and restoration actions, and the extension of a protected area (Arbelaez and Sagre 2015). The strategy also included a quite novel scheme combining offsetting for ecosystems, species and ecosystem services (Jacob et al. 2016). In both cases described above, biodiversity hotspots and native populations dependent on natural resources are being impacted and despite increasing anecdotal evidence that offsetting has both severely intensified inequalities and used to greenwash socially environmentally destructive projects and enable funding for the companies' operation, publicly available data and assessments are scarce (Crezee 2017; Christians et al. 2018; Betancur and Villa 2016; Teherán 2014).

Crucially, more and more similar examples are emerging almost every day crossing geographical boundaries and divides between the Global South and North. One such case that I want to describe here is the tar sands and pipelines in Alberta, Canada. The tar sands, also known as bitumen deposits, are the second largest source of oil globally and are considered an emblematic case of the planetary spread of urbanization (Brenner 2014). Tar sands have been mainly located in remote and sparsely populated areas that have significantly expanded due to the migration of workers. Even though they initially relied on temporary labour with work camps built close to the sites, settlements have gradually grown in size and complexity and become more permanent (Ferguson 2014). Their constant expansion has also increased the need for infrastructure projects (Hallstrom et al. 2015). Over the last decade, offsets have emerged as a leading solution to the major ecological consequences of the tar sands boom (Weinhold 2011) by various actors, including the oil and gas industry, provincial governments, conservation NGOs and some First Nations (Dyer et al. 2008), and have been widely used to re-brand Canadian tar sands as non-environmentally destructive (Hackett 2016). These have included voluntary and quasi-voluntary offsetting programs that have been implemented over the last decade under the Alberta Conservation Association's Habitat Conservation Initiative (Poulton 2014), such as the Shell Canada-True North Forest project that started in 2011.[17] These programs involved the purchase of boreal forest habitat for conservation and recreation purposes to offset habitat disturbance from the tar sands mining operations and included detailed offsetting methodologies for achieving No Net Loss through habitat creation and restoration activities (Poulton 2014). Importantly, evidence suggests that severe dispossession and displacement have occurred at both the extraction and the offset sites with major

[17] See https://www.shell.ca/en_ca/sustainability/environment/land-conservation.html#vanity-aHR0cHM6Ly93d3cuc2hlbGwuY2EvZW4vZW52aXJvbm1lbnQtc29jaWV0eS9lbnZpcm9ubWVudC10cGtnL3RydWUtbm9ydGguaHRtbA.

impacts on local communities (Hackett 2015, 2016). There are also several individual offsets required by regulators, including three cases where the National Energy Board approved pipeline development in caribou habitat contingent upon the provision of habitat compensation (Poulton 2014). These include the recent (2018) White Spruce project which involves the construction of two crude oil pipelines crossing Fort McKay's territory where inhabitants hold treaty and aboriginal rights, including rights to hunt, fish, trap, and gather culturally important natural resources for social, cultural, consumption, recreation, and general enjoyment purposes (TransCanada Pipelines Limited 2018). Part of the White Spruce project is located within lands designated federally as critical caribou habitat, a cultural keystone species, and involves an offsetting strategy to compensate for caribou habitat's destruction, as required by the Alberta Energy Regulator in 2018 (ibid.). This is a case of profound importance not only for Canada but also for the whole of North America where pipelines are altering landscapes, primarily in indigenous, rural, poor and people-of-color communities (Winter 2018; Ordner 2018).

The convergence of offsetting and urbanization is also increasingly observable in Europe. I will mention here only two recent cases that show this well. The first is the Trans-Adriatic-Pipeline (TAP), a large-scale gas pipeline of 878 km which connects with the Trans Anatolian Pipeline at the Greek-Turkish border, then crossing Northern Greece, Albania and the Adriatic Sea before coming ashore in Southern Italy to connect to the Italian gas network (TAP 2015). Construction started in 2015 in Albania and TAP is expected to be ready for operation by 2020. TAP involves the construction of new roads, compressors, and metering stations, which are expected to promote economic development in the affected localities and attract large numbers of workers with projections talking about more than 10.000 jobs in Greece and Albania (Danchev et al. 2013; Oxford Economics 2013). The pipeline passes through a wide range of land use types impacting around 21,000 lands parcels utilized by local households and affecting important ecosystems, including Natura 2000 sites, and approximately 45,000 landowners and land users living along its corridor (CEE 2017; TAP 2013a, b, c). The project has attracted intense scrutiny by stakeholders, the press, and civil society and has been subject to a massive number of complaints from affected communities and NGOs, particularly in Greece and Italy (European Investment Bank 2017). Biodiversity offsetting has been selected as the key tool to address the residual impacts of the pipeline's construction and consists on an ambitious strategy aiming to offset its adverse impacts on people, their rights, livelihoods, culture, and the environment (TAP 2016).

The second case is the expansion of Heathrow airport in the UK (Airport Expansion Consultation 2018) that is expected to finish in 2025 and create up to 40,000 jobs locally (McDonough 2015). It involves the creation of a third runaway (approved in 2018), as well as airport-related development, supporting facilities and transport infrastructure (Airport Expansion Consultation 2018) and it will affect approximately 973,000 households due to the intensification of urbanization (including increased noise, traffic and air pollution), uneven urban development, and expropriation and demolition of local villages and houses (Airport Expansion Consultation 2018; Department of Transport 2017a; Aviation Environment Federation 2018). The expansion will also have severe impacts on ecosystems and species, including areas of community importance under the Habitats and Birds Directives, Ramsar sites, and Sites of Special Scientific Interest (Department for Transport 2017b). Heathrow Airport Limited has prepared a detailed offsetting strategy that, inter alia, includes a peatland restoration project to offset carbon emissions (Heathrow Airport Ltd 2018). Heathrow's expansion has a long history of conflicts with local communities and is currently the subject of major controversies.[18]

The list goes on including the Shuakhevi hydropower plant in Georgia, promoted since 2017 from the European Bank for Reconstruction and Development as the world's first carbon neutral hydropower project, the BRUA Gas Pipeline, crossing Bulgaria-Romania-Hungary-Austria, and several other cases of European and international significance showing. This rapidly growing list of offsetting projects is the clearest manifestation of the fact that despite the scarce evaluations of their impacts and the extremely limited research on the combined effects of their convergence with extended urbanization, offsetting mechanisms

[18]See for example: http://stopheathrowexpansion.co.uk/our-history/.

are spreading across the globe altering socionatural metabolisms, places and livelihoods, and actively remaking the earth in line with the patterns of capitalist growth.

References

Adams, W. M. (2017). Geographies of conservation I: De-extinction and precision conservation. *Progress in Human Geography, 41,* 534–545.

Adams, W. M., & Hutton, J. (2007). People, parks and poverty: Political ecology and biodiversity conservation. *Conservation and Society, 5*(2), 147–183.

Agrawal, A., & Redford, K. (2009). Conservation and displacement: An overview. *Conservation and Society, 7,* 1–10.

Airport Expansion Consultation. (2018). *The Case for Heathrow Expansion.* London: Airport Expansion Consultation.

Altvater, E. (1990). The foundations of life (nature) and the maintenance of life (work). *International Journal of Political Economy, 20,* 10–34.

Altvater, E. (1993). *The future of the market: An essay on the regulation of money and nature after the collapse of "actually existing socialism"*. London: Verso.

Andreucci, D., García-Lamarca, M., Wedekind, J., & Swyngedouw, E. (2017). 'Value Grabbing': A political ecology of rent. *Capitalism Nature Socialism, 28,* 28–47.

Angelo, H., & Wachsmuth, D. (2015). Urbanizing urban political ecology: A critique of methodological cityism. *International Journal of Urban and Regional Research, 39,* 16–27.

Apostolopoulou, E. (2019). Beyond post-politics: Offsetting, depoliticization and contestation in a community struggle against executive housing. *Transactions of the Institute of British Geographers.* https://doi.org/10.1111/tran.12354.

Apostolopoulou, E., & Adams, W. M. (2015). Neoliberal capitalism and conservation in the post-crisis era: The dialectics of 'green' and 'un-green' grabbing in Greece and the UK. *Antipode, 47,* 15–35.

Apostolopoulou, E., & Adams, W. M. (2019). Cutting nature to fit: Urbanization, neoliberalism and biodiversity offsetting in England. *Geoforum, 98,* 214–225.

Apostolopoulou, E., Adams, W. M., & Greco, E. (2018). Biodiversity offsetting and the production of "equivalent natures": A Marxist critique. *ACME: An International Journal for Critical Geographies, 17*(3), 861–892.

Apostolopoulou, E., & Cortes-Vazquez, J. (Eds.). (2018). *The right to nature: An academic-activist dialogue on environmental justice, social movements and neoliberal natures.* Abingdon: Routledge-Earthscan.

Apostolopoulou, E., & Pantis, J. D. (2009). Conceptual gaps in the national strategy for the implementation of the European Natura 2000 conservation policy in Greece. *Biological Conservation, 142,* 221–237.

Arbelaez, D. M. L., & Sagre, J. D. Q. (2015). Compensaciones de biodiversidad: experiencias en Latinoamérica y aplicación en el context colombiano. *Gestión Y Ambiente, 18,* 159–177.

Arboleda, M. (2016a). In the nature of the non-city: Expanded infrastructural networks and the political ecology of planetary urbanisation. *Antipode, 48,* 233–251.

Arboleda, M. (2016b). Spaces of extraction, metropolitan explosions. *International Journal of Urban and Regional Research, 40,* 96–112.

Arsel, M., Hogenboom, B., & Pellegrini, L. (2016). The extractive imperative in Latin America. *The Extractive Industries and Society, 3,* 880–887.

Asafu-Adjaye, J., Blomqvist, L., Brand, S., Brook, B., De Fries, R., Ellis, E., et al. (2015). *An ecomodernist manifesto.* Available at: http://www.ecomodernism.org/manifesto.

Aviation Environment Federation. (2018). *Campaigning against unsustainable expansion.* Available at: https://www.aef.org.uk/campaigning-against-unsustainable-expansion/.

Bayon, R., Fox, J., & Carroll, N. (2008). *Conservation and biodiversity banking: A guide to setting up and running biodiversity credit trading systems.* London: Earthscan.

BBOP (Business and Biodiversity Offsets Programme). (2009a). *Biodiversity offset design handbook* (Appendices). Washington, DC: BBOP.

BBOP (Business and Biodiversity Offsets Programme). (2009b). *BBOP pilot project case study. Akyem Gold Mining project, Eastern Region, Ghana.* Washington, DC: BBOP.

Bebbington, A., & Humphreys-Bebbington, D. (2018). Mining, movements and sustainable development: Concepts for a framework. *Sustainable Development, 26,* 441–449.

Becker, B. K. (1988). Significado atual da fronteira: Uma interpretação geopolítica a partir da Amazônia brasileira. In C. Aubertin (Ed.), *Fronteiras* (pp. 60–89). Brasilia: Universidade de Brasilia (UNB)/ORSTOM.

Bennett, G., Gallant, M., & ten Kate, K. (2017). *State of Biodiversity Mitigation 2017*. Markets and Compensation for Global Infrastructure Development: Ecosystem Marketplace, Forest Trends, Washington D.C.

Betancur, A. C., & Villa, W. (2016). *Mining and indigenous peoples in Colombia*. IWGIA.

Bitsakis, E. (Μπιτσάκης, E. I.). (1997). Οικολογία: μια μαρξιστική προσέγγιση. *Ουτοπία, 26*, 23–52 (in Greek).

Blaikie, P., & Brookfield, H. (1987). *Land degradation and society*. London: Methuen.

Bond, P. (2012). Emissions trading, new enclosures and eco-social contestation. *Antipode, 44*, 684–701.

Borras, S. M., Jr., Franco, J. C., Gómez, S., Kay, C., & Spoor, M. (2012). Land grabbing in Latin America and the Caribbean. *The Journal of Peasant Studies, 39*, 845–872.

Bracking, S., Bond, P., Brockington, D., Büscher, B., Igoe, J. J., Sullivan, S., & Woodhouse, P. (2014). *Human, non-human and environmental value systems: An impossible frontier?* (LSCV Working Paper Series No. 1). Leverhulme Centre for the Study of Value, School of Environment, Education and Development, The University of Manchester.

Brenner, N. (Ed.). (2014). *Implosions/explosions: Towards a study of planetary urbanization*. Berlin: Jovis.

Brenner, N., & Katsikis, N. (2014). Is the mediterranean urban? In N. Brenner (Ed.), *Implosions/explosions: Towards a study of planetary urbanization* (pp. 428–459). Berlin: Jovis.

Brenner, N., & Schmid, C. (2011). Planetary urbanization. In M. Gandy (Ed.), *Urban constellations* (pp. 10–13). Berlin: Jovis.

Brenner, N., & Schmid, C. (2014). Planetary urbanization. In N. Brenner (Ed.), *Implosions/explosions: Towards a study of planetary urbanization* (pp. 160–163). Berlin: Jovis.

Brenner, N., & Schmid, C. (2015). Towards a new epistemology of the urban? *City, 19*, 151–182.

Bridge, G. (2011). Resource geographies 1: Making carbon economies, old and new. *Progress in Human Geography, 35*, 820–834.

Brockington, D., & Igoe, J. (2006). Eviction for conservation: A global overview. *Conservation and Society, 4*(3), 424–470.

Burchardt, H. J., & Dietz, K. (2014). (Neo-)extractivism, a new challenge for development theory from Latin America. *Third World Quarterly, 35*, 468–486.

Burkett, P. (1999). *Marx and nature: A red and green perspective*. New York: St. Martin's Press.

Burkett, P. (2003). The value problem in ecological economics lessons from the physiocrats and Marx. *Organization & Environment, 16,* 137–167.

Burkett, P. (2006). *Marxism and ecological economics: Toward a red and green political economy* (Vol. 11). Historical Materialism Book Series. Boston: Brill.

CAF. (2013). *Carretera Interoceánica Sur del Perú: Retos e Innovación*. Bogota: CAF.

Çaglar, A., & Schiller, N. G. (2018). *Migrants and city-making: Dispossession, displacement, and urban regeneration*. Durham: Duke University Press.

Callon, M. (2007). What does it mean to say that economics is performative? In D. MacKenzie, F. Muneisa, & L. Sui (Eds.), *Do economists make markets? On the performativity of economics* (pp. 311–357). Princeton: Princeton University Press.

Callon, M., & Muniesa, F. (2003). Les marchés économiques comme dispositifs collectifs de calcul. *Réseaux, 21,* 189–233.

Carbones del Cerrejón Limited y Conservación Internacional. (2012). *Propuesta modelo de compensación ambiental aplicado a la intervención por los permisos de aprovechamiento forestal*. Departamento de Gestión Ambiental. Carbones del Cerrejón Limited.

CEE Bankwatch Network. (2017). *Land lost but not forgotten: Impacts of the trans-Adriatic Pipeline on the land and livelihoods of farmers*. Available at: https://bankwatch.org/publication/land-lost-but-not-forgotten.

Christians, C., Mears, L., Whitmore, A., & Rhoades, H. (2018). Self-determination as resistance: Re-asserting control over natural resources in Colombia. In E. Apostolopoulou & J. Cortes-Vazquez (Eds.), *The right to nature: Environmental justice, social movements and neoliberal natures* (pp. 17–28). Abingdon: Routledge-Earthscan.

Clark, B., & York, R. (2005). Dialectical materialism and nature: An alternative to economism and deep ecology. *Organization & Environment, 18,* 318–337.

Commonwealth of Australia. (2012). *Environment protection and Biodiversity Conservation Act 1999 environmental offsets policy*. Department of the Environment, Canberra, ACT, Australia. Available at: http://www.environment.gov.au/system/files/resources/12630bb4-2c10-4c8e-815f-2d7862bf87e7/files/offsets-policy_2.pdf.

Conway, M., Rayment, M., White, A., & Berman, S. (2013). *Exploring potential demand for and supply of habitat banking in the EU and appropriate*

design elements for a habitat banking scheme (Final Report Submitted to DG Environment). UK: ICF GHK and BIO Intelligence Service.
Corbera, E., & Brown, K. (2010). Offsetting benefits? Analysing access to forest carbon. *Environment and Planning a, 42,* 1739–1761.
Cortes-Vazquez, J. A., & Apostolopoulou, E. (2019). Against neoliberal natures: Environmental movements, radical practice and "the right to nature". *Geoforum, 98,* 202–205.
Crezee, B. (2017). *Preocupantes efectos de una carretera: la interoceánica y la deforestación que causa.* Mongabay. Available at: https://es.mongabay.com/2017/10/preocupantes-efectos-una-carretera-la-interoceanica-la-deforestacion-causa/.
Cruz, J., Cenzano, A., & Supa, H., et al. (2008). *Impactos ambientales y sociales de la carretera Interoceánica* (Parliamentary Report).
d'Amour, C. B., Reitsma, F., Baiocchi, G., et al. (2017). Future urban land expansion and implications for global croplands. *Proceedings of the National Academy of Sciences, 114,* 8939–8944.
Danchev, S., Paratsiokas, N., & Tsakanikas, A. (2013). *Economic impact from the construction and operation of the trans Adriatic pipeline on Greek territory.* Athens: Foundation for Economic and Industrial Research.
Dauguet, B. (2015). Biodiversity offsetting as a commodification process: A French case study as a concrete example. *Biological Conservation, 192,* 533–540.
DEC. (2006). *BioBanking: A biodiversity offsets and banking scheme: Conserving and restoring biodiversity in NSW* (Working Paper). Sydney: Department of Environment and Conservation NSW.
DEC. (2007). *BioBanking: Biodiversity banking and offsets scheme: Scheme overview.* Sydney: Department of Environment and Climate Change NSW.
Defra (Department for Environment, Food and Rural Affairs). (2013). *Biodiversity offsetting in England green paper*. Department for Environment, Food and Rural Affairs. London, UK. Available at: https://consult.defra.gov.uk/biodiversity/biodiversity_offsetting/supporting_documents/20130903Biodiversity%20offsetting%20green%20paper.pdf.
Dempsey, J., & Suarez, D. C. (2016). Arrested development? The promises and paradoxes of "selling nature to save it". *Annals of the American Association of Geographers, 106*(3), 653–671.
Department for Transport. (2017a). *Revised draft airports national policy statement: New runway capacity and infrastructure at airports in the South East of England*. Presented to Parliament pursuant to Section 9(2) of the Planning Act 2008. London: Department for Transport.

Department for Transport. (2017b). *Appraisal of sustainability: Draft airports national policy statement*. Consultation draft. Exeter: Department for Transport and WSP, Parsons Brinckerhoff.

Department of Defense and EPA (Environmental Protection Agency). (2008). *Compensatory Mitigation for Losses of Aquatic Resources; Final Rule*. U.S. Army Corps of Engineers, DoD and Environmental Protection Agency.

Dickinson, S., & Berner, P. O. (2010). Ambatovy project: Mining in a challenging biodiversity setting in Madagascar. In S. M. Goodman & V. Mass (Eds.), *Biodiversity, exploration, and conservation of the natural habitats associated with the Ambatovy project*. Malagasy Nature 3 (pp. 2–13). Antananarivo: Association Vahatra.

Dourojeanni, M., Barandarián, A., & Dourojeanni, D. (2010). Amazonía peruana en 2021: Explotación de recursos naturales e infraestructura ¿Qué está pasando? ¿Qué es lo que significa para el futuro? *Bois et Forets des Tropiques, 305*, 77–82.

Dyer, S., Grant, J., Lesack, T., et al. (2008). *Catching up: Conservation and biodiversity offsets in Alberta's boreal forests*. Ottawa: Canadian Boreal Initiative.

EFTEC and IEEP. (2010). *The use of market-based instruments for biodiversity protection—The case of Habitat Banking*. Technical Report for European Commission DG Environment, EFTEC and IEEP, London.

Ekers, M., & Loftus, A. (2013). Revitalizing the production of nature thesis: A Gramscian turn? *Progress in Human Geography, 37*, 234–252.

Elnaschie, S. (2013). *Rio Tinto: Compensation manipulation in Southeast Madagascar*. Available at: http://www.ejolt.org/2013/11/rio-tinto-compensation-manipulation-in-southeast-madagascar/.

Engels, F. (1844). *The condition of the working-class in England*. Moscow: Progress Publishers.

Engels, F. (1940). *Dialectics of nature*. New York: International Publishers.

Engels, F. (2017 [1878]). *Anti-dühring*. São Paulo: Boitempo Editorial.

Escalas, A., Ayllón, R., Escobedo Grandez, J., et al. (2017). *Estudios de caso para el desarrollo de la compensación ambiental en el Perú: Madre de Dios*. Lima: CSF.

Escobar, A. (1999). After nature: Steps to an antiessentialist political ecology. *Current Anthropology, 40*, 1–30.

European Investment Bank. (2017). *Environmental and social data sheet TAP—Trans Adriatic Pipeline*. Available at: http://www.eib.org/attachments/registers/66312481.pdf.

Fairhead, J., Leach, M., & Scoones, I. (2012). Green Grabbing: A new appropriation of nature? *The Journal of Peasant Studies, 39,* 237–261.
Felli, R. (2014). On climate rent. *Historical Materialism, 22,* 251–280.
Ferguson, N. (2014). From coal pits to tar sands: Examining labour migration between the Athabasca oil sands and an Atlantic Canadian region. *Just Labour, 17–18,* 106–118.
Fine, B., & Lapavitsas, C. (2000). Markets and money in social theory: What role for economics? *Economy and Society, 29,* 357–382.
Fish and Wildlife Service. (2003). *Federal guidance for the establishment, use, and operation of conservation banks.* Washington, DC: United States Department of the Interior.
Foster, J. B. (1999). Marx's theory of metabolic rift: Classical foundations for environmental sociology. *American Journal of Sociology, 105,* 366–405.
Foster, J. B. (2000). *Marx's ecology: Materialism and nature.* New York: Monthly Review Press.
Foster, J. B. (2002). *Ecology against capitalism.* New York: Monthly Review Press.
Foster, J. B., Clark, B., & York, R. (2011). *The ecological rift: Capitalism's war on the earth.* New York: NYU Press.
Gandy, M. (2003). *Concrete and clay: Reworking nature in New York City.* Cambridge: MIT Press.
Greco, E. (2015). *Value or rent? Leverhulme centre for the study of value* (Working Paper No. 12). University of Manchester.
Greco, E., & Apostolopoulou, E. (2019). Value, rent and nature: The centrality of class. *Dialogues in Human Geography, 10*(1).
Grimm, N. B., Faeth, S. H., Golubiewski, N. E., et al. (2008). Global change and the ecology of cities. *Science, 319,* 756–760.
Hackett, R. (2015). Offsetting dispossession? Terrestrial conservation offsets and First Nation treaty rights in Alberta, Canada. *Geoforum, 60,* 62–71.
Hackett, R. (2016). 'Shell games', displacement and the reordering of boreal landscapes in Alberta, Canada. *Area, 48,* 153–160.
Hahs, A. K., McDonnell, M. J., McCarthy, M. A., et al. (2009). A global synthesis of plant extinction rates in urban areas. *Ecology Letters, 12,* 1165–1173.
Hallstrom, L., Stonechild, J., & Reist, W. (2015). Alberta. In S. Markey, S. Breen, A. Lauzon, et al. (Eds.), *State of rural Canada report* (pp. 17–24). CRRF: Canada.
Harvey, D. (1985). *The Urbanization of Capital.* Baltimore: John Hopkins University Press.

Harvey, D. (1996). *Nature, justice and the geography of difference*. Oxford: Blackwell.

Harvey, D. (2006). *The limits to capital*. London and New York: Verso.

Harvey, D. (2010a). *A companion to Marx's capital*. London and New York: Verso.

Harvey, D. (2010b). *The enigma of capital: And the crises of capitalism*. London: Profile Books.

Harvey, D. (2012). *Rebel cities: From the right to the city to the urban revolution*. London and New York: Verso.

Harvey, D. (2014). *Seventeen contradictions and the end of capitalism*. London, UK: Profile Books.

Heathrow Airport Ltd. (2018). *Heathrow 2.0*. Carbon Neutral Growth Roadmap. Available at: https://your.heathrow.com/wp-content/uploads/2018/12/Heathrow-Carbon-neutral-growth-online.pdf.

Heynen, N. (2014). Urban political ecology I: The urban century. *Progress in Human Geography, 38*, 598–604.

Heynen, N., Kaika, M., & Swyngedouw, E. (Eds.). (2006). *In the nature of cities: Urban political ecology and the politics of urban metabolism*. London and New York: Routledge.

Huber, M. (2009). Energizing historical materialism: Fossil fuels, space and the capitalist mode of production. *Geoforum, 40*, 105–115.

Huber, M. (2017). Value, nature and labor: A defense of Marx. *Capitalism Nature Socialism, 28*, 39–52.

Ibañez, D., & Katsikis, N. (2014). Editorial. In D. Ibañez & N. Katsikis (Eds.), *New geographies 06: Grounding metabolism* (pp. 2–9). Cambridge: Harvard University Press.

Igoe, J., Neves, K., & Brockington, D. (2010). A spectacular eco-tour around the historic bloc: Theorising the convergence of biodiversity conservation and capitalist expansion. *Antipode, 42*, 486–512.

International Monetary Fund (IMF). (2011). *Regional economic outlook, Western Hemisphere*. Washington, DC: IMF.

Jacob, C., Vaissiere, A. C., Bas, A., et al. (2016). Investigating the inclusion of ecosystem services in biodiversity offsetting. *Ecosystem Services, 21*, 92–102.

Jones, P. (2009). Saving the planet or selling off the atmosphere? Emissions trading, capital accumulation and the carbon rent. *Marxist Interventions, 1*, 9–22.

Kaika, M. (2005). *City of flows*. New York and London: Routledge.

Kaika, M., & Swyngedouw, E. (2014, May 14). Radical urban political-ecological imaginaries: Planetary urbanization and politicizing nature. *Derive Journal*.

Kallis, G., & Swyngedouw, E. (2018). Do bees produce value? A conversation between an ecological economist and a Marxist geographer. *Capitalism, Nature, Socialism, 29,* 36–50.

Katz, C. (1998). Whose nature, whose culture?: Private productions of space and the 'preservation' of nature. In B. Braun & N. Castree (Eds.), *Remaking reality: Nature at the millennium* (pp. 45–62). London and New York: Routledge.

Keil, R. (2003). Progress report: Urban political ecology. *Urban Geography, 24,* 723–738.

Keller, R., & Turek, M. (1998). *American Indians and National Parks*. Tucson: University of Arizona Press.

Kelly, A. B. (2011). Conservation practice as primitive accumulation. *Journal of Peasant Studies, 38,* 683–701.

Kemf, E. (1993). *Indigenous peoples and protected areas: The law of mother Earth*. London: Earthscan.

Lapeyre, R., Froger, G., & Hrabanski, M. (2015). Biodiversity offsets as market-based instruments for ecosystem services? From discourses to practices. *Ecosystem Services, 15,* 125–133.

Latour, B. (2005). *Reassembling the social: An introduction to actor-network theory*. Oxford: Oxford University Press.

Lefebvre, H. (1970). *La révolution urbaine* (Vol. 216). Paris: Gallimard.

Lefebvre, H. (1991). *The production of space*. Oxford: Blackwell.

Loftus, A. (2012). *Everyday environmentalism: Creating an urban political ecology*. Minneapolis: University of Minnesota Press.

Lohmann, L. (2012). Financialization, commodification and carbon: The contradictions of neoliberal climate policy. *Socialist Register, 48,* 85–107.

MacKenzie, D. (2009). Making things the same: Gases, emission rights and the politics of carbon markets. *Accounting, Organizations and Society, 34,* 440–455.

Magdoff, F. (2002). Capitalism's twin crises: Economic and environmental. *Monthly Review*, 54. Available at: https://www.monthlyreview.org/0902magdoff.htm.

Malm, A. (2016). *Fossil capital: The rise of steam power and the roots of global warming*. Brooklyn: Verso Books.

Martinez-Alier, J. (1995). Political ecology, distributional conflicts and economic incommensurability. *New Left Review, 211,* 70–88.

Marx, K. (1887). *Capital: A critique of political economy* (Vol. I). Moscow: Progress Publishers. Available at: https://www.marxists.org/archive/marx/works/1867-c1/.

Marx, K. (1894). *Capital: A critique of political economy* (Vol. III). New York: International Publishers. Available at: https://www.marxists.org/archive/marx/works/1894-c3/.

Marx, K. (1964). *The economic and philosophic manuscripts of 1844*. New York: International Publishers.

Marx, K. (1973 [1857]). *Grundrisse*. London: Penguin.

Marx, K. (1990). *Capital: A critique of political economy* (Vol. I, Ben Fowkes, Trans.). London: Penguin Classics.

Marx, K., & Engels, F. (1970). *The German ideology* (Vol. 1). New York: International Publishers.

Massey, D. (1984). *Spatial divisions of labour: Social structures and the geographies of production*. Basingstoke: Macmillan.

Massey, D. (2005). *For space*. London: Sage.

McAfee, K. (2012). The contradictory logic of global ecosystem services markets. *Development and Change, 43,* 105–131.

McDonough, E. (2015). *Global flows, local conflicts and the challenge of urban governance: Managing the city-airport interface in London, UK*. BIVEC/GIBET Transport Research Days 2015: Current Issues in Transportation Research.

Mercer and Inter-American Development Bank (IDB). (2017). *Crossing the bridge to sustainable infrastructure investing*. Inter-American Development Bank. Available at: https://publications.iadb.org/en/publication/12717/crossing-bridge-sustainable-infrastructure-investing-exploring-ways-make-it.

Merrifield, A. (2013). The urban question under planetary urbanization. *International Journal of Urban and Regional Research, 37,* 909–922.

O'Connor, J. (1998). *Natural causes*. New York: Guilford.

Ordner, J. (2018). Petro-politics and local natural resource protection: Grassroots opposition to the keystone XL pipeline in Nebraska. In E. Apostolopoulou & J. Cortes-Vazquez (Eds.), *The right to nature: Environmental justice, social movements and neoliberal natures* (pp. 29–42). Abingdon: Routledge-Earthscan.

Oxford Economics. (2013). *The economic impact of the trans-Adriatic pipeline on Albania: A report for TAP AG*.

Padoch, C., Brondizio, E., Costa, S., et al. (2008). Urban forest and rural cities: Multi-sited households, consumption patterns, and forest resources in Amazonia. *Ecology and Society, 13*(2), 2.

Peet, R., & Watts, M. (2000). *Liberation ecologies: Environment, development, social movements*. London: Routledge.

Perreault, T. (2013). Dispossession by accumulation? Mining, water and the nature of enclosure on the Bolivian Altiplano. *Antipode, 45*, 1050–1069.

Pierce, J., Martin, D., & Murphy, J. (2011). Relational place making: The networked politics of place. *Transactions of the Institute of British Geographers, 36*, 54–70.

Poulton, D. W. (2014). *Biodiversity offsets, a primer for Canada*. Sustainable Prosperity and Institute of the Environment, Ottawa.

Purcell, M. (2014). Possible worlds: Henri Lefebvre and the right to the city. *Journal of Urban Affairs, 36*, 141–154.

Purcell, T. F., Loftus, A., & March, H. (2019). Value–rent–finance. *Progress in Human Geography*. https://doi.org/10.1177/0309132519838064.

QMM (QIT Madagascar Minerals). (2009, March). *A mine at the rescue of the unique biodiversity of the littoral zone of Fort-Dauphin*. QIT Madagascar Minerals SA Press kit. Available at: http://www.riotintomadagascar.com/english/media.asp.

Re:Common. (2019). *Turning forests into hotels: The true cost of biodiversity offsetting in Uganda*. Available at: https://www.recommon.org/eng/turning-forests-into-hotels-the-true-cost-of-biodiversity-offsetting-in-uganda/.

Robbins, P. (2012). *Lawn people: How grasses, weeds, and chemicals make us who we are*. Philadelphia: Temple University Press.

Robertson, M. (2012). Measurement and alienation: Making a world of ecosystem services. *Transactions of the Institute of British Geographers, 37*, 386–401.

Ruddick, S., Peake, L., Tanyildiz, G. S., et al. (2017). Planetary urbanization: An urban theory for our time? *Environment and Planning D: Society and Space, 36*, 387–404.

Schmid, C., Karaman, O., Hanakata, N. C., et al. (2018). Towards a new vocabulary of urbanisation processes: A comparative approach. *Urban Studies, 55*, 19–52.

Seagle, C. (2012). Inverting the impacts: Mining, conservation and sustainability claims near the Rio Tinto/QMM ilmenite mine in Southeast Madagascar. *Journal of Peasant Studies, 39*, 447–477.

Seto, K. C., Fragkias, M., Güneralp, B., et al. (2011). A meta-analysis of global urban land expansion. *PLoS ONE, 6*(8), e23777.

Seto, K. C., Güneralp, B., & Hutyra, L. R. (2012). Global forecasts of urban expansion to 2030 and direct impacts on biodiversity and carbon pools. *Proceedings of the National Academy of Sciences, 109*, 16083–16088.

Smith, N. (2005). Foreword. In N. Heynen, M. Kaika, & E. Swyngedouw (Eds.), *In the nature of cities: Urban political ecology and the politics of urban metabolism* (pp. xi–xv). Abingdon: Routledge.

Smith, N. (2006a). Foreword. In N. Heynen, M. Kaika, & E. Swyngedouw (Eds.), *In the nature of cities: Urban political ecology and the politics of urban metabolism*. London and New York: Routledge.

Smith, N. (2006b). Nature as accumulation strategy. In L. Panitch & C. Leys (Eds.), *Socialist register 2007: Coming to terms with nature* (pp. 16–36). London: Merlin.

Smith, N. (2010). *Uneven development* (3rd ed.). New York: Verso.

Soja, E. W. (1980). The socio-spatial dialectic. *Annals of the Association of American Geographers, 70,* 207–225.

Soja, E. W., & Kanai, J. (2006). The urbanization of the world. In N. Brenner (Ed.), *Implosions/explosions: Towards a study of planetary urbanization* (pp. 142–159). Berlin: Jovis.

Solly, R. (2016). *A struggle between a Tethered Donkey and a Tiger*. London Mining Network. Available at: http://londonminingnetwork.org/2016/08/cerrejon-coal-a-struggle-between-a-tethered-donkey-and-a-tiger/.

Sullivan, S. (2013). After the green rush? Biodiversity offsets, uranium power, and the 'calculus of casualties' in greening growth. *Human Geography, 6,* 80–101.

Sweezy, P. M. (2004). Capitalism and the environment. *Monthly Review, 56,* 86–93.

Swyngedouw, E. (1996). The city as hybrid: On nature, society, and cyborg urbanization. *Capitalism Nature Socialism, 7,* 65–80.

Swyngedouw, E. (2012). Rent and landed property. In B. Fine & A. Saad-Filho (Eds.), *The Elgar companion to Marxist economics* (pp. 310–315). Northhampton: Edward Elgar.

Swyngedouw, E., & Heynen, N. C. (2003). Urban political ecology, justice and the politics of scale. *Antipode, 35,* 898–918.

TAP (Trans Adriatic Pipeline). (2013a). *ESIA Albania*. Available at: https://www.tap-ag.com/resource-library/reference-documents/esia-documents/esia-albania-in-english.

TAP (Trans Adriatic Pipeline). (2013b). *ESIA Greece*. Available at: https://www.tap-ag.com/resource-library/reference-documents/esia-documents/esia-greece-in-english.

TAP (Trans Adriatic Pipeline). (2013c). *ESIA Italy*. Available at: https://www.tap-ag.com/resource-library/reference-documents/esia-documents/esia-italy-in-english.

TAP (Trans Adriatic Pipeline). (2015). *Environmental, socioeconomic, and cultural heritage overview*. Available at: https://www.tap-ag.com/resource-library/reference-documents/project-finance-disclosure/environmental-social-and-cultural-heritage-overview.

TAP (Trans Adriatic Pipeline). (2016). *Biodiversity offset strategy*. CAL00-C5577-640-Y-TRY-0001. RSK.

Teherán, S. (2014). Impactos socioterritoriales de la explotación minera en los departamentos del Cesar y La Guajira. In G. Rudas & D. Hawkins (Eds.), *La minería de carbón a gran escala en Colombia: Impactos económicos, sociales, laborales, ambientales y territoriales* (pp. 44–67). Friedrich-Ebert-Stiftung: Bogotá.

Temper, L., Demaria, F., Scheidel, A., et al. (2018). The Global Environmental Justice Atlas (EJAtlas): Ecological distribution conflicts as forces for sustainability. *Sustainability Science, 13*, 573–584.

TransCanada Pipelines Limited. (2018). *Applications for the white spruce pipeline project Fort McKay area*. Alberta Energy Regulator. Available at: https://www.aer.ca/documents/decisions/2018/2018ABAER001.pdf.

United Nations (UN). (2018). *World urbanization prospects* (The 2018 Revision). New York: UN.

Vlachou, A. (2005). Environmental regulation: A value-theoretic and class-based analysis. *Cambridge Journal of Economics, 29*, 577–599.

Vergara, K., Figallo, M., & Glave, M. (2014). Infraestructura en la Amazonía peruana: Una propuesta para proyectar cambios en la cobertura boscosa en la carretera Pucallpa-Cruzeiro do Sul. In R. Barrantes & M. Glave (Eds.), *Amazonía peruana y desarrollo económico* (pp. 161–208). Lima: IEP and GRADE.

Vicente, A., Martin, N., Slee, D., et al. (2011). Carbón para el mundo, retroceso para La Guajira. In *PBI Colombia, Boletín informativo no 18* (pp. 32–36). Colombia: PBI.

Villarroya, A., Barros, A. C., & Kiesecker, J. (2014). Policy development for environmental licensing and biodiversity offsets in Latin America. *PLoS ONE, 9*, e107144.

Wachsmuth, D. (2014). City as ideology: Reconciling the explosion of the city form with the tenacity of the city concept. *Environment and Planning D: Society and Space, 32*, 75–90.

Weinhold, B. (2011). Alberta's oil sands: Hard evidence, missing data, new promises. *Environmental Health Perspectives, 119*, A126–A131.

Winter, A. (2018). Illegal camping in "stolen native land". In E. Apostolopoulou & J. Cortes-Vazquez (Eds.), *The right to nature: Environmental justice, social movements and neoliberal natures* (pp. 263–276). Abingdon: Routledge-Earthscan.

World Bank. (2009). *Global Economic Prospects*. Washington, DC: World Bank.

WRM & Re:Common. (2016). *Rio Tinto in Madagascar: A mining destroying the unique biodiversity of the littoral zone of Fort Dauphin*. WRM and Re:Common.

5

Biodiversity Offsetting in England: Deepening the Neoliberal Production of Socionatures

E. Apostolopoulou, *Nature Swapped and Nature Lost*,
https://doi.org/10.1007/978-3-030-46788-3_5

It was a town of unnatural red and black like the painted face of a savage. It was a town of machinery and tall chimneys, out of which interminable serpents of smoke trailed themselves for ever and ever, and never got uncoiled. It had a black canal in it, and a river that ran purple with ill-smelling dye, and vast piles of building full of windows where there was a rattling and a trembling all day long, and where the piston of the steam-engine worked monotonously up and down, like the head of an elephant in a state of melancholy madness. It contained several large streets all very like one another [...] inhabited by people equally like one another, who all went in and out at the same hours, with the same sound upon the same pavements, to do the same work, and to whom every day was the same as yesterday and tomorrow, and every year the counterpart of the last and the next.

—Charles Dickens (1854)

5.1 Introduction

Let's now move to the exploration of the emergence and evolution of biodiversity offsetting in a specific country, the UK and explore two major aspects of offsetting that are deeply intertwined: how its emergence relates to the deepening of neoliberal conservation (this chapter) and to governmental attempts to facilitate economic development, often in the form of urban development (next chapter). This chapter, as well as Chapters 6, 7, and 8, largely draws on empirical data obtained through ethnographic research across England from the summer of 2014 until the autumn of 2019. During these five years I conducted more than 80 interviews with governmental officials, environmental administrations and public bodies, including Natural England and Defra, members of several environmental NGOs, conservation brokers (including the Environment Bank), representatives of industries, as well as various economists, regulators, conservation scientists and environmentalists. Interviewees came from very different backgrounds, but they were all involved in the designation and implementation of offsetting in the UK. I have also participated in various community meetings and conducted more than 70 interviews and group discussions with local authorities, environmental administrations, private sector organizations, businesses, conservation NGOs and local community groups in seven field areas. The community groups that participated in my research included residents participating in Hands off Thaxted group, community groups against the Coventry Warwickshire Gateway, community groups opposed to the Lodge Hill housing development, STOP HS2, local authorities participating in 51m, Save Gosforth Wildlife Campaign, West Moor Residents Association, and community groups in Lambeth.

I would like to briefly explain here the logic behind the selection of these case studies. When I started this research, even though offsetting was from time to time at the center of national debates, mainly in high profile cases like the Lodge Hill housing development and the HS2 train-line, it was not clear what exactly was happening in the pilot areas or how widespread its implementation was beyond the pilots. I, therefore, decided to include in my research areas that were part of the official Defra pilots, and preferably, cases for which there was already some information available. I also wanted to select pilots where the Environment Bank, the first conservation broker in the UK, was involved as a partner from the beginning, mainly because its establishment was something new that had the potential to change the way conservation was done in England. I, therefore, included in my research the Essex pilot and the Warwickshire, Coventry and Solihull pilot. To shed light on the links between offsetting and extended urbanization but also to both the social impacts of offsetting and social resistance against its implementation, I also included cases of infrastructure and housing development where opposition to offsetting was strong and has significantly influenced the planning process, and this is how I decided to include the cases of the housing development in Lodge Hill, the case of the HS2 railway, the case of the Whitehouse Farm housing development in North Tyneside and the case of the Coventry and Warwickshire Gateway. Finally, I decided to also include the Thameslink project for two main reasons: firstly, because it was an important case of a large-scale urban development project where offsetting has been voluntary proposed, and, secondly, because it was one of the cases that have been promoted as a success story of offsetting. The period when this research started also offered me a unique opportunity to evaluate the impacts of a policy that had just emerged and started to institutionalize in the UK. This conjuncture made many people willing to share their thoughts and criticisms on offsetting that, at the time, had provoked much-polarized commentary, outrage from many local communities and had divided conservationists across the country (see also Sullivan and Hannis 2015). Despite being a policy with strong depoliticizing effects (Apostolopoulou 2019) offsetting has managed to trigger heated debates primarily due to its deeply controversial message, namely that the loss of nature due to development can be compensated and even lead to No Let Loss (or Net Gain) by simply (re)creating nature somewhere else.

In what follows, I explore the evolution of market environmentalism in England with a focus on the case of biodiversity offsetting and I pay attention to the role of the financial crash of 2008, a key moment in recent British, and more broadly European, history.

5.2 Nature Conservation in the UK After the 2008 Financial Crash: Consolidating the Hegemony of Market Environmentalism

The publication of the Millennium Ecosystem Assessment (MEA) in 2005 was a key moment for conservation policy in the UK. Following the MEA, Defra published several documents (Defra 2007a, b) under the Labor government that included specific priorities in moving toward an ecosystems approach and brought forward an understanding of ecosystems as environmental assets that, like other capital assets, provide a flow of services over time which if consumed in a sustainable manner, then capital can be kept intact (Defra 2007a). The fact that environmental damage has a potential to undermine long-term economic prosperity has also been clearly acknowledged by both the government and the Treasury (HM Treasury 2006). As an employee in Defra pointed out during our discussion:

> For Defra attention on ecosystem services and natural capital goes back to 2005 and the Millennium Assessment. Defra started looking at ecosystem services and the ecosystem approach and this gradually led to the commissioning of the funding of the UK National Ecosystem Assessment. And then to the publication of the White Paper. So, the pathway is several years. The thinking was, and still is, that this approach tries to make the case why ecosystem services and biodiversity loss do matter for the economy. Along with that there is an ongoing effort to see how this new approach applies in practice, and this links to monetary evaluation. These ideas are now at the core of policy-making and I don't see any signs that this would change anytime soon either here or internationally.

To test the new approach, Natural England launched three Uplands Ecosystem Service Pilots at the end of 2009 in Cumbria, Yorkshire, and South West England[1] in collaboration with various partners to demonstrate how the provision of ecosystem services, including water quality,

[1] For the evaluation of these pilots see: http://publications.naturalengland.org.uk/publication/4915928315985920?category=10006.

flooding and carbon storage, could be turned into "genuine business opportunities" by adopting a Payments for Ecosystem Services (PES) approach (Defra 2010) and enhance natural capital for multiple benefits. In June 2011, the UK National Ecosystem Assessment (UK NEA 2011) was published showing, inter alia, that more than 30% of the services provided by the natural environment were in decline. UK was the first country that undertook a complete assessment of ecosystem services, exploring, inter alia, their economic value and contribution to the economy. The NEA has been followed by the Lawton Report (Lawton et al. 2010) that highlighted the major habitat fragmentation that was occurring across the country and its impacts to climate and demographic change. During the same period, the Natural Environment White Paper entitled "The natural choice: securing the value of nature" (HM Government 2011) has been also published emphasizing the need for "innovative" approaches for considering the role of nature in supporting economic growth.

The White Paper marked the hegemonic position of market environmentalism and the increasing penetration of neoliberal logic in UK conservation. It signaled a widespread neoliberal restructuring of environmental and conservation policy driven, at that time, by the Conservative–Liberal Democrat coalition government, consolidating and deepening the previous neoliberal restructuring. The key aim of the White Paper (HM Government 2011, p. 3) was to propose specific ways through which the government would "mainstream the value of nature across society," widely understood as its economic value as it has been obvious in the discussions I had with people involved in the designation of these policies in England. As several interviewees explained to me there were three key policy goals that stemmed from the White Paper: firstly, an emphasis to the creation of a green economy in which economic growth and "healthy" natural resources sustain each other, and government works closely with businesses to understand the economic value of nature, secondly, an emphasis to the local scale, and, thirdly, an enhancement of the third sector's role. The White Paper strongly influenced the governmental agenda and the Coalition government promoted several policies for the conservation sector flowing from these positions. One of the most prominent ways through which the neoliberal logic was consolidated was the heightened role given to the private sector: new policies favored the inclusion of various businesses in conservation in the specific ways that are outlined below.

Firstly, the White Paper announced the creation of a business-led Ecosystem Markets Task Force (sic) which was expected to review "the opportunities for UK business from expanding green goods, services, products, investment vehicles, and markets which value and protect nature's services" with the goal to "gain the maximum benefit for UK businesses from new market opportunities which protect and improve natural capital" (HM Government 2011, p. 40). The Task Force was made up of business leaders and experts, appointed by the Defra Secretary of State and included representatives of Jaguar Land Rover, Unilever UK and Ireland, Aldersgate Group, United Utilities, Johnson Matthey Fuel Cells, and the Environment Bank.[2] It had an explicitly market-driven agenda and the responsibility to advise the Government on the scope of market development and value creation for businesses linked to ecosystem services and, therefore, it was explicit that its advice had to be relevant to a business audience. Secondly, the government established the Natural Capital Committee,[3] an independent advisory committee consisting of seven experts, which initially included representatives of the Business and Biodiversity Offsets Programme (BBOP) and consultants. Its key role was to provide advice to the government on the sustainable use and economic valuation of natural capital, including forests, rivers, land, minerals, and oceans. The Committee was renewed in 2016 with the key role to provide advice on the development of a 25 year plan for the environment.[4] As we read in the relevant announcement, to achieve this goal the Committee considered necessary to make use of appropriate knowledge, tools, and techniques to ensure that natural capital can be consistently assessed, valued and accounted for in decision-making and economic planning. Importantly, this also meant developing national

[2] See http://webarchive.nationalarchives.gov.uk/20140305215952/http://www.defra.gov.uk/ecosystem-markets/about/membership/.

[3] See https://www.gov.uk/government/groups/natural-capital-committee.

[4] See https://assets.publishing.service.gov.uk/government/uploads/system/uploads/attachment_data/file/517123/ncc-terms-of-reference.pdf.

natural capital accounts and work with businesses to develop and apply corporate natural capital accounts recognizing that much of England's natural capital is privately owned (HM Government 2016, p. 2[5]). Thirdly, the government further enabled business participation in environmental governance by endowing the Foresight Group, a private asset manager, with 80m pounds to invest in green infrastructure projects. As we read in its website the Foresight Group is "the leading independent infrastructure and private equity investment manager," and assets under its management have grown by more than 50% in 2013 to over £1bn.[6] Fourthly, the White Paper announced a competitive process for the selection of Nature Improvement Areas (NIA) projects. NIAs were established to create "joined up and resilient ecological networks at a landscape scale" and they would be run by partnerships of local authorities, communities and landowners, the private sector and conservation organizations with funding provided by Defra and Natural England.[7] The Government provided £7.5 million to support the creation of 12 initial NIAs as seed funds: partners were expected to pool resources and draw together funding from various sources, including the National Lottery, environmental charities, business, local authorities, and communities (Adams et al. 2014). Along with NIAs, Local Nature Partnerships (LNPs) have been also established to strengthen action "at the right scale" and mirror Local Enterprise Partnerships (LEPs). LEPs and LNPs were supposed to play complementary roles in creating the necessary conditions for "thriving local enterprise, innovation and inward investment" (HM Government 2011, p. 20), contributing to the wider goal of a green economy by promoting green growth locally. LNPs echoed the key conclusions of the UK NEA that emphasized the need for a landscape-scale approach to conservation. In 2012, Defra published an overview of

[5]See https://assets.publishing.service.gov.uk/government/uploads/system/uploads/attachment_data/file/517123/ncc-terms-of-reference.pdf.

[6]See http://www.foresightgroup.eu/news/foresight-group-assets-under-management-exceed-1-billion/.

[7]See https://www.gov.uk/government/publications/nature-improvement-areas-improved-ecological-networks/nature-improvement-areas-about-the-programme.

their role emphasizing that to be effective LNPs would need to be "self-sustaining strategic partnerships" of local organizations, businesses, and people (Defra 2012, p. 1).

Both NIAs and LNPs are state-funded conservation partnerships that make business/private sector involvement in conservation compulsory. By being localist partnerships between local councils, NGOs, and businesses, they epitomize the neoliberal aspect of the rescaling of biodiversity governance: as competitive funding schemes, they work at vertical scale construction and outward business/NGO inclusion fostering competition and entrepreneurial tactics in localities (Apostolopoulou et al. 2014). It is important to emphasize here that fiscal austerity along with competitive funding and novel monitoring frameworks acted as powerful mechanisms at the state's disposal to canalize and discipline conservation into pragmatic market logic signaling a post-political (see Allmendinger and Haughton 2012) turn in UK conservation. Actors with competing agendas were expected to work together to create "shared visions" for their areas and the ability of these partnerships to reach consensus and resolve or sideline conflicts played a key role in their funding by the Government. NIAs and LNPs were quite indicative examples of Defra's new approach. As a representative of Natural England explained to me:

> The White Paper changed the trajectory of environmental policy, it talks to localism, it talks to Big Society, and so it can be a defining moment. Local Nature Partnerships and Nature Improvement Areas are good examples of the on-the-ground implementation of White Paper's logic; they show that partnerships are the way forward, conservation cannot survive in the 21st century without having the private sector as an ally.

Finally, the White Paper also referred to the need to develop Payments for Ecosystem Services schemes and test a voluntary approach to biodiversity offsetting. From 2012 to 2015, Defra commissioned 16 PES pilot studies over three competitive rounds across England and Wales in partnership with actors from the civil society, NGOs, private, and public sectors (Defra 2016) whereas six offsetting pilots[8] ran from 2012 to 2014 (see next section).

It is important to point out here that the shift toward neoliberal conservation did not firstly emerge when the Coalition Government came to power. The first explorations of a market-based approach started under the Labor government and, as several interviewees explained to me, even though the emphasis on economic growth had been stronger since 2010, there has been no fundamental difference between the policies of different governments. What some people saw as different was the language used and the admission that the "win-win-win" discourse of New Labor had failed. This was reflected in the Coalition's Big Society,[9] the flagship policy of the 2010 UK Conservative Party general election manifesto and part of the legislative program of the Conservative–Liberal Democrat coalition agreement.[10] Big Society was launched in 2010 by using slogans like "put more power and opportunity into people's hands" and "we are all in this together" (p. 1), emphasizing volunteering, philanthropy, charitable giving, and social enterprises (p. 2). Within months it became an Oxford dictionary word of the year. However, launching the Big Society coincided with large-scale cuts in public expenditure programs and the widespread adoption of austerity politics: in 2010, Ipsos-Mori poll found that 57% of respondents agreed that the Big Society was just an excuse to save money by cutting public services. David Cameron tried to respond to these allegations by arguing that cuts were temporary and had been enacted from economic necessity.

[8]For the official evaluation report of the offsetting pilots see here: http://sciencesearch.defra.gov.uk/Default.aspx?Menu=Menu&Module=More&Location=None&Completed=0&ProjectID=18229.

[9]The Big Society was a political ideology developed in the early twenty-first century. It proposed "integrating the free market with a theory of social solidarity based on hierarchy and voluntarism" and conceptually it was drawing on "a mix of conservative communitarianism and libertarian paternalism." The roots of the Big Society ideology "can be traced back to the 1990s, and to early attempts to develop a non-Thatcherite, or post-Thatcherite, brand of UK conservatism, such as David Willetts' Civic Conservatism and the revival of Red Toryism. Some commentators have seen the Big Society as invoking Edmund Burke's idea of civil society, putting it into the sphere of one-nation conservatism (https://en.wikipedia.org/wiki/Big_Society).

[10]"Cameron and Clegg set out 'big society' policy ideas". BBC News, 18 May 2010. Available at: http://news.bbc.co.uk/2/hi/uk_news/politics/8688860.stm.

However, three years later, in 2013, it became evident that the government had no intention of resuming spending once the structural deficit had been eliminated. This led critics to conclude that the Big Society was primarily intended as a mechanism for reducing the size of the state and for replacing what was left of the welfare state with a rapidly expanding philanthropic sector. Big Society has now almost disappeared from the public discourse and from governmental rhetoric, "like an embarrassing fashion item, worn once and pushed to the back of the cupboard" whereas people seem less able to influence local decisions than they did 25 years ago.[11]

5.3 Biodiversity Offsetting in the UK and Neoliberal Conservation

> Biodiversity offsetting can support improvement to the planning systems in particular how it tackles the impacts of development on nature.
>
> It has the potential to be quicker: a simple, standard framework for evaluating the impacts of development on biodiversity can speed up assessment.
>
> Allowing any required compensation to be bought "off-the-shelf" from a market removes the need for negotiation on what will be provided.
>
> Defra (2013, p. 5)

The evolution of offsetting offers a good example of both the continuity of policy between different governments and the intensification of the neoliberal restructuring of environmental policies since 2008. Offsetting was officially introduced in the 2011 White Paper but, as mentioned above, the first explorations of the concept started under the Labor government elected in 2007 as part of a wider discussion on the possibility of market creation in biodiversity (Lockhart 2015; Defra 2007a, b; Treweek et al. 2009). A key moment in the evolution of offsetting

[11] See https://www.theguardian.com/society/2015/jan/20/the-big-society-civil-exchange-audit-shows-coalition-contempt-and-hypocrisy; http://www.huffingtonpost.co.uk/caroline-slocock/big-society_b_6505902.html.

has been the establishment of the first national broker, namely the Environment Bank, in 2008 by Professor David Hill.[12] The profile of the Environment Bank's founder shows well its links with the state and the government whereas evidence suggests that the Conservative Party had strongly supported its establishment (Apostolopoulou and Adams 2019). The Environment Bank is a private compensation brokering and consultation company working to broker biodiversity compensation agreements for developers and landowners and typifies a market-based, third party approach to offsetting and mitigation schemes.[13] As we read on their website[14]:

> As an impartial adviser, we calculate the environmental impact of development proposals using approved Government metrics. Our ecological experts then match a developer's compensation requirement with sites put forward by landowners and conservationists who have conservation credits available to sell. Credits are sold in exchange for the creation or enhancement of habitats, generating biodiversity gain.

One year after the establishment of the Bank, Defra commissioned a scoping study for the design and use of biodiversity offsets (Treweek et al. 2009). According to the study, the situation was leading itself to offset development, firstly, because the UK Biodiversity Action Plan was focusing on targeted management of individual habitat types and species offering a basis for defining tradable biodiversity credits and, secondly, because many local authorities were drastically under-resourced with respect to the Biodiversity Duty[15] and thus would significantly benefit from increased investment (coming from offsets) to address "the hidden costs of development" (Treweek et al. 2009, p. 4). Defra has also published a discussion material on offsetting in December 2010. More specific policy proposals were brought by the previous Coalition government in 2011 (HM Government 2011): offsetting was expected

[12]See http://www.environmentbank.com/team-biographies.php.

[13]See http://ec.europa.eu/environment/biodiversity/business/assets/pdf/workstream2/ebl_en.pdf.

[14]See http://www.environmentbank.com/about.php.

[15]Under the Natural Environment Research Council-NERC, UK's leading public funder of independent research, training and innovation in environmental science.

to provide a straightforward approach to assessing development impacts, creating greater certainty and leading offset suppliers to establish an offsets market, increasing the number and quality of biodiverse habitats in England.

Biodiversity offsetting had a prominent role in all the policy documents mentioned in the previous section. Offsetting was suggested as a means to achieve habitat restoration and deliver a more "landscape scale" approach to conservation in the Lawton review. The latter identified key principles that should be followed in the establishment of offsets that were largely taken up by Defra in developing the pilots. It recommended the establishment of an ecological network of "more, bigger, better and joined" areas of wildlife habitat (Lawton et al. 2010, p. 3) and the creation of "Ecological Restoration Zones" offering the scientific case for the White Paper's (2012) language of physical extension and connectivity of conservation sites. As Adams et al. (2014) argue this, inter alia, manifested the growing policy interest for restoring "wilderness" in UK's severely transformed landscapes. The interrelationship between the neoliberal market-based reframing of biodiversity and the new expansive intent in UK conservation (Adams et al. 2014) underpinned the ecological logic behind offsetting. The policy has been promoted as simultaneously having the potential to attract private investments for conservation and pool a number of offsets required for separate small developments into a larger habitat block without imposing additional burdens on developers (Lawton et al. 2010). The benefits of offsetting for landscape-scale conservation were also emphasized in the *ThinkBIG* report (England Biodiversity Group 2011) that considered offsets capable to provide a funding mechanism for landscape-scale conservation via the sale of credits to developers. Offsetting was, thus, being promoted not only as a pro-development policy but also as an innovative and economically efficient conservation tool. As an interviewee from Defra argued:

> Offsetting should complement existing habitat designations and help to expand and restore the ecological network in England. Used in a strategic way offset sites can provide the necessary resources to deliver more, better, bigger and joined up networks of habitat.

The 2011 White Paper signaled Defra's commitment to offsetting and piloting offsets was considered as one of the key environmental reforms: "through reforms of the planning system, we will take a strategic approach to planning for nature within and across local scales. This approach will guide development to the best locations, encourage greener design and enable development to enhance natural networks [...] We will establish a new, voluntary approach to biodiversity offsets and test our approach in pilot areas" (HM Government 2011, p. 3). In 2011, Defra also published several other reports on offsetting and, as mentioned in the previous section, in 2012, the Coalition Government launched an offsetting pilot scheme consisted of six voluntary pilot areas in England: Devon (comprising three sub-pilots North, South and East Devon), Doncaster, Essex, Greater Norwich, Nottinghamshire, Warwickshire, Coventry and Solihull.[16] In these areas, as in the ecosystem services pilots, in line with the post-2008 neoliberal rescaling of biodiversity governance (Apostolopoulou et al. 2014), groupings of Local Planning Authorities (LPAs), and a range of other partners, including NGOs and businesses, formed public–private partnerships to pilot voluntary offsetting and test the Defra metric (see also Collingwood Environmental Planning Limited 2013). Simultaneously, in many areas across England that were not officially declared as pilots the potential of offsetting to compensate the impacts of—mainly controversial—development projects was being tested on the ground. 2012 was also the year where the Environment Bank launched the Environmental Markets Exchange (EME, initially known as Conservation Credits Exchange—CCE). Its goal was to provide "a 'one-stop-shop' for people to register their offsetting receptor sites and have the credit value of their site measured".[17] The Bank believed that the EME would rapidly lead to the creation of an offsetting market in the UK by providing a national resource for identifying sites to receive offset funds and that, in turn, a large number of

[16]In 2016, Defra published an official evaluation of the offset pilots that is available here: http://randd.defra.gov.uk/Default.aspx?Menu=Menu&Module=More&Location=None&ProjectID=18229&FromSearch=Y&Publisher=1&SearchText=WC1051&SortString=ProjectCode&SortOrder=Asc&Paging=10#Description.

[17]See http://www.environmentbank.com/documents/EnvironmentBankNewsletterSpring2012_000.pdf.

available sites ready to sell credits would foster companies and developers to buy conservation credits. The Environment Bank formed a partnership with AB Agri (the agricultural division of Associated British Foods) to increase the number of potential offset receptor sites (from among their 30,000 British farmers in 2014) and with Shell Foundation in 2011 to pilot the use of credits.[18] The plan that the supporters of offsetting initially had in mind was that offsets would be produced in a way to match supply and demand. As a Defra employee explained to me if offsetting was to be successful it had to produce "as much nature" as needed:

> While you build such a system it's quite hard to get the supply and demand matched and you would have to be careful to avoid both sides of the problems, and not having people going around and looking for an offset which doesn't exist. But equally not to encourage offset providers to be flooding the market with things that are not required.

In 2013, the Green Paper on offsetting was published defining offsets as "conservation activities that are designed to give biodiversity gain to compensate for residual losses" (Defra 2013, p. 3). The proposed system was a variation of the habitat hectares approach (BBOP 2009; see also Chapter 3) and has been largely based on the work done for the Defra scoping study (Treweek et al. 2009). A section of the Green paper was entitled "Why the Government is interested in biodiversity offsetting" offering one of the most lucid accounts of the governmental discourse on offsetting. As we read in the relevant section: "The Government is determined to succeed in the global race by creating growth and delivering lasting prosperity. At the same time the Government wants this generation to be the first which leaves the natural environment of England in a better state than it inherited" (Defra 2013).

It is important to note here that offsets, along with payments for ecosystem services, have eventually become a criterion for landscape-scale conservation funding, NIAs and LNPs. As a matter of fact, according

[18]See http://www.shellfoundation.org/Our-Focus/Partner-Profiles/Environment-Bank/Summary.

to Defra,[19] one of the key objectives of LNPs was to promote interest in, and uptake of, offsetting. Indeed, LNPs were from the beginning expected to work with local authorities in the pilots to find potential offset providers and encourage developers and local businesses to participate. Importantly, linking offsetting to landscape-scale conservation and local partnerships was also a goal of many conservation NGOs. Thus, for example, in the Wildlife Trusts webpage we read:

> Any biodiversity offsetting scheme should have a clear purpose, set within a framework for delivering ecological networks. The Wildlife Trusts believe Government needs to set out a vision to create a National Ecological Network for England, which should be informed by ecological network strategies produced by Local Nature Partnerships (or other similar strategic partnerships of a broad range of local organizations, businesses and people who want to help bring about improvements in their local natural environment). Within these strategies, ecological network maps would show the existing natural assets (our statutory and non-statutory sites and priority habitats) and important areas for re-establishing strategic footholds for nature, e.g., buffers around protected sites and Nature Improvement Areas (NIAs) across the length and breadth of England.[20]

Similarly, RSPB in a publication regarding the Scottish Borders biodiversity offsets argues: "The Council has, together with stakeholders, developed a biodiversity offset scheme that accounts for the residual environmental impacts of renewable energy and mineral development on black grouse, and on blanket bog and other upland habitats. The implementation of eight schemes has mainstreamed biodiversity into the planning process by seeking biodiversity benefits at the landscape scale, while simultaneously benefiting ecosystem services. [...] Careful negotiation by the partner NGOs with farmers and landowners has balanced their needs with those of biodiversity and flood protection gains [...] By working with local partners to develop an offset mechanism, the

---

[19] See https://assets.publishing.service.gov.uk/government/uploads/system/uploads/attachment_data/file/192580/local-nature-partnerships-overview120402.pdf.

[20] See http://www.wildlifetrusts.org/biodiversityoffsetting. I have to point out here that this web link has stopped working in July 2018, so it seems that it is not available online anymore.

Council has ensured that, where compensation is the only option, appropriate habitat compensatory measures can be secured"[21] & "...when well planned and executed, compensation in the form of offsite biodiversity enhancement may be more beneficial for the impacted wildlife than seeking to maintain existing, isolated habitats and species on proposed development sites [...] Offsets have drawn in, through the Scottish Rural Development Programme, more than £3 million in extra resources for habitat enhancement and management."

5.4 The Defra Offset Metric: The Triumph of Reductionism

> The science behind the metric obviously needs improving and refinement and we worry whether that's much of a priority for the government.
>
> Member of a biodiversity offset pilot steering committee

A key aspect of the Green Paper on offsetting has been the introduction of the offset metric, the so-called Defra metric. The proposed metric was based on habitat types meaning that development sites should be mapped and divided into habitat parcels (Defra 2013). It also quantified the value of habitats on the basis of three criteria: (i) the distinctiveness of the habitat (assessed as low, medium or high) which should, inter alia, reflect the rarity of the habitat at different scales, and the degree to which it supports species rarely found in other habitats, (ii) the quality of the habitat (assessed as poor, moderate, or good), based on Natural England's Higher level stewardship: Farm Environment Plan (FEP) Manual, and (iii) the area of the habitat in hectares. According to the guidelines given in the Green Paper after assessing habitats against these factors, their value should be calculated in terms of *biodiversity units* using Table 5.1 (Defra 2013, p. 10).

The Defra metric was originally designed to complement the UK Biodiversity Action Plan. The way the Action Plan was structured

---

[21] See https://www.cieem.net/data/files/Resource_Library/News/Planning_Naturally.pdf.

Table 5.1 The Defra biodiversity offset metric

Value of 1 ha in biodiversity units		Habitat distinctiveness		
		Low (2)	Medium (4)	High (6)
Habitat quality	Good (3)	6	12	18
	Moderate (2)	4	8	12
	Poor (1)	2	4	6

implied a No Net Loss target only for irreplaceable vulnerable biodiversity and then a zone within which No Net Loss would be delivered in various ways, for example through increasing the area under conservation status and the type of biodiversity represented or through improving the way it was managed and its condition. The scoping study (Treweek et al. 2009) reflected this approach by arguing that these three axes needed to be considered and that the metric should reflect intrinsic properties, like diversity itself, and condition as driven by management. As an interviewee who contributed to the designation of the metric explained to me:

> So that was what made it different from other approaches, like conventional sort of quality times area measures, which were often too subjective and obscuring too much detail and maybe not clarifying the actual properties that you would be trying to manage to improve. It has the potential to work well but what hadn't been done at the point where Defra adopted that metric was that they hadn't developed a scale for condition scoring. Having researched it quite a lot with Natural England and others we did feel that it was possible to pre-assign vegetation types to distinctiveness categories just based on what everybody already knows and that's what they did, they pre-assigned everything to one of the distinctiveness categories.

According to the original plan the initial metric was supposed to be replaced by a more tailored approach and complemented by a condition scoring method. However, Defra decided to issue the metric without having a condition scoring method. During my fieldwork many interviewees mentioned that this has made it very difficult to assign timeframes for achieving different levels of improvement and to convert levels of improvement and condition into credits. They argued that without such a framework it has been almost impossible to work out how many credits should be released per year over time to get to the outcome that they had agreed to deliver. As an ecologist working in a local council where one of the pilots has been established and a member of a local conservation NGO, respectively, argued:

> The metric doesn't work for some habitats that obviously don't fit neatly into the UK BAP action planning approach. So, we 've always struggled with brownfield sites. There is a priority habitat type called Open Mosaic Habitats of Previously Developed Land I think, but the definition is so vague that it's very hard to run the metrics on that and how do you assess condition, there is no recognized condition assessment. And here we have some sites that are regularly threatened from development, kind of brownfield, very interesting biologically. Much better research is needed, because a lot of the definitions and condition assessments were taken from the Farm Environment Plan Handbook, so they are very focused on the farm landscape and do not work well for urban habitats. And of course, they do not work for species, I think that's pretty obvious.

And:

> The main problem is trying to categorize something that is a continuum. Because the quality of habitat and its distinctiveness is a continuum and we often disagree on that. So, to put an absolute figure on that is debatable. I think they need to at least give us the option to adjust that locally, to give some scope for planning authorities to decide their own metrics because there are important differences between areas. Some habitats are rarer, some are more common. So, a national approach does not work well I think.

In practice, in many cases the way the metric has been used was quite simplistic and reductionist (see also Apostolopoulou 2019; Carver and Sullivan 2017) and the absence of field-testing and of the necessary research to underpin the matrix clearly contributed to that. The excerpt that follows from a discussion I had with an interviewee who has been involved in the designation of offsetting in the UK is indicative of those weaknesses:

Interviewee: Well number one, at one end the scoping report was a very preliminary version done through voluntary work. The feeling at the time was that given the politics of the situation if everybody didn't agree to sign up to a consistent framework it would have been chaos because it would have actually made it very difficult to develop any approach that would work with the planning system or that could be built into the way that development is planned and implemented. The available time was very short. It was developed incredibly fast but what's a shame is that there's been no follow-up through funded research to underpin it and test the metric itself. So, all the pilots were concerned more with how to integrate offsets into the planning system and attract developers rather than actually look at the metric.

Interviewer: You mentioned before the politics of the time, could you be more specific?

Interviewee: One of the things was the government then and especially now even more so, this government, are very reluctant to regulate things like this. So they wanted a system that could be stand up to scrutiny even if not regulated. The metric allows you to kind of do that because it provides a consistent basis for designing and implementing things. That means it can be tested and it's always possible to unpick what decisions have been made and how the offset has been determined. That's all good but the lack of willingness to regulate means that you need to be simple, I 'd say probably simplistic, really pragmatic and not add significantly to the costs of environmental assessment. So yes, I think it's even more the case now that everybody agrees that for offsetting to work you need a regulated system and yet the government is incredibly unwilling to engage in that. And I think they are missing the opportunity that it might bring to actually open up markets and generate supply-demand relationships. But obviously we do not have a thriving system and judging from the amount of housing development that is being proposed, development is being kind of fast tracked through now.

Interviewer: So, in principle do you think that it's possible to design metrics that will permit this kind of exchangeability between a development and an offset site?

Interviewee: Well the one thing I would say is that you cannot have one metric that deals with every aspect of nature and everybody knows that. The Defra offset metric intended to be a minimum requirement that would allow you to determine an appropriate area, size and context for an offset. But if you then had specific habitat or species population requirements you would have to tailor those additionally and look at the spatial context for impacts and gains through offsets. The key weaknesses now are around spatial aspects and species population aspects but to fix that would require a certain amount of investment in good work that doesn't appear to be likely at the moment. So yes, the Defra metric as it stands obviously isn't appropriate for every detail of an impact, it's not complete and it wasn't ever intended to be.

It is important to emphasize here that the lack of the necessary follow-up research indicated the approach that the government had in mind. As evidence from my fieldwork suggests, there was not only a clear unwillingness to spend more money on designing an offsetting policy but there was also an expectation that offsetting would act as a new funding mechanism for conservation and green infrastructure planning. This has been accompanied by a commitment to keep the metric as simple as possible to make it attractive to developers. The following discussion I had with two conservationists that participated in one of the pilot's steering committees is indicative of the restrictions and barriers that this approach has been creating in practice:

Interviewee 1: The Defra metric is a very simple metric that has been used without considering the complexities or the details that are needed. It is more like a first stand or something, it can be used to say yes this is roughly where you are looking at, but the actual assessment should be much more detailed. As it is now it leads to rather crude measurements.

Interviewee 2: The other thing is that we have no experience in implementing it. So, we went to one of the meetings and we said: we need to be trained, the steering group should be trained as assessors for the metric. They didn't want to do that. Because obviously…

Interviewee 1: The university, somebody could do that.

Interviewee 2: Yes, exactly, but as I said they didn't want to do it because for them it was useless. The whole point of offsetting wasn't to train people to a new method of calculating impacts, they just wanted to find a way to make things easier for developers and create a space for consultants and brokers. And there is another issue in the way things are being run: they have somebody dependent to assess the site and then they have somebody else to apply a metric or whatever, it doesn't seem quite right. There wasn't a kind of a procedure to say, ok, you've got this site, this is what you need to ask somebody to properly assess the impacts of the development.

Interviewee 1: If the developers are doing the assessments there is an obvious conflict of interest.

Interviewee 2: And the other thing we suggested is that if the steering group is trained then certain ecologists can become registered or accredited by the steering group to make sure that they can appropriately assess the habitat and calculate the number of credits. Otherwise someone/anyone comes across and do it, and we don't know how true it is and how wrong it is.

Interviewee 1: But we never really got any final conclusions from this meeting. We haven't seen a report yet. Nobody has. It provided nothing like final outcomes, that everybody could see around and say: yes, I am quite satisfied with that… and that was the end of it.

Importantly, for the supporters of offsetting the simplicity of the metric was one of its key strengths and its deficiencies were often acknowledged but recognized as a necessary compromise to achieve offsetting's primary goal, namely render the calculation of environmental compensation easier, faster and, ideally, also cheaper, streamlining the provision of environmental compensation. It was widely believed that if the metric was more complicated or sophisticated the policy would fail since no developer would be convinced to implement it. As an interviewee from Defra argued:

The idea of the metric is one of the big successes of biodiversity offsetting. We like that and we try to get the message across the people. The strength of it is that it is clear and simple. With the metric you can go to a local council they've been told to accept a planning application and you can tell them 'you lost 25 units and you gained 25'. There were of course several people during the consultation that suggested that it must be improved by adding more dimensions, more criteria etc. I understand the logic, but I am not sure if this is a good idea. Because as I said the great strength is that the metric is understandable, simple and transparent. If you start making it more complicated and more subtle you start losing the simplicity.

The goal to keep the metric as simple as possible inevitably made it highly reductionist. This was not only a weakness of the Defra metric, which despite the problems mentioned above was based on one of the most sophisticated approaches that were available at that time, namely the habitat hectares method, but also, as we have seen in Chapter 3, a result of the irreconcilable contradictions that characterize offsetting's designing principles. The Defra metric, as all offset metrics, followed a strictly quantitative approach for the calculation of ecological losses and gains. Thus, for example, the risk of failure or the time needed to restore or recreate a habitat, or the location of the offset, could be represented through numbers in order to allow the calculation of losses and gains in terms of a number of units or credits. In the case of restoration, this has been possible by classifying restoration and recreation activities in four bands from low to very high difficulty. For low difficulty sites there is no requirement to increase the area while for very high difficulty sites there is a requirement to improve 10 times as much area to generate the same number of units. Similarly, the time lag between ecological losses in the development site and the creation of equivalent gains in the offset site can be addressed by applying a 3.5% discount rate as set out in HM Treasury's Green Book (Defra 2013). Following this logic, literally anything can be quantified in order to be included in offsetting's calculations and this is how people have been gradually implementing the metric on the ground: multipliers have been used to incorporate into the calculation of units the risk of non-delivering the offset, the uncertainty involved in grounding prospective gains on idealized target conditions that would be reached in several years' time, the time lag between the loss of habitats in the development site and the creation of new habitats in the offset site, or even the loss of irreplaceable habitats, like ancient woodlands. This offers an important counter-argument to those who argue that the offset metric despite its deficiencies could improve transparency by providing a consistent way of calculating losses and gains. Representing through numbers and scores issues that obviously cannot be quantified in a meaningful way has often obscured the discussion by creating empty categories that render mathematical calculations idealistic and any meaningful public debate practically impossible. One of the most characteristic examples that shows where the above logic can lead is the case of the offset that has been offset (!) in Thaxted. As an interviewee from the relevant local council explained to me:

> A site next door was developed where there was less valuable habitat and reptiles and the mitigation for them was dealt with on what became the next development site. So that has been considered in the final offset, the offset has been offset if you like, but I think both sides agreed that it wasn't a kind of issue really as long as they accounted for it in their final calculation of the required biodiversity units which they did.

A final comment is important here. Both the way the Defra metric was designed and the policy was launched in England made clear that offsetting was expected to achieve a quantitative balance between nature lost due to economic development and nature saved through offsets so that the stock of natural capital would be kept constant. However, the many weaknesses that have been mentioned above along with the patchy evidence base, including the absence of comprehensive and up-to-date information on habitat condition and spatial data for habitats and species for most areas, created major concerns about the actual ability of the policy to achieve any meaningful equivalence even to those who were only expecting from offsetting to streamline the delivery of compensation for biodiversity loss due to economic development.

5.5 Offsetting, Depoliticization and the Neoliberal Rescaling of Governance: It Ain't Technical

Offsetting has been relying from the very beginning on a scientific technocracy portrayed as objective and neutral, on expert conferences, like the "To No Net Loss of Biodiversity and Beyond' conference in

London,[22] on best practice guidance (see, e.g., BBOP 2012), and on various case-study manuals (see, e.g., the website of the Environment Bank[23]), and has given increased roles to intermediaries, advocates, and experts as shapers of and conduits for policy messages. The supporters of the policy have been claiming that this approach was consistent with its key aim, namely the streamlining of the way that environmental compensation was being calculated and of the ecological science on which these calculations were based. Offsetting has been a product and a trigger of the privatization of policy expertise, including consultants and conservation brokers, and of delivery systems following a widespread outsourcing of state functions in the post-2008 era and a deference to market solutions, along with a reciprocal exploitation of governmental markets by management consultants and contract-service companies (Apostolopoulou and Adams 2015; Apostolopoulou et al. 2014; see also Peck et al. 2012). As we have seen in previous sections, these developments manifest the broader shift towards neoliberal conservation in England particularly the last decade. It is indicative, as many interviewees told me, that before conservation brokers came onboard things were already moving towards a situation where consultants were doing most ecological assessments partly due to the decreasing numbers of local council ecologists and the limited state funding for the environment.

The emphasis on the role of experts, primarily on consultants and brokers, within the wider context of conservation austerity was used to support the claim that offsetting could transfer costly and time-consuming responsibilities from the public to the private sector. The role of conservation brokers has been explicitly related to the new opportunities that offsetting was offering to private landowners and developers who could now expect to have on their side an ally from the conservation sector (namely the brokers). Brokers were also expected to ensure that compensation would be both ecologically sound and attractive to developers. By being the intermediate, brokers were expected to facilitate the provision of compensation for the developers by finding the offset site for them while providing an independent assessment to ensure

[22]See http://bbop.forest-trends.org/events/no-net-loss/.

[23]See http://www.environmentbank.com.

that any agreement would make economic sense from the developers' point of view. Conservation brokers in the UK made a systematic effort to prove their valuable role that was only partially successful. On the one hand, many interviewees seemed to believe that having a third party involved in the process would act as a safeguard ensuring a smooth and cost-effective planning process using the typical argumentation in favor of the superiority of the private in comparison to the public sector. This is well illustrated in the following two excerpts from the discussions I had with two brokers:

> The idea behind the emergence of conservation brokerage companies is they act as the fundholder, a local authority and a developer would have their application process, etc. etc. and they would have a number of credits that they need to offset. At that point the developer would sign an agreement with the broker to say we're going to offset 25 woodland credits. The broker then goes out and identifies a site and a land manager and sort of brings the whole process together without the local authority staff member having to spend time looking for that site which it's a real drain on resources. Then the broker also acts as a fundholder and pays an annual basis to ensure that the offset is happening. So, we have a third party that can do almost all the work faster, more efficiently and without requiring important input from local authorities.

And:

> We can ensure that the offset will be in favor of the developer in several ways. We can provide the offset to a developer more quickly. We can also provide protection to the local landowner because we can check the cost of a specific deal. So, if, for example, they think that offset credits are costed too low we will revisit the cost with them. We can also try to assess the future opportunities for the site, and some landowners appreciate this a lot. So, the key idea is once we present an offset to the developer and the local authority all the checks have been made and we can confirm that this is a good offsetting scheme. And we are not going to present it to you if it is not appropriate for your developments. But we can also protect the landowner from the developer saying that: I don't want to pay that, make it cheaper for me. Because we can't have 10 years down the

line and the provider doesn't have enough money to do all the work as required. That is a situation we need to avoid.

On the other hand, others were very cautious of the fact that the increased role of brokers and consultants made the offsetting process highly technocratic and less transparent changing significantly the terms of the public debate, inter alia, by restricting the ability of those who were not familiar with offsetting, including the majority of local communities, to challenge claims of no net loss or net gain. Indeed, during our discussions, brokers, and consultants but also several state officials and members of conservation NGOs were framing offsetting as primarily a technical issue where experts were better positioned to judge whether its outcome would be beneficial or not. This framing of offsetting has often contributed to the depoliticization of conservation–development conflicts: supporters of offsetting used the policy as stratagem to shift the public debate from the impacts and scope of controversial development projects to the narrower question of appropriate compensation, in an attempt to foreclose public debate (Apostolopoulou 2019; Apostolopoulou and Adams 2017a, 2019; Spash 2015). It is interesting to point out here that the "intellectual terrorism" of technicians, technocrats, and the wider epistemology of a purely technical order which allows only certain things to be said has been criticized since the 1970s from Lefebvre (1976, p. 33). Similarly, in the 1990s, Harvey (1996, pp. 374–375), has aptly described how the translation of environmental problems into the domain of expert discourse entailed an application "of bureaucratic-technocratic rationality under the dual influence of the state and corporations." As Harvey (1996) explains democratic politics have been viewed as obstructing "rational" and "sensible" regulatory activities, and concerns for environmental justice have been kept "strictly subservient to concerns for economic efficiency, continuous growth and capital accumulation".[24]

[24]See also in Chapter 3 the reference to Jasanoff's work.

The intensification of neoliberal policies since 2008 and their increasing association with TINA dogma has further reinforced these trends: in mainstream conservation discourse, there has been a clear tendency to circumscribe political contradictions and conflicts by narrowly defined boundaries and to see them as manageable by experts and legitimated through consultation processes in which possible outcomes are largely pre-defined (see also Wilson and Swyngedouw 2014). Thus, despite the widespread acknowledgment that development cannot proceed without considering its environmental impacts, mainstream solutions to environmental loss, like offsetting, never question economic growth and the need to support capital accumulation and official governance modes are always structured around technocratic management and problem-focused decision-making (Swyngedouw 2010, 2015). In the case of offsetting accepting the inevitability of continuous economic growth and development, a priori reframes conservation as a pragmatic search for the least bad outcome separating the practice of conservation from social-environmental movements.

It is important to point out here many interviewees saw the role of the Environment Bank as one of the most problematic aspects of offsetting and as an indication of the real goal of the policy. In particular, the establishment of a national broker signaled for many the outsourcing of conservation to the private sector and a further withdrawal of the state from its responsibilities with profound implications for the transparency, legitimacy and accountability of the planning process. As a conservationist working at the local level and a former employee in a conservation brokerage company respectively argued:

> I am unsure how independent these companies are, I guess they aren't, we 've seen that with consultancies and environmental impact assessments so we know how the private sector can be used. The state isn't perfect but there are specific processes that you can follow to challenge decisions on the public domain, there is the environmental legislation, what happens with these companies? Who can challenge them and how? So far, we have very limited access to their calculations and their agreements with landowners and authorities. And who controls them, how can we ensure that they are accountable? Huge question marks.

And:

> I do not trust these companies. It's too risky to make conservation a commercial agreement, and this is what you do when you give such a key role to brokers. And who can guarantee where this may lead in the future? I do have serious concerns that we can end up with a kind of Bob's biodiversity offsetting broker setting up and just delivering projects that aren't properly costed, aren't looked at by an ecologist, aren't managed in the long-term, aren't consulted with anybody.

Several interviewees also strongly criticized the way that offset calculations had been run and found problematic not only the decisive role of brokers and consultants but also the fact that the whole process was highly fragmented since different people were responsible for different parts of the calculations. The following excerpts from the discussions I had with an interviewee involved in the delivery of an offset for the impacts of a large-scale infrastructure project and a conservation broker respectively are indicative:

> *Interviewee*: I'm not an ecologist so I don't conduct the surveys to collect information, but I do the Defra metric calculations. One of the consultancies have done the fieldwork for this project and I fed in the information, what tree types would be planted, the mapping of the site on the ground, the actual hectarage, all that information has been provided by them and then I've done the calculations to make sure that they're meeting the required number of credits.
>
> *Interviewer*: So, you did the metric calculations?
>
> *Interviewee*: Yes, I just didn't do the fieldwork, the actual on the ground surveys.

And:

> *Interviewee*: Once you have a detailed survey, the completion of the metric isn't a demanding job. You have probably seen somewhere this 20 minutes, that came out to be misinterpreted. If you've got an ecological assessment that has assessed the condition of the habitat, taking that information and putting it into the calculator, doesn't take very long. You

don't survey the site in 20 minutes, but you can put the information from the survey into the calculator in 20 minutes.

Interviewer: So, are you doing the surveys?

Interviewee: No, I am not, this is not our job actually.

Interviewer: Have you visited any of these sites that are included in this excel sheet?

Interviewee: I haven't, but the consultants have, and they have sent me all the information I need so it's not really necessary and most of these are not very special in any case."

5.6 Offsetting and Habitat Banking: Buying Biodiversity "Off-the-Shelf"?

> The democracy of Essex is maybe the land prices that are too high. Are Wales going to be offsetting for developments in Essex because it's cheaper to do it there?
>
> Local council employee

As we have seen earlier in this chapter there has been a strong link between the emergence of offsetting and the consolidation of market environmentalism in England. Indeed, the key logic behind offsetting is a reflection of the core tenets of market environmentalism, namely, that the primary reason of environmental degradation is that the environment is being treated as an externality in conventional decision-making processes. This has been explicitly stated in the impact assessment published in 2013 (HM Government 2013, p. 1) where the introduction of a regulated biodiversity offsetting market was considered necessary because currently "habitat in England is being lost, fragmented and degraded as biodiversity -as an 'unpriced externality'- suffers from market failure," Offsetting's supporters by building on this line of argumentation claimed that the policy could support the inclusion of biodiversity in the economic calculus of development costs. But they went further than that by also claiming that offsetting had the potential to offer a new "win-win-win" vision for UK conservation by speeding up the planning process, providing more money for conservation and offering a new source of income to those who would be involved in the exchange of offset credits. The role of brokers was decisive in advertising offsetting to landowners by emphasizing the economic benefits that their involvement in the delivery of offset sites and credits trading could offer them:

> The idea that any piece of land that could be valuable for wildlife has a sort of market value, that's quite useful. If a landowner feels his woodland has value, he might manage and look after it. To a big farmer pieces of woodland are a liability really; they are of very little value. So, we [conservation brokers] are trying to show them that if there was an exchange and a market and that piece of woodland could be registered that could be quite useful, in many ways more useful than the actual offsetting metric itself really.

The initial idea about how offsetting was supposed to work was that landowners would register their land in conservation banks, creating the basis for "buying biodiversity off-the-shelf" to borrow the expression of the Defra Green Paper (2013, p. 5). This was expected to benefit developers since they would be able to provide the required compensation by purchasing the necessary number of credits from a bank. The multiple benefits of this "win-win-win" vision have been emphatically explained to me by a local officer in one of the pilots and by a Defra employee respectively:

> A national offsetting scheme, as robust as the one we've got in here, it does make efficiencies, because it reduces the time spent on negotiations. So, when compensation was previously required there was a lot of discussion on how much is required, what management is suitable etc. And that can take a lot of time. Whereas with the metric even if there is a discussion, there is a framework to base on, and with a functioning offset market, like the one the Environment Bank is trying to establish, because they have networks, they have sites of their own register, they have contacts with local land agents, they have a partnership with AB Sustain, and AB Sustain have a large network of landowners. The idea is that when somebody says I need a site they can go to the Bank and they will find a site for them. So, if you've got a bank setup you can buy the required credits within a week. And so that will not only make the whole process much

> easier but will clearly speed up the planning process as well while giving money from development to conservation and opening a new market for offset providers.

And:

> There is clear evidence from the United States that there was a definite save in the time to take the decision. In our own impact assessment, we estimated it was only a week. Of course, it depends on how developed the system is. If you are imagining that in a sort of 20 to 30 years' time you've got a fully operational system that is working as a well-oiled machine, you can see that there will be a considerable saving. You won't get that early on when each offset must be looked out individually and negotiate it. But in the future, you can have massive gains from a well-organized offset market.

Conservation banking was also attractive to many NGOs that saw in this a new opportunity for supporting conservation activities and primarily landscape conservation projects. However, their experience with the implementation of offsetting in England gradually made them quite uncertain on whether this would happen and how exactly it would benefit them. The following excerpt from my discussion with a member of a local conservation NGO is indicative:

> As an organization we think that habitat banking is a good idea. If we could get the capital funding to start buying areas and land and bank them we would. But we ain't got money to do it. Developers might be able to buy land and bank it, but conservation organizations can't. But if you think about it logically it would make a lot of sense. What is happening now is that, for example, there is a development with protected species like reptiles, so developers and their consultants start ringing NGOs asking for sites to put their reptiles on. If there was a better system, we might already have sites ready and developed in advance, before any specific development project. [...] In an ideal world we would have a system of land banking, we would create grassland and improve the quality of that grassland so we could accept things like reptiles from development sites. We were expecting that offsetting could lead there, this was supposed to be the idea initially, but it doesn't seem to work

like that. Which is not surprising, this country always struggled with the appropriate planning for biodiversity.

The above observations give a good sense of how offsetting was initially expected to work in the UK. As it became clear during my fieldwork a strong link between offsetting and banking was desirable to many not only because it was expected to save considerable time and to convince landowners to be involved in conservation in an era of prolonged austerity, but also because it would mean that compensation credits would be available for selling in advance of a project. This would mean that the time-consuming and, often, uncertain negotiations that currently take place between developers, their consultants and lawyers, local councils and councilors, local communities, NGOs and other interested parties would be replaced by a trading of offset credits. Under that scenario, a developer would have no other responsibility than giving a specific amount of money to buy the required credits from a bank and claim that she/he has provided the necessary compensation. In practice, however, at least so far, things have not been developed in the way that those who conceived and designed the policy were hoping to and, even though there are still ongoing attempts to establish banks, for example in Warwickshire, linking offsetting to habitat banking has not been as successful as expected. One of the reasons is that, as experience with the offset pilots gradually showed, involving landowners (including farmers), has not been as straightforward as initially thought:

So potentially you can convince people to restore their land, for example their quarry, and then offer up all these units in a conservation bank and gain income from this. With farmers, however, this is not easy. One of the problems we've found when looking for offset sites is that farmers in Essex are very unwilling to offer up their land because they get a lot of money from subsidies, the wheat prices have been high in the last two years. So, you've got to offer them quite a good deal, particularly if you want an offset to last longer than 10 years, they are very reluctant to go beyond 10 years. They can also get almost £1,000 a hectare for solar farms now, so you're competing against wheat prices and the price that they're getting for solar farms.

Moreover, several conservationists were not convinced that banking would ever work as a "well-oiled machine" in favor of both development and conservation. One of the key concerns was that giving a price to nature, through the monetary valuation of both its destruction and its conservation, would transform the process of providing environmental compensation to a market exchange with all associated risks:

> The main concern we have is that if something is valued in cash terms it means that somebody gives a load of cash and then that's it, end of the story. In most cases nothing happens afterwards, it's highly possible that you won't get any biodiversity benefits. We think the losses must be linked specifically to an offset project [...] But the danger is that someone gives someone a million pounds to do something and they sit on that money. In the meantime, the development happened, and you have this loss of continuity of habitat. We have discussed a lot about it because you destroy the habitats where a species goes, and these are not recreated. That was a quite complex discussion. We did respond on that matter to the national consultation basically saying we can't reduce the biodiversity impacts to pounds and pence or just to credit. And with banking this is the key idea.

Another important issue had to do with how the price of offset credits would be calculated and what it would reflect. State officials argued that this is something for the offset providers to decide without state intervention. As they explained to me, landowners involved in the provision of credits are providing a service to developers and, therefore, the price of a credit should take into account anything that the landowner "would not avoid anyway," including legal fees, the cost of ecological surveys, the costs for designing and reviewing the management plan and conduct new surveys when needed (since we refer to schemes that would last up to 30 years), the initial capital works, ongoing maintenance, management fees, etc. As a Natural England employee pointed out the costs should also include land use change caused by the creation of the bank:

> If, for example, we talk about a farmer who must change how he uses the site, and that has reduced his income, compensation for that should be in there. They must look to future duties, they are doing long-term

> things, and so they should think how the cost would change in the future. Potentially, they've been able to do something else with this land and increase their main income so we cannot ignore this. To create a bank will take this land out of production, for grazing or whatever, and they would keep that land free for the next 30 years. And they've got to make sure they've got enough money to run a viable project. They should not imagine that as a financial loss. It is not reasonable, and it is not fair, so those kinds of constraints need to be taken into consideration when designing an offset and when deciding its price.

It has to be pointed out here that offsetting's supporters, in line with the mainstream argumentation in favor of the economic valuation of nature, often adopted a rather simplistic market logic during our discussions arguing that putting a price to nature would increase the recognition of its value and hence would reduce or even prevent its destruction. Even if we ignore the major theoretical problems of this approach (which is largely based on neoclassical economics) and we only consider empirical evidence we will rather clearly see that subjecting non-human nature to the vagaries of the market has not been successful so far but rather the opposite. Prices of mitigation and offset credits have been highly variable and have been influenced by all kinds of factors, including variations in restoration costs, land prices, supply and demand, speculative action by landowners and financial crises (see, e.g., Madsen et al. 2011; Muradian and Rival 2012). In Europe, credit prices have been ranging from 30.000 euros to 1.2 million euros per hectare (Conway et al. 2013) whereas in New South Wales BioBanking only in 2016 prices ranged from $1 for Darwinia biflora to $9293 for gray box-forest red gum grassy woodland on flats of the Cumberland Plain, in Sydney Basin Bioregion (Bennett et al. 2017, p. 32). Similarly, in Queensland, the price of offset credits ranged from $10,395 for freshwater wetlands to $178,571 for seagrass (ibid). The crucial point here is that conservation activities by being transformed into an exchange of credits, namely of priced ecological assets, are directly dependent on the function of the market and on price fluctuations (see also Chapter 4, here, and particularly the last paragraph of Sect. 4.3). This reframes a genuine concern for the use value of nature, expressed in the need to prevent environmental degradation, into a matter of market prices and potentially profitable exchanges, a question of economic value "that is entirely inimical to the original concern" (Smith 2010, p. 249).

Given the slow development of conservation banking in the UK and the challenges of creating a functional national system, many interviewees emphasized that state regulation would be necessary for offsetting to work and that in the absence of a regulated and mandatory system no market will ever be created. However, in my discussions with state officials it became evident that the government had a clear preference for a voluntary scheme and no intention to be directly involved in regulating the price of offset credits. The following quote from a discussion I had with Defra employees is indicative:

> *Interviewee*: The proposal is that it will be entirely a matter between the developer and the provider of the offset to decide the price. So, we wouldn't put a sort of value on it. I think that there is a wide range and so fundamental differences/variations; so when for example we are looking to do an offset in London there are various options, and to deliver a unit of biodiversity –in our jargon- that's going to cost 2.000 pounds in one site and 20.000 on another, so, you know, a huge range. And what will it cost to deliver the same benefit.
>
> *Interviewer*: That means that the price will also reflect the value of land.
>
> *Interviewee*: Yes, definitely. And this is something we have in mind from the beginning. So, someone could argue that offsets will mainly be where land is cheaper – and this is something we should carefully explore.

All the above empirical observations corroborate the argument made in Chapter 4 that biodiversity offsetting should be primarily seen as an instance of rent. This is well reflected in the key role that land prices have played in both the location of offset sites and the price of credits. Moreover, as also explained in Chapter 4, offsetting in most cases changes land uses and affects land access, and can impact the level of rent and the profits generated through landownership, often furthering processes of land speculation. For example, in the case of the Whitehouse Farm in

North Tyneside that will be described in the next chapter even though there was an initial agreement with a landowner for the provision of an offset site when the landowner discovered that Bellway Homes was involved in the purchase he tripled the price of the land. The fact that offsetting can have both positive and negative roles to play in relation to capital accumulation and the way this differs according to the interests of different sections of capital has been vividly reflected in the offset pilots in England where some landowners and developers perceived offsets as a new tax, others as a new land use, and others as an opportunity for speculation over land prices. As a Defra employee pointed out:

> Responses from developers are very mixed. Some, mainly small developers, don't want it. Another group, they don't basically really mind what the rules are as long as the rules are the same to everybody. So, if it is requirement, they will do it if that will ensure fair competition. Another group, the X [the group is not identified to ensure anonymity], the chief extractor, the ecosystems market taskforce, and some major industries clearly support offsetting. And there are also some companies, I have in mind a housing company, that market their homes on the grounds of user friendly homes, and that they are on a nice setting and so on, so they are happy to offset voluntarily, this is part of their green profile and they will fund it.

A final observation about the nature of offset credits is important here. In offset trading, the production of equivalent natures is instrumental as we saw in Chapter 3 in order to show that the whole process is not a chaotic exchange of non-comparable things. The debates on whether in-kind or out-of-kind exchanges are appropriate reveal such worries, but hardly grasp the problems arising from enabling the trading of ecological losses and gains across time and space. Even if offsetting involved only in-kind exchanges, this would imply that a certain number of a particular species, for example of nightingales threatened or lost by large-scale housing developments at Lodge Hill (see next chapter), would be equivalent to the same number of nightingales "conserved" in a different place, for example in Essex. This equivalence assumes that nightingales can exist outside of the specific ecosystem and reflects an understanding of species and habitats as unrelated to place, culture and people. This implies that nature can be measured and produced as any other product. In out-of-kind exchanges the crudeness and the reductionism of the offsetting process is even more obvious since the exchanged credits do not even refer to the same species or habitat(s). In both cases, nature is stripped of its specificity (sensu Smith 2006) and the socioecological origins of offset credits are obscured with profound implications for the involved socionatures. Offsetting in that sense reproduces the idea of a flat world and, as pointed out in Chapter 3, produces and reproduces non-human nature as reproducible.

5.6.1 Can Money Capture Nature's Destruction? A Brief Note on the Pervasive Influence of the Logic of Cost–Benefit Analysis

Cost–benefit analysis dates back to infrastructure appraisal efforts in France in the second half of the nineteenth century when a series of books on its use have been published following the work of a group of engineers at the École Nationale des Ponts et Chaussées, and particularly the work of Jules Dupuit on the utility of public road improvement (Thoft-Christensen 2012). One of the most striking examples of the penetration of the cost–benefit logic in public life since the early twentieth century can be found in an article published in 1910 in the New York Times headlined "What the baby is worth as a National Asset".[25] The article provides an estimation of the value of an eight-pound baby at birth as \$362 a pound and attempts to offer specific figures of children's "earning capacity". The calculations were based on the work of Irving Fisher, a Professor of Political Economy at Yale University, who was trying to calculate, based on cost–benefit analysis, the economic value of what he called "the baby crop" to prove the financial benefit of saving children's lives. Cost–benefit analysis was popularized after 1945

[25] See https://www.nytimes.com/1910/01/30/archives/what-the-baby-is-worth-as-a-national-asset-last-years-crop-reached.html.

by the World Bank especially through its application in dam construction and despite several criticisms since the 1960s, it is now recognized as the major appraisal technique for public investments and policy.[26] Cost–benefit analysis is a tool that is assumed to support political decisions by using a set of simplified assumptions (Arler 2006), essentially by adding up total money costs and benefits regardless of who receives them. It lies at the core of environmental economics and aims to determine whether a policy promotes economic efficiency. Importantly, as Coplan (2017, p. 283) explains, there is a wealth transfer implicit in any cost–benefit analysis: "simply put, that it will cost a business enterprise $2 to prevent $1 of harm to an uncompensated member of the public cannot ever justify failing to prevent the injury." To show the implications of the latter Coplan uses one of the most exemplar cases of regulatory cost–benefit standard setting, the drinking water standard for arsenic according to which the Environmental Protection Agency (EPA) in the United States accepted that a number of water-consumer deaths and illnesses due to lung and bladder cancers was necessary to avoid imposing an excessive compliance cost on public drinking water enterprises (ibid).

As Coplan (2017) further explains cost–benefit analysis is based on a flawed libertarian baseline that entails a resistance to environmental legislation and public rights and a preference for unrestricted limits on the environmentally destructive activities of a property owner. The cost–benefit logic clearly underpins the polluter pays principle and consequently both biodiversity and carbon offsetting policies. This exposes the naivety of the arguments of offsetting's supporters who suggest that putting a price to environmental compensation will increase the recognition of the value of nature and hence will reduce its destruction. As evidence from England shows offsetting was considered the preferable policy option when it made economic sense for developers and it was abandoned when it did not (Apostolopoulou 2019; Bormpoudakis et al. 2019; Apostolopoulou and Adams 2019).

[26]See http://www.oecd.org/environment/tools-evaluation/36190261.pdf.

5.7 Framing the Social as Irrelevant: Obscuring the Unevenness and Inequality Behind Offsetting's Equivalence

> You cannot quantify a landscape, you cannot quantify a view.
>
> Local activist

> One of the big things people coming down on us was the associated benefits of nature, the benefit to individuals of having the natural world close to them. The metric cannot quite capture that. This is still an open question.
>
> Defra Employee

The social impacts of offsetting constitute one of the most controversial aspects of the policy in England and, particularly, the way that offsetting, primarily through calculating ecological losses and gains by using the Defra metric, excluded from consideration the social and cultural aspects of biodiversity loss. This was one of the issues that sharply divided interviewees along two major categories: on the one hand, there were those who argued that the policy is a conservation tool which aims to accurately measure biodiversity loss and that as such it should focus only on strictly defined ecological criteria. On the other hand, there were those who argued that the above approach is a convention since biodiversity cannot be separated from its social and cultural aspects and, thus, any calculation of losses and gains based only on ecological criteria will inevitably not only highly reductionist but would also cause unevenness and inequality. There were, of course, also other opinions that did not fall exactly within these two categories. There were those, for example, who argued that issues of culture, place, and history are relevant but are currently dealt separately in the planning system and, thus, they are covered by other regulations:

> I know there are obviously links between the landscape and biodiversity; people's enjoyment of the site and biodiversity, but cultural or historical

issues are dealt separately; there is a planning policy to do with heritage and open space. So some will say that such a debate is covered properly. In our case offsetting mainly focuses on whether it is possible to offset the loss of species using the Defra metric, but there are a lot of other discussions about traffic, privacy, amenity, and landscape. In theory all these are dealt with in other ways but how effective it will be to deal with those issues if you move the offset site somewhere else? You are effectively separating the issue out. And I'd say the same about banking. We can stick land together and manage it as a whole. It is feasible and it may work. But what this site will be? Because now a site is supposed to be conserved for biodiversity but at the same time you conserve that open space for local people, the heritage of the site, and cultural aspects. How these aspects will be covered by a conservation bank located in a different area?

And:

I think the problem is we were trying to distinguish between different roles that a particular piece of land might play. We were only latching onto the biodiversity side. For us that is the bit not carried in the planning system. In real life when you go to the place at the end of your street to walk your dog or play football or admire the green there or whatever, you don't distinguish between its role as green infrastructure that is already covered by quite precise rules. So, I think the question for us is, is there something, is there a gap, so we just latch on the biodiversity value because all the other aspects are recognized in the planning system. Of course, then you end up separating things that are actually interrelated with all sorts of problems that such separations create.

And:

In a way issues of place or culture should not be part of this policy, I mean the policy is about biodiversity loss so there should be other policies for green spaces and recreational enjoyment. The two might overlap but offsetting offers compensation for biodiversity loss. But people have raised it, particularly in urban areas they feel they might loose a valuable piece of open space and then the offset will be located close to a rural community,

> so that could be quite significant. But that's a failure of open space policy really, not a failure of biodiversity policy.

The above opinions reflect the reductionist understanding of non-human nature that pervades environmental and planning legislation and policy and which has been primarily defended by state officials, consultants and legal representatives of developers. On the other hand, in all the case studies included in my fieldwork, local communities did not separate biodiversity concerns from other issues arguing that the double process of destroying the development site and creating an offset site somewhere else entailed the loss of rights for the involved communities in both places. Importantly, many advocates of offsetting when faced with escalating criticism about the social and spatial injustices of trading ecological losses and gains across space and time argued that development in England mainly occurs in private land and, therefore, issues of public access are not as relevant as the public may think:

> There is an argument about the human use of the site, that people are using it for public purposes. But most developments are on private land, so they don't have public access anyway. So, can you take nature away from people? Because some people have used this as an argument against offsetting. But if there is no public access to the development site why should I need public access to the offset site? And as I said before most development sites are just farms and fields effectively. And if the site has a high level of public use because of footpaths or whatever, that's a separate planning issue that has nothing to do with offsetting. In my perspective, issues of public access should not be considered. If such issues exist and locals are worried about losing a site, then this is a secondary planning issue.

Of course, such arguments are never as neutral as their advocates try to argue whereas the overemphasis on formalistic issues in most cases is used to disorientate the public discussion or lock disagreement in legalistic aspects. Offsetting's extreme reductionism was thus serving specific political purposes almost in all the cases I visited in England. The policy's ignorance for the social reflected the governmental priority for economic growth at any cost and the underestimation for the social impacts of land use change. It is important to understand here that a class analysis is the only capable to show that the main reason that the social aspects of offsetting are framed as irrelevant is because they are antithetical to the interests of the housing or the infrastructure industry. Offsetting attempts to mask the unequal power relations that drive the production of new social-ecological configurations by portraying the remaking of socionatures as a matter of expert knowledge whose goal is to offer compensation to an unspecified subject (Defra 2013). The policy's supporters in most cases remain characteristically silent on who gains and who loses from the trading of environmental goods and bads that offsetting enables. However, there is increasing evidence that offsetting mostly facilitates the interests of specific sections of capital to the detriment of the social majority who tends to lose its access to land and resources with major implications for livelihoods and the quality of everyday life (Apostolopoulou 2019; Re:Common 2019; Seagle 2012; WRM and Re:Common 2016). Offsetting's focus on ecological criteria was used in England to legitimize this view as a concern for a scientifically sound assessment of environmental impacts that will not be affected by "populist demands," a triumph of technocrats who have the rationality and knowledge to decide what is worth to be compensated for:

> My concern is that if you add other issues the number you get in the end is biased by things that are not biodiversity related. So, it will skew the biodiversity portion of that metric and I am very worried if that happens. If the site has historical significance or cultural significance that is a planning issue in of itself. I don't think it is appropriate to include this in offsetting metrics. We are talking about biodiversity value here. And to trample different things into a metric it will become more difficult to use and less applicable. I am even worried of including species into the metric for the same reason. We have to safeguard scientific assessments from populist arguments; and this is exactly what offsetting is offering.

Ironically, the contradictions of the extreme reductionism of the Defra metric and of the offsetting's disinterest for the social became clear when the issue of ecosystem services entered the discussion, a contested term that has been highly criticized for bringing about the monetary valuation of nature:

> When we asked them about the cultural and the social value of the area they are destroying, and we used their language, the language of ecosystem services because this is what they are - isn't that right?- they said that we can perhaps ask for a bigger offset, they would then use a multiplier and include those values but that right now the Defra metric measures biodiversity and not ecosystem services. (Local activist)

This brings at the forefront the tension between offsetting for the loss of biodiversity and for the loss of ecosystem services, an issue that has received relatively limited attention so far even though it is quite apocalyptic of the conundrums of extreme reductionism. During my fieldwork, I realized that this was an uncomfortable issue to discuss for many supporters of offsetting who were otherwise also warm supporters of ecosystem services but were finding hard to combine them. Some interviewees involved in the designation of the Defra metric told me that even though concerns about how to combine offsetting for biodiversity loss with offsetting for the loss of ecosystem services are legitimate it is not "appropriate" to conflate the two terms. The argumentation was that since there is not a direct overlap between biodiversity and ecosystem services, the two cannot be considered equivalent and, therefore, they cannot be measured in the same way. According to this logic, offsetting and ecosystem services should be dealt with separately and by using a different type of metric that would, inter alia, be able to quantify the ability of users to access a service. However, I received no clear responses on what this would mean in practice since, for example in England, the metric currently focuses exclusively on biodiversity. Does this mean that Defra despite having adopted an ecosystem services approach, as explained in the earlier sections of this chapter, is accepting that there is no need to compensate for the loss of ecosystem services? This means accepting that offsetting can lead to the loss of ecosystem services since these are not being taken into consideration. The following quote from a discussion I had with an interviewee who have been involved in the designation of the metric shows well the limits of the reductionist representation of non-human nature either in the form of ecosystem services or biodiversity units:

> Where an offset is located would have a bearing on people's ability to use and access a service and therefore the benefits that they get and that might be entirely different from where you might want to locate something to maximize or to get the right biodiversity outcome. It would be very challenging to do them together. That doesn't mean there might not be overlaps but often looking at, say, green infrastructure to benefit biodiversity you might need to exclude people but to give the service to the people for recreation and amenity and access to green space you might need to allow the people to access it. I don't think it's impossible to offset impacts on ecosystem services, but I don't think you can do it using something like the Defra metric as it stands.

Interestingly, one of the key arguments that interviewees used to explain to me that the Defra metric was inappropriate for offsetting the loss of ecosystem services was its inability to capture cultural aspects, the sense of place and local ties between people and biodiversity. Interviewees who supported the ecosystem services approach perceived all these issues as examples of ecosystem services that are not being properly accounted for:

> *Interviewee*: Biodiversity is providing people with a sense of place or cultural significance, that is an ecosystem service and that needs to be offset and that might be in the same place as where you offset intrinsic biodiversity impacts on intrinsic biodiversity values, or it might be something entirely different. People need to distinguish clearly between biodiversity per se and the services it gives to people. Where an offset is located should be all about ecosystem functioning, biodiversity, species populations and where ecological equivalence is achieved, not just putting it nearby because people use that area, that's a different value that we should be offsetting.
>
> *Interviewer:* Yes, but this means, borrowing your language, that we will end up with many trade-offs and who and how will decide?
>
> *Interviewee*: There are certainly many trade-offs. This is partly why in the States they have different kinds of credits and different kinds of banks

and offsets. You might have species specific ones or multiple species ones or habitat ones or ecosystem ones like wetland ones. We haven't really sorted that out in the UK at this point.

5.8 Interregnum: A Discussion with a Designer of Offset Metrics on Ecosystem Services, Offsetting and the Social Implications of the Economic Valuation of Nature

Interviewer: Let's start by discussing how offsetting relates to the broader shift towards the economic valuation of nature.

Interviewee: Well, it certainly does but we took a deliberate decision to avoid valuation per se because, from my own extensive experience, I can see that it's fraught with pitfalls. So, you need a rationale for prioritising based on people's dependence on a service and the alternatives that they have available and the extent to which they consider it to be important for their livelihoods and well-being. And if it is a priority then it should be maintained and again there will be a cost involved in maintaining it, but the decision about how much it's valued so that you can decide whether you should maintain it or not I think is a whole different argument and is just a never-ending debate that will never be resolved. I think money isn't really the sensible way to value nature because it doesn't capture well the less tangible benefits that people still think are very important to them. So, we need to find some way of capturing those values and benefits that isn't entirely based on money, on financial valuation because I just think it doesn't work.

Interviewer: Do you fear that payments for ecosystem services or biodiversity offsetting may facilitate this kind of valuation of nature in economic terms?

Interviewee: They definitely appear to, but I think the reason they do is because of the failure to grasp the No Net Loss concept and if there isn't a policy or regulation that makes that very clear then yes I see the fear. So, I'm in favour of these things only if they are appropriately used as tools towards a defined policy outcome. Using them on a case-by-case basis without a well-defined policy framework can be dangerous and they can be abused, and they are being abused and there's lots of evidence. But I'm not sure what the other solutions are. I'm not seeing other powerful solutions being suggested by anybody.

Interviewer: So, you think that ecosystem services and biodiversity offsetting if implemented properly do not have to lead necessarily to the monetary valuation of nature?

Interviewee: I would say yes but only for ecosystem services. I mean as a practitioner doing environmental assessments ecosystem services perspective does bring you to different conclusions than conventional approaches looking at livelihood restoration only or looking at ecosystems only.

Interviewer: Could you give me an example?

Interviewee: Well for example the sense of place attachment that people have which is often driven by ecosystems and biodiversity can cause a lot of social unrest that can seriously hold up projects. If you don't get people accepting the changes they're going to have in their landscape and their environment and if they feel that their whole sense of community, belonging and sense of place is going to be destroyed, they're unlikely to accept your project and I think project proponents underestimate the power of those kinds of emotions that people have to their cost and often have to accommodate big delays. But also, for example, they may not be very good at predicting their own needs for services that they depend on. So, developments often depend quite heavily on natural resources and ecosystem services, and treat them as free goods that can be taken for granted so I think having the ecosystem services perspective has maybe made people a little bit more aware of their dependencies on the environment. Because if, for example, you're developing a project in a landscape that is essentially rural with high levels of poverty and people have a very high dependency on subsistence farming to live, then the consequences of destroying natural resources can be really serious, millions of people can be affected to the point of malnutrition, all sorts of human rights being compromised. The social aspects are not a peripheral issue and if you don't recognise them and you don't manage them then...

Interviewer: Do you consider this more relevant to the Global South context or would you say the same about the UK for example?

Interviewee: It's less obvious in the UK but, on the other hand, look at flooding impacts.

Interviewer: That's a good example.

Interviewee: Yeah, I mean the financial cost of that was absolutely enormous and some of that was exacerbated by failure to recognise ecosystem services provided by farmland managed in a certain way. I think it's generally accepted that upper catchment land management does have an impact on sedimentation downstream and the need for dredging and the costs of flood evacuation. But again, as I said this is completely different from the discussion on profitability etc. That's mostly overseas so more in Uganda and places like that. So, the way ecosystem services seem to have been conceptualised here to me seems to be quite high level and supply oriented and largely ignoring the social aspects.

Interviewer: On the other hand, it is also quite challenging to figure out how exact prices will be defined either for offset credits or for ecosystem services and even more how these prices will capture the social aspects – if we accept that this is even possible.

Interviewee: Well I don't know if anybody has looked at these, at least seriously. In the Somerset Levels, for example, various decisions were made in the decades leading up to those floods last year that have to do with the level of management, upper catchment management, with whether pumps would be maintained. So, they had Dutch pumps that they got rid of and then they had to buy them back in. So, some of the costs are quite tangible.

Interviewer: Yes, of course, but only some of the costs.

Interviewee: Exactly, to come up with a global cost that captures everything, including the social impacts, I don't think it's possible. To know whether good stewardship would have been a better alternative than what happened, it's sad to say but theoretically it ought to be possible. The Environment Agency's ability to dredge in the years preceding the flood was heavily influenced by the way Defra costed benefits from flood protection. So, they have established methods for deciding what flood control schemes they will support based on costed benefits in terms of loss of life or damage to property and those are heavily skewed in favour of properties, numbers of properties, so that meant that schemes in areas with lower levels of habitation never get through. But that doesn't mean they're not important in terms of the overall picture and managing the environment properly. And the consequences of that cost–benefit approach that pervades everything in Britain we now see that can be detrimental.

References

Adams, W. M., Hodge, I. D., & Sandbrook, L. (2014). New spaces for nature: The re-territorialisation of biodiversity conservation under neoliberalism in the UK. *Transactions of the Institute of British Geographers, 39,* 574–588.

Allmendinger, P., & Haughton, G. (2012). Post-political spatial planning in England: A crisis of consensus? *Transactions of the Institute of British Geographers, 37,* 89–103.

Apostolopoulou, E. (2019). Beyond post-politics: Offsetting, depoliticization and contestation in a community struggle against executive housing. *Transactions of the Institute of British Geographers.* https://doi.org/10.1111/tran.12354.

Apostolopoulou, E., & Adams, W. M. (2015). Neoliberal capitalism and conservation in the post-crisis era: The dialectics of 'green' and 'un-green' grabbing in Greece and the UK. *Antipode, 47,* 15–35.

Apostolopoulou, E., & Adams, W. M. (2017a). Biodiversity offsetting and conservation: Reframing nature to save it. *Oryx, 51,* 23–31.

Apostolopoulou, E., & Adams, W. M. (2017b). Biodiversity offsetting and the reframing of conservation: A reply to ten Kate & von Hase and Dempsey & Collard. *Oryx, 51,* 40–42.

Apostolopoulou, E., & Adams, W. M. (2019). Cutting nature to fit: Urbanization, neoliberalism and biodiversity offsetting in England. *Geoforum, 98,* 214–225.

Apostolopoulou, E., Bormpoudakis, D., Paloniemi, R., Cent, J., Grodzińska-Jurczak, M., Pietrzyk-Kaszyńska, A., et al. (2014). Governance rescaling and the neoliberalization of nature: The case of biodiversity conservation in four EU countries. *International Journal of Sustainable Development and World Ecology, 21,* 481–494.

Arler, F. (2006). *Ethics and cost-benefit analysis* (Research Report 4). Aalborg, Denmark: Department of Development and Planning, Aalborg University.

BBOP (Business and Biodiversity Offsets Programme). (2009). *Biodiversity offset design handbook* (Appendices). Washington, DC: BBOP.

BBOP (Business and Biodiversity Offsets Programme). (2012). *Resource paper: No net loss and loss-gain calculations in biodiversity offsets.* Washington, DC: BBOP.

Bennett, G., Gallant, M., & ten Kate, K. (2017). *State of biodiversity mitigation 2017.* Markets and Compensation for Global Infrastructure Development. Ecosystem Marketplace, Forest Trends, Washington DC.

Bormpoudakis, D., Tzanopoulos, J., & Apostolopoulou, E. (2019). The rise and fall of biodiversity offsetting in the Lodge Hill large-scale housing development, South East England. *Environment and Planning E: Nature and Space*. https://doi.org/10.1177/2514848619884890.

Carver, L., & Sullivan, S. (2017). How economic contexts shape calculations of yield in biodiversity offsetting. *Conservation Biology, 31*, 1053–1065.

Collingwood Environmental Planning Limited. (2013). *Evaluation of the biodiversity offsetting pilot phase, WC 1051*. Summary of Interim Report Collingwood Environmental Planning Limited in partnership with The Institute for European Environmental Policy (IEEP).

Conway, M., Rayment, M., White, A., & Berman, S. (2013). *Exploring potential demand for and supply of habitat banking in the EU and appropriate design elements for a habitat banking scheme* (Final Report Submitted to DG Environment). UK: ICF GHK and BIO Intelligence Service.

Coplan, K. S. (2017). The missing element of environmental cost-benefit analysis: Compensation for the loss of regulatory benefits. *Georgetown International Environmental Law Review, 30*, 281.

Defra (Department for Environment, Food and Rural Affairs). (2007a). *An introductory guide to valuing ecosystem services*. Department for Environment, Food and Rural Affairs. London, UK. Available at: https://assets.publishing.service.gov.uk/government/uploads/system/uploads/attachment_data/file/69192/pb12852-eco-valuing-071205.pdf.

Defra (Department for Environment, Food and Rural Affairs). (2007b). *Securing a healthy natural environment: An action plan for embedding an ecosystems approach*. Department for Environment, Food and Rural Affairs. London, UK. Available at: https://ecosystemsknowledge.net/sites/default/files/wp-content/uploads/Defra%20eco-actionplan.pdf.

Defra (Department for Environment, Food and Rural Affairs). (2010). *Payments for ecosystem services: A short introduction*. Department for Environment, Food and Rural Affairs. London, UK. Available at: http://www.fwr.org/WQreg/Appendices/payments-ecosystem.pdf.

Defra (Department for Environment, Food and Rural Affairs). (2012). *An overview of the local nature partnership role*. Department for Environment, Food and Rural Affairs. London, UK. Available at: https://assets.publishing.service.gov.uk/government/uploads/system/uploads/attachment_data/file/192580/local-nature-partnerships-overview120402.pdf.

Defra (Department for Environment, Food and Rural Affairs). (2013). *Biodiversity offsetting in England green paper*. Department for Environment, Food and Rural Affairs. London, UK. Available at: https://consult.defra.

gov.uk/biodiversity/biodiversity_offsetting/supporting_documents/20130903Biodiversity%20offsetting%20green%20paper.pdf.

Defra (Department for Environment, Food and Rural Affairs). (2016). *Defra's payments for ecosystem services pilot projects 2012–15: Review of key findings*. Department for Environment, Food and Rural Affairs. London, UK. Available at: https://assets.publishing.service.gov.uk/government/uploads/system/uploads/attachment_data/file/578005/pes-pilot-review-key-findings-2016.pdf.

England Biodiversity Group. (2011). *ThinkBIG: How and why landscape-scale conservation benefits wildlife, people, and the wider economy*. Available at: http://publications.naturalengland.org.uk/publication/30047.

Harvey, D. (1996). *Nature, justice and the geography of difference*. Oxford: Blackwell.

HM Government. (2011). *The natural choice: Securing the value of nature*. Available at: https://assets.publishing.service.gov.uk/government/uploads/system/uploads/attachment_data/file/228842/8082.pdf.

HM Government. (2013). *Biodiversity offsetting* (p. 1126). Defra: Impact Assessment.

HM Government. (2016). *Natural Capital Committee: Terms of reference*. Available at: https://assets.publishing.service.gov.uk/government/uploads/system/uploads/attachment_data/file/517123/ncc-terms-of-reference.pdf.

HM Treasury. (2006). *Long-term opportunities and challenges for the UK: Analysis for the 2007 comprehensive spending review*. London, UK.

Lawton, J. H., Brotherton, P. N. M., Brown, V. K., Elphick, C., Fitter, A. H., Forshaw, J., et al. (2010). *Making space for nature: A review of England's wildlife sites and ecological network*. Report to Defra.

Lefebvre, H. (1976). Reflections on the politics of space. *Antipode, 8*, 30–37 (M. J. Enders, Trans.).

Lockhart, A. (2015). Developing an offsetting programme: Tensions, dilemmas and difficulties in biodiversity market-making in England. *Environmental Conservation, 42*, 335–344.

Madsen, B., Carroll, N., Kandy, D., & Bennett, G. (2011). *State of biodiversity markets report: Offset and compensation programs worldwide*. Washington, DC: Forest Trends.

Muradian, R., & Rival, L. (2012). Between markets and hierarchies: The challenge of governing ecosystem services. *Ecosystem Services, 1*, 93–100.

Peck, J., Theodore, N., & Brenner, N. (2012). Neoliberalism resurgent? Market rule after the great recession. *The South Atlantic Quarterly, 111*, 2.

Re:Common. (2019). *Turning forests into hotels: The true cost of biodiversity offsetting in Uganda*. Available at: https://www.recommon.org/eng/turning-forests-into-hotels-the-true-cost-of-biodiversity-offsetting-in-uganda/.

Seagle, C. (2012). Inverting the impacts: Mining, conservation and sustainability claims near the Rio Tinto/QMM ilmenite mine in Southeast Madagascar. *Journal of Peasant Studies, 39*, 447–477.

Smith, N. (2006). Nature as accumulation strategy. In L. Panitch & C. Leys (Eds.), *Socialist register 2007: Coming to terms with nature* (pp. 16–36). London: Merlin.

Smith, N. (2010). *Uneven development* (3rd ed.). New York: Verso.

Spash, C. L. (2015). Bulldozing biodiversity: The economics of offsets and trading-in nature. *Biological Conservation, 192*, 541–551.

Sullivan, S., & Hannis, M. (2015). Nets and frames, losses and gains: Value struggles in engagements with biodiversity offsetting policy in England. *Ecosystem Services, 15*, 162–173.

Swyngedouw, E. (2010). Impossible sustainability and the post-political condition. In M. Cerreta, M. Concillio, & G. Monno (Eds.), *Making strategies in spatial planning* (pp. 185–205). The Netherlands: Springer.

Swyngedouw, E. (2015). Politicizing urban political ecologies. In T. Perreault, G. Bridge, & J. McCarthy (Eds.), *The Routledge handbook of political ecology* (p. 609). London and New York: Routledge.

Thoft-Christensen, P. (2012). Infrastructures and life-cycle cost-benefit analysis. *Structure and Infrastructure Engineering, 8*, 507–516.

Treweek, J. with contributions from: Ten Kate, K., Butcher, B., Venn, O., Garland, L., Wells, M., Moran, D., Thompson, S. (2009). *Scoping study for the design and use of biodiversity offsets in an English Context*. Final Report to Defra (Contract NE 0801). Available at: https://www.cbd.int/financial/offsets/unitedkingdom-scoping.pdf.

UK NEA. (2011). *The UK national ecosystem assessment: Technical report*. Cambridge: UNEP-WCMC.

Wilson, J., & Swyngedouw, E. (Eds.). (2014). *The post-political and its discontents: Spaces of depoliticization, spectres of radical politics*. Edinburgh, UK: Edinburgh University Press.

WRM & Re:Common. (2016). *Rio Tinto in Madagascar: A mining destroying the unique biodiversity of the littoral zone of Fort Dauphin*. WRM and Re:Common.

6

Offsetting, Urbanization and the Neoliberalization of Space in Post-crisis England

E. Apostolopoulou, *Nature Swapped and Nature Lost*,
https://doi.org/10.1007/978-3-030-46788-3_6

When one observes how here in London alone a greater quantity of manure than is produced by the whole kingdom of Saxony is poured away every day into the sea with an expenditure of enormous sums, and what colossal structures are necessary in order to prevent this manure from poisoning the whole of London, then the utopia of abolishing the distinction between town and country is given a remarkably practical basis.

—Friedrich Engels (1872)

6.1 The Neoliberal Restructuring of Planning and the War on Red Tape

> If the idea that space is political is abandoned, a double criticism, itself political (both in theory and in practice), arises: the criticism of the right and the criticism of the left. Right-wing criticism is largely a criticism of the bureaucracy and of governmental intervention, insofar as this intervention constrains private initiative, that is to say capital investment. Leftist criticism is equally aimed at the bureaucracy and governmental intervention, but its criticism is that governmental intervention does not take into account, or else inadequately considers, in the planning process, the people and the social customs, in other words, the people's urban life styles.
>
> Lefebvre (1976)

In the aftermath of the 2008 financial crash, securing land for construction was identified as a key governmental priority. In the Coalition Agreement published on 20 May 2010 (HM Government 2010), the incoming government made clear its aim to support and facilitate housebuilding and large infrastructure projects (or megaprojects[1]), with indicative examples the HS2 and the Thameslink Programme (Apostolopoulou and Adams 2019). The housing market was considered as one of the biggest casualties of the economic crisis: from 2009 to 2010, 115,000 new-builds were completed in England, fewer than during any other year in peace time since the 1920s and nearly a quarter of a million homes have remained empty for more than 6 months. To respond to the so-called housing crisis the government made one of its key priorities to rectify the situation "where lenders couldn't lend, so builders couldn't build and buyers couldn't buy" by, inter alia, removing "unnecessarily complex regulations".[2] The latter has been translated in practice in a range of actions aimed at reducing environmental and planning regulations. These included a commitment to cut red tape, introducing a one-in, one-out rule whereby no new regulation would be brought in

[1]Megaprojects are commonly understood to be projects that cost at least a billion dollars.

[2]See https://www.gov.uk/government/publications/2010-to-2015-government-policy-house-building/2010-to-2015-government-policy-house-building#background.

without other regulation being cut by a greater amount (Apostolopoulou and Adams 2015). In April 2011, the Coalition Government launched the Red Tape Challenge[3] inviting the public to help cutting regulations. In September 2012, the implementation plan for the Red Tape Challenge for the Environment has been published mentioning that environmental legislation, including the Habitats and Birds Directives, would be under review with a focus on its impact on proposed developments and "with a view to reducing burdens on business" (Defra 2012, p. 10). The implementation plan included the removal and merging of various regulations, including environmental permitting and regulations for the protection of the landscape.

The intentions of the Coalition Government to simplify and deregulate the planning system have been also evident in the National Planning Policy Framework (NPPF) that was launched in 2012 (NPPF 2012). The new NPPF included a "presumption in favor of sustainable development" which would run as "a golden thread" through decision-making (NPPF 2012, pp. 3, 4, 13, 28, 37, 46). According to the governmental rhetoric, the new NPPF was primarily introduced "as a way of cutting back on red tape and endless planning documents to focus on what people care about: local roads, schools and homes that meet their needs".[4] This was accompanied by the abolition of Regional Strategies and a reinforcement of the importance of councils' Local Plans.[5] My fieldwork across England confirmed that the NPPF has been translated on the ground as a clear encouragement of economic development and growth as the only pragmatic and efficient ways to deal with economic recession and austerity. As the Minister of State for Housing and Planning argued in 2013: "We enabled a locally controlled, plan-led approach by abolishing the top down regional strategies and by replacing over 1300 pages of central government guidance with the 52-page NPPF. These changes have already achieved significant results. Local Plans adopted since the NPPF was published allocate substantially more housing than

[3] See https://www.gov.uk/government/news/red-tape-challenge.

[4] See https://www.gov.uk/government/news/prime-minister-councils-must-deliver-local-plans-for-new-homes-by-2017.

[5] See https://www.gov.uk/government/news/new-step-for-localism-as-every-regional-plan-has-gone.

those adopted before the NPPF, and 261,000 homes were granted planning permission in the year to March - the highest annual total since before the 2008 economic crash".[6]

In 2013, in an assessment of planned and potential infrastructure investment until 2020 and beyond, the government has announced that the capital value of permitted large infrastructure projects was more than £50 million. The overall value was estimated at £375 billion of investment, updated to £383 billion in the summer of 2014. Most of this planned investment was in the energy and transport sectors. About two thirds of these investments were expected to be financed from the private sector,[7] a fifth from public sources and the rest from a mix of public and private finance (HM Treasury 2013). Even though the government made clear that its role varied between sectors, it was acknowledged that a degree of government involvement was necessary since large infrastructure projects, due to their important land use implications, would involve securing planning permission, a matter that relates directly to governmental policy.

In 2014, the at the time Prime Minister David Cameron announced that since the beginning of the "war on red tape" 800 regulations had been abolished or simplified.[8] He emphasized that the government's priority was to make it "vastly easier and cheaper" for businesses to meet environmental obligations and promised that, by March 2015, Defra would have further slashed 80,000 pages of environmental guidance saving businesses around £100 million per year. The government was determined to help housebuilders by cutting down 100 standards to less than 10, saving them around £60 million per year.[9] In a similar tone, the at

[6]See https://www.parliament.uk/documents/commons-vote-office/July%202015/21%20July/8-Communities-and-Local-Government-Local-Plans.pdf.

[7]It has to be pointed out here that the UK is rather unusual among advanced economies in the extent to which it relies on the private sector to finance and provide infrastructure (https://www.instituteforgovernment.org.uk/sites/default/files/publications/Political%20economy%20of%20infrastructure%20in%20the%20UK%20final%20v1.pdf).

[8]See https://www.gov.uk/government/news/supporting-business-david-cameron-announces-new-plans.

[9]See http://webarchive.nationalarchives.gov.uk/20150423101309/; http://www.redtapechallenge.cabinetoffice.gov.uk/themehome/pm-speech-2/.

the time Business Minister Matthew Hancock, argued that the government's war on red tape had saved business £10 billion from 2010 to 2014 and announced plans for delivering £20 billion of savings by 2020 "by scrapping even more pointless regulations"[10] (sic).

The new Conservative Government elected in 2016, continued and further deepened the widespread deregulation of legislation by moving to a "one-in, three-out" rule for new legislation to help deliver the government's commitment to cut a further £10 billion of red tape.[11] Defra has once again been one of the key targets in order to deal with "delays and unnecessary burdens associated with the permitting system" which "cost legitimate businesses millions of pounds each year". To get an idea of what the Conservative Government had in mind (and still has at the time of writing) it is useful to look at the "Cutting Red Tape Review of House Building" that was published in February 2017 (HM Government 2017) with the aim to boost housing and increase the number of new buildings in line with the governmental priorities set out in the White Paper on Housing. The review identified six areas as the cause of burdens including planning and environment. Regarding planning, one of the key complaints of the sector was the "too many unnecessary planning conditions" that were imposed, and which could take too long to be discharged, including pre-commencement planning conditions. Local planning authorities have been also criticized for often lacking the necessary expertise to assess or validate whether the required conditions have been met and should be discharged. It has been highlighted that this lack of expertise may force developers to consult experts or bodies, such as Natural England or the Environment Agency, causing further delays and additional costs. Finally, it was reported that agreements necessary under Section 106 of the Town and Country Act 1990 have been subject to lengthy delays. Housebuilders blamed the lack of staffing resources in Council Legal Departments as the key cause of delays while also citing the involvement of parish, district and county councils as a contributing factor (HM Government 2017).

[10]See http://webarchive.nationalarchives.gov.uk/20150418202319/; https://www.gov.uk/government/news/hancock-red-tape-drive-saves-business-a-record-10-billion.

[11]See https://www.gov.uk/government/news/government-going-further-to-cut-red-tape-by-10-billion.

Regulations related to protected species have been also heavily criticized for causing significant costs and delays to housing developers. Special reference was made to delays in protected species mitigation which has been blamed for causing "a knock-on effect," creating further delays for housebuilders. The fact that costs related to specific protected species could be burdensome was also pointed out with housebuilders reporting that the cost of dealing with great crested newts in 2013 was at an average £2261.55 per newt relocated and that a sum of £500,000 had been spent only on one site where five newts were found (HM Government 2017). The government's decision to deregulate the environmental and planning legislation was accompanied by an emphasis to the positive role of the market and the need to actively support a parallel process of market-friendly reregulation. A prominent example of governmental intensions has been the introduction of market-based mechanisms for the delivery of environmental compensation in the form of offsetting.

The governmental rhetoric against state regulation and in favor of the market was not of course new to British politics. Since the Thatcher government, the market has been rhetorically framed as the guarantor of freedom, efficient allocation of land and effective choice, and the state's role was to remove barriers that interfere with market forces. It is interesting to see what the Secretary of State for the Environment, Michael Heseltine, was arguing in 1981: "I do not look to the planning profession to recreate an economic base. But I would ask you to think and act with resolve as you create some of the necessary preconditions and remove some of the constraints".[12] In line with this logic, in 1985, in "Lifting the Burden," published by the Department of the Environment, it was argued that "too many people in central and local government spend too much of their time regulating the activities of others" and that the planning system "imposes costs on the economy and constraints on enterprise that are not always justified by any real public benefit" (Department of the Environment 1985). The conclusion was that there is a need to simplify the system and improve its efficiency and to accept a presumption in favor of development. As Blowers (1987) explains such arguments were echoed in "Building Business, Not Barriers," published by

---

[12]Quoted in Cloke (1999, p. 281).

the Department of Employment in 1986, which recommended a reduction in controls, simplified procedures, and a positive approach to economic development. These have been promulgated in two familiar ways: by widespread deregulation and by an explicit support for developers' interests.

6.2 The Convergence of Offsetting and Urbanization: Fetishizing Economic Growth

> We 've been told quite categorically that we need to be more solution-focused, not to put disproportionate barriers to development, block housing, saying yes rather than no, helping the country in this difficult condition to achieve a broader set of goals than just conserving biodiversity. Biodiversity offsetting reflects this exact approach.
>
> Natural England Employee

> We will get a license to trash with offsetting, but we won't get the kind of power to create. It seems that we will still be destroying something, and we won't really have the resources to create something equal or better. We will just do the minimum.
>
> Conservationist, Member of the Steering Committee in one of the offset pilots

As already mentioned, the introduction of biodiversity offsetting in the UK was part of a wider planning reform in the post-2008 era that was aiming at supporting and facilitating economic development. The Conservative-Liberal Democrat Coalition Government made clear from the very beginning that offsetting was not meant to introduce more environmental red tape but to be a pro-development policy. Thus, even though past failure to achieve satisfactory environmental compensation for the residual impacts of development projects was mentioned in the White Paper to support the introduction of offsetting, it was made clear that this was not the primary driver of the policy. Indeed, according to the White Paper (HM Government 2011) the planning system has been too "costly," "bureaucratic" and "with excessive central control" (p. 21) and the adoption of more "innovative" and "flexible" (p. 22) approaches to nature conservation, such as biodiversity offsets and Nature Improvement Areas (see Chapter 5), could simplify the system and streamline development while bringing economic benefits to developers. Offsetting has, thus, been introduced as a policy that could offer a "quicker and more certain" mechanism for dealing with biodiversity in the planning system while increasing land availability for development and net developable area (HM Government 2013, pp. 2–3). As we read in the White Paper, offsetting "could make the process of managing the impacts of development on biodiversity simple for all involved, by providing a straightforward and cost-effective way to assess the impact of a development and to agree the requirements for compensation" (p. 22).

The ultimate goal, as stated in the White Paper, was to protect nature's value through the planning system (HM Government 2011), translated to the policy goal of finding a way to deliver "the homes, business, infrastructure and thriving local places that the country needs" while "protecting and enhancing the natural and historic environment" (ibid., p. 21). Offsetting was thus widely understood as a policy that could render economic development compatible with environmental protection by, inter alia, addressing the increasing environmental impacts of urbanization (IEEP 2016). This explicit link between offsetting and urbanization was not unique to the UK but it has been also obvious in the broader EU guidelines for Environmental Impact Assessments (EIA). According to the latter, projects that include urban expansion, infrastructure, and industrial activity (e.g., energy, extractive industries) are subject to offset requirements whereas for the residual impacts from projects and activities related to agriculture, forestry, or fishery there are no offsetting requirements (Tucker et al. 2014).

Offsetting's introduction in the UK planning system as a policy that was compatible with the new presumption in favor of sustainable development has been supported using three key arguments (HM Government 2013). Firstly, the policy was considered capable to make the delivery of biodiversity requirements in the planning system quicker, cheaper, and more certain for developers. Secondly, it was expected that it would lead to a net gain of biodiversity by ensuring that biodiversity units lost at a development site would be equally matched by the number of units conserved at the offset site. Thirdly, in line with governmental commitments to developers not to increase net burdens on them, offsetting was designed in a way to avoid additional regulations and costs to businesses. Offsetting was expected to bring economic benefits for businesses that would use it to bring forward additional development and enable the relocation of onsite compensation (£43 million) and conservation benefits by introducing a new way to source environmental compensation.

The strong link between offsetting and economic development, primarily urban development as mentioned above, has been consolidated in the Green paper (Defra 2013) where it was argued that according to existing evidence the cumulative costs of complying with biodiversity policy strongly affect the viability of development projects and often delay or even prevent projects from materialization. It was also stated that both compensation and mitigation have often been inadequate, inter alia, because they have been using land that would be more suitable for development. The introduction of offsetting along with a set of other governmental measures, including the reduction of paperwork required with a planning application and the Habitats and Birds Directives Implementation Review, was expected to reduce costs and delays in the planning process related to ensuring compliance with environmental regulations. The Green Paper also included the following important statements. Firstly, the introduction of the Defra offset metric was accompanied by the promise that it would provide a simple and standard framework for evaluating development impacts speeding up assessments and saving developers' time. Secondly, it was argued that offering would offer a standard framework that would give developers more certainty since it would make clear what exactly the planning system would require from them. Thirdly, it was emphasized that offsetting by enabling off-site compensation would also save developers money by facilitating the provision of compensatory habitats on less-expensive land. Off-site compensation was expected to allow a larger development footprint and unblock development that based on the mitigation hierarchy has been considered unfeasible because developers could not provide the required compensation onsite or negotiate appropriate off-site solutions. Finally, it was stated that the introduction of a national standardized system would ensure consistency across the country, simplifying the planning process and benefiting developers that would not have to grapple with varying local approaches. The Green Paper made clear that offsetting would be introduced nationally as a mandatory policy only if it succeeded in fulfilling the above promises, namely, making the delivery of biodiversity requirements quicker, cheaper and more certain for developers while avoiding additional costs and offering them reputational benefits. This line of argumentation was criticized by several interviewees who even though were sympathetic to the idea of introducing a new, consistent tool for quantifying ecological losses due to economic development, were disappointed by the governmental framing of the policy. Others had more substantial disagreements with offsetting's underlying logic and neoliberal origins. As an environmental activist said to me:

> The way the Green Paper introduced offsetting made clear that the intention was to fast-track development. That was the Conservative government's version of sustainability that we all know. However, this only makes things worse; offsetting no matter how it would be introduced it's clearly inappropriate. What we need is not a neoliberal-inspired market-based instrument but a full, in-depth discussion about land use change, this is a big discussion and that is why planning should be a slow and complicated process and also that is why planning is contested. Offsetting can speed up planning policies and we should worry about it and not feel better, there is a reason planning processes take a long time and permanently changing land use by using ecological arguments and reductionist metrics is a technocratic approach to a political problem.

It is important to mention here that the parallel establishment of the Environment Bank significantly influenced the introduction and implementation of offsetting in the UK. In line with governmental rhetoric, the Bank tried to convince developers to be involved in offsetting because it could offer them a more streamlined planning approval process (Environment Bank 2014a). The certainty that offsetting would facilitate planning permission was rather obvious in the Bank's reassurance to developers that any upfront costs would be factored into residual land values that would be substantially uplifted "as a result of a planning permit" (ibid.).

Before closing this section, it is important to briefly refer to Section 106 (S106) agreements, the main instruments that local authorities had at their disposal before offsetting to apply restrictions on developments and address impacts on local communities. S160 agreements were introduced through the Town and County Planning Act (1990) enabling private landowners and local authorities to enter into legally binding agreements. For example, to be granted a planning permission, a developer might be required to dedicate a piece of land to the local authority and pay a charge for its future maintenance through a S106 agreement. Additionally, a developer may pass over funding to the local authority, or any other body acting for it, to maintain green space or recreational facilities lost due to development. As Rotherham (2015) argues S106 agreements were launched at a period of active development opportunities leading many local authorities to see them as a potential source of income that by facilitating development could lever funding. Such systems were, thus, criticized for potentially compromising the planning role of local authorities by the incentive of lucrative cash donations (ibid.). Biodiversity offsetting has a similar logic, in the sense that it often requires from developers to compensate the local authority for the environmental impacts of their development. Offsetting can be combined with S106 agreements but also brings important changes to the previous system as it has been explained in the previous sections. An observation that several interviewees from the local administration, including employees with a long-term experience in planning, made during our discussions was that in many cases where offsetting had been implemented, there was an upgraded role for ecologists, consultants, and brokers that favored a more technocratic approach: consultants and ecologists have been proven more amenable to a site away from the application site, whereas councilors would have probably been more concerned about the location of the offset site and the need to place it close to the development site in order to show that compensation would favor the local community.[13] Importantly, interviewees also mentioned several examples where offsetting's existence as the last element in the

[13]This does not mean that they would have managed to locate the offset site locally or that their interest would have been genuine but that these concerns would have been stronger in the previous system and less easily ignored.

mitigation hierarchy (see paragraph 118 of the NPPF 2012) has led to an underuse—or even complete dismissal—of its earlier stages facilitating planning permissions that would have been otherwise refused.

6.3 The Selection of Offset Pilots: Deepening the Urban/Rural and Nature/Society Divides

As expected, the selection of the offset pilots has been based on their potential to prove governmental promises by offering some concrete success stories that would convince more developers and more landowners to be involved in offsetting. An indicative example is the Essex pilot where the County Council conducted a feasibility study of offsetting in order to achieve its inclusion in the pilots. The Council worked with the Environment Bank at the outset proposing a broker-led offsetting scheme, emphasizing offsetting's benefits for developers in the broader area. As an employee in the Essex County Council explained to me:

> One of the reasons Essex was included in the pilots was that the County Council proposed a broker-led scheme. We tried to show that offsetting would save developers money by simplifying the process and reducing the meetings with the planning authority and the need for legal advice and ecological consultants. We strongly encouraged developers, particularly of big schemes, to use Defra's metric in their environmental/ecological impact assessment, and the Environment Bank's calculator to work out their impacts and the required compensation. We published some of our case studies where money had been wasted, where offsetting would bring benefits, well these were quite sensitive so we could never really put them in the public domain. However, during the pilot there weren't that many developments, obviously due to economic recession, that actually caused a significant enough impact to require compensation. So the right scheme never came up really.

Experience from the pilots gradually convinced also those who were more optimistic about offsetting's potential to support conservation goals that the actual driving force behind its introduction during the economic crisis was to facilitate or, ideally, boost economic development. The win-win rhetoric that has initially surrounded the policy started to fade out spreading disappointment to several conservationists who had seen to offsetting a beneficial compromise. The following quote from my discussion with a conservationist that was a member of the offsetting steering committee in one of the pilots is indicative of this gradual disillusionment:

> At first, we thought that offsetting could be good if developers believed in it, and if conservationists believed in it. In a sense it is a license to trash, but it is also a license to create elsewhere. And while we concrete everything, we can still do a huge amount about wildlife, new sites. The techniques are there. We thought that we could use offsetting to link several small sites together to create bigger sites and restore them for variable land, and then we could have massive biodiversity benefits. But our experience with the pilot is very disappointing. We see that we get a license to trash, but we don't get the kind of power to create. This should be offsetting's key idea: you might destroy something here, but you are also creating something somewhere else. But I do not think that this is what drives this policy in the UK. So, my worry is that we will get some poor system implemented rather than a better system [...] Whatever the results are now, I think that offsetting will become part of the planning system. The government wants it, and many developers want it, because it simplifies things for them. So, I think that offsetting will be happening a lot in the future and we will just get the minimum.

On the other hand, there were other cases, like the Warwickshire pilot, where offsetting's win-win discourse proved more enduring. In my discussions with local offsetting officers and representatives of the local council, which was pro-offsetting, it became obvious that there was an expectation that through offsetting they would get better compensation while, simultaneously, creating a more attractive planning process which would encourage developers to invest in the region:

> I think the government saw a way of doing something that would promote growth and reduce the burden on developers while delivering positive gains for the environment. There is too much regulation around protected species for developers and a lot of money wasted which could be used on the ground to do more beneficial things.

This pro-offsetting stance was met with strong opposition from local communities and activists who saw in it an open invitation to developers to select the region for their business due to its pro-development regulatory environment. As a representative of the local committee against the construction of the Coventry Gateway told me:

> We found quite worrying that although it was only a pilot scheme the county ecologists said that in the future most of our biodiversity enhancement will come through offsetting which we thought it was a little bit presumptuous. It effectively showed to developers what they should do to gain planning permission, opening the way for the exploitation of our area. So, instead of dealing with environmental impacts it seems that offsetting is being used to actively unblock development. But where this could lead in the future? Will we be creating development-friendly zones across the country attracting businesses as it happens with special economic zones?

6.4 Austerity Localism, Housing and the Presumption in Favor of "Sustainable" Development

> This would have been a poor development under any regime, biodiversity offsetting just made things worse. The district didn't have a local plan in place. They should have thought more strategically about their housing allocations, but then they'd probably say they've had budget cuts and local plans are quite expensive to get to public hearing and signed off by a planning inspector. But in the current political climate if you do not have a local plan then it's hard to reject a planning application especially

> one that claims to build more executive houses while achieving no net loss of biodiversity.
>
> Local council employee

A significant aspect of the post-2008 planning reform was the radical devolution of power. This was one of the key goals of the 2010 Coalition Government (HM Government 2010; Maclennan and O'Sullivan 2013) and has been accompanied by the desire to create a tight link between localism and growth. The NPPF significantly reinforced the plan-led English system that is driven by Local Plans claiming that "a combination of greater autonomy and specific incentives will unleash a desire to enable development" (Cowell 2013, p. 36) by enabling people to "specify what kind of development they want to see in their area" (Conservative Party 2010, p. 73). Local Government Secretary Eric Pickles made the governmental intentions clear in one of his announcements in March 2013 when he argued that "this government is committed to localism and greater local decision-making in planning […] We are committed to decentralizing as much power as possible and these important and popular planning reforms will bring a significant shift in power to local people." The deregulation of the planning system through the NPPF was thus accompanied by its further decentralization rendering planning a key arena, as Hannis and Sullivan point out (2012, p. 22), "for the spectacles of localism and marketization." Importantly, these reforms took place within the broader context of austerity politics that, as Gray and Barford (2018) explain, has reshaped the relationship between central and local government, shrinking the capacity of the latter, by, inter alia, "dumping" the fiscal crisis onto the local state (see also Peck 2012), increasing inequality between local governments and exacerbating territorial injustice.

In their informative periodization of the planning system, Allmendinger and Haughton (2013) explain that the dominant paradigm in the post-2010 period combined pro-market localism, with deregulation of controls and targets, loosening of policy cascade and hierarchy. They furthermore show that the transition in the English planning system from spatial planning to localism, constituted a form of, and contributed to, neoliberal spatial governance: the flip side of allowing communities to play a greater role in identifying and addressing local needs has been a significantly reduced role for the state "in favor of a plurality of localist interventions" which, and this is crucial, despite the rhetoric, "is tightly circumscribed" (ibid., p. 7). This has been evident in the post-2008 era offering one of the most clear manifestations of the increasing contradiction between democracy and exclusion under capitalism in crisis: the governmental rhetoric on community participation and the autonomy of the local level has been translated on the ground to a form of *austerity localism*. This meant that only those who had the resources, expertise and social capital to become involved in the provision of various services and facilities were empowered to shape local policies (Apostolopoulou et al. 2014; Featherstone et al. 2012). Several communities and activists have strongly criticized the tightly circumscribed plurality of localist interventions citing their disempowering experience from participating in public hearings and consultation processes. As local activists in Newcastle explained to me most decisions were largely predetermined in the hearings they participated and their role in the process was essentially to legitimate these decisions and offer evidence of a consensual process:

> The way I describe these public debates it's like watching a six-hour play where you have audience participation. And then when it is your time to speak you understand that the game is rigged because already a lot of your best arguments are ruled out, they are outside the terms of reference that a planning inspector is working, so you can mention them until you are blue in the face and people know they are very relevant, and your community will support you, but it won't cut any ice because the rules of the game are excluding critical things. And the inspector is just smiling away to what you are saying and listening and nodding, but he is not writing anything down.

Indeed, as an ecologist working for a local council said to me referring to a public hearing:

> The case went to enquiry, but we kind of settled out of court if you like and we came to an agreement. But they still had to go through the formality of the enquiry, to give members of the public and Parish Council

the chance to kind of express their concerns. But we more or less knew the outcome, we came to an agreement in the end.

Before continuing it is important to make here a reference to the role of Local Plans. According to the NPPF (2012) the presumption in favor of sustainable development means that local planning authorities should positively seek opportunities to meet the objectively assessed development needs of their area with enough flexibility to adapt to rapid change. However, as the NPPF further explains, where there is no Local Plan or it is out-of-date, planning permission should be granted to ensure that developments would be promoted "without delay" and, thus, that the presumption in favor of development would be applied in practice. Following the same logic, neighborhood plans are also required to be in line with Local Plans and to actively explore development opportunities (NPPF 2012). In practice, this meant that the new presumption in favor of sustainable development has been translated to an uncritical support for any development proposal that had support either from the national or the local administration.

Housing offers an emblematic example of what the above reforms meant on the ground. According to the NPPF local authorities have to objectively assess housing needs by following online guidance, namely, the National Planning Practice Guidance (NPPG[14]) and deliver a five-year supply of housing land to meet the set targets (NPPF 2012). The NPPF (2012, p. 13) clarifies that in case where a local planning authority cannot demonstrate a five-year supply of deliverable housing sites then the relevant policies should not be considered up-to-date and housing applications should be judged "in the context of the presumption." The transformation of housing to a big business (Harvey 2014) had led to a situation where whether housing needs are being objectively and rationally assessed or driven by the willingness to support the profits of the housing industry, including big estate agents, downplaying other much more crucial investments for local communities, such as public health and education, is highly questionable. But even if governmental housing targets are often arbitrary and inflated, they bind local authorities

[14] See http://planningguidance.communities.gov.uk.

to meet them (CPRE 2015). When these increased and unsustainable housing targets are not met, the result is a disproportionate power to developers in identifying sites and a loosening of local planning control. Considering the need for more housing as an exceptional circumstance to allow exceptions to existing regulations has been a common practice in England. This has often facilitated or even enabled the location of housing sites in Greenfield areas or within the Green Belt leading to countryside being lost to urban development.[15] According to a recent satellite survey[16] only between 2006 and 2012, 22,000 hectares of green space, an area of land twice the size of Liverpool, has been converted to artificial surfaces, mostly housing. This included the loss of more than 7000 hectares of forest, 14,000 hectares of farmland, and 1000 hectares of wetland (Mathiesen 2015).

Biodiversity offsetting was widely perceived as a policy instrument that would fit neatly into the above context. Several interviewees explained to me with specific examples how offsetting has been linked to the new local plans and the land release that the government had been enabling in the post-2008 era, including more recent plans for a further release of greenbelt sites.[17] Crucially, offsetting has been particularly attractive not only to specific sections of capital, including the housing industry, but also to a much broader audience in the context of prolonged austerity. Offsetting was shifting the focus from avoidance and minimization to the provision of the adequate compensation offering an effective way to foreclose the public debate on the scope of controversial development projects and unblock development. The fact that the policy was favoring the role of industries, experts and consultants and was downplaying communities (Apostolopoulou and Adams 2017a, 2019; Apostolopoulou 2019) contributed to the framing of economic growth as an unquestionable national priority to fight recession and austerity. As Montgomerie[18] puts it: "austerity changed the landscape of the UK, but

[15]See http://www.cpre.org.uk/magazine/opinion/item/3845.

[16]See https://www2.le.ac.uk/offices/press/press-releases/2015/june/state-of-our-countryside-land-use-map-of-the-united-kingdom-reveals-large-scale-changes-in-environment.

[17]See https://www.theguardian.com/commentisfree/2017/sep/22/green-belt-housing-crisis-planning-policy.

[18]Available at: https://blogs.lse.ac.uk/politicsandpolicy/economic-storytelling-about-debt/.

also Europe, since the 2008 economic crisis. Austerity is the new normal and we are getting used to it. No one believed Phillip Hammon's 2018 claim that austerity was over, and everyone expects an emergency budget the moment the UK leaves the European Union. Austerity has become more than a public policy platform: it is a mood, a depression, a chronic case of financial melancholia. Austerity is, above all, an experiment. This means that its policies are in motion and can be contested, but also that its results and consequences are not yet fully know."

6.5 Offsetting, Urbanization, and the Right to Nature as a Social Right: Community Struggles Against Large-Scale Infrastructure and Speculative Housing Development in Austerity England

> Urbanisation and capital accumulation go hand in hand and that is one of the aspects of Marxist thought that has been historically underdeveloped. [...] Questions of daily life in environments constructed for purposes of capital accumulation is a big issue and a source of contradiction and conflict. This is emphasised politically by the pursuit of the right to the city: e.g., class struggle in and over the qualities of urban life. Many of the major social movements in recent decades have been over such questions.
> Harvey (2019)[19]

The way biodiversity offsetting intertwines with urban development in post-crisis England showing a broader strategic choice to align conservation and urbanization patterns becomes more evident if we look at specific cases. In what follows, I will focus on three cases. The first two refer to housing developments in South East and North East England where offsetting has been proposed after the initial refusal of planning permission and the third case is the HS2 trainline. All cases show well some of the key dimensions of offsetting that this book tries to tackle:

[19] See https://thewire.in/economy/david-harvey-marxist-scholar-neo-liberalism.

from the controversial and reductionist calculations of ecological losses and gains in terms of priced credits and the ignorance for the ways offsetting relates to social inequality exposing the policy's class implications to offsetting's key role in granting planning permission and reordering the landscape according to urban patterns.

6.5.1 Contesting Executive Housing and the Loss of Open Space in North Tyneside[20]

> This is not biodiversity offsetting, it is biodiversity upsetting. Seriously now this is the last open space that this community has, they are consciously destroying it for profit and their argument is that they will protect another place to counterbalance its destruction. For me, this is one of the most absurd things I 've heard for years. You cannot offset Whitehouse Farm –what does this even mean? - because it won't be Whitehouse Farm –you 'll either loose it or save it.
>
> Local activist

> This is arable farmland, I think the offsetting they are providing is more than enough for the loss of this area. I know some people may say that the community is losing a lot and I sympathize, but this is compensation for biodiversity loss, so I do not think that the community's opinion is quite relevant to this particular instance.
>
> Local conservationist

Whitehouse Farm is approximately 7 km from the city center of Newcastle and 1.5 km from the center of Killingworth and consists mainly of arable farmland. Whitehouse Farm is in the constituency of North Tyneside that is bounded by Newcastle upon Tyne to the west, the North Sea to the east, the River Tyne to the south and Northumberland to the north.[21] The North East region of England has been historically

[20]For a detailed analysis of the Whitehouse Farm case see Apostolopoulou, E. (2019). Beyond post-politics: Offsetting, Depoliticization and contestation in a community struggle against Executive Housing. Transactions of the Institute of British Geographers, https://doi.org/10.1111/tran.12354.

[21]See https://en.wikipedia.org/wiki/North_Tyneside.

associated with shipbuilding, coal mining and heavy engineering (Nayak 2003). North Tyneside followed a similar trajectory and has traditionally been a center of heavy industry and coal export. As in many other areas in Britain, the deindustrialization of the 1980s led to high unemployment in the wider area rendering North Tyneside one of the most deprived areas in England (NTC 2015a). Deindustrialization was accompanied by extensive deregulations and expansion of property markets marking a gradual transition to a service economy (Madanipour 2009). This has been a key aspect of rural gentrification (Stockdale 2010), a process which as Darling (2005, p. 1016) points out is characterized by "the 'colonization' of the British countryside by an exurban or suburban, middle class of homeowners who seek to create a lifestyle organized around the consumption of 'nature' and/or rurality, and the subsequent displacement of working-class rural residents as a result of rising local land and housing prices." A key aspect of rural gentrification is the emergence of a "post-productive" countryside (Marsden and Sonnino 2008), in which service-oriented accumulation regimes, including residential real-estate development, gradually replace industrial and agricultural production (Darling 2005), reflecting a broader restructuring of space driven by the needs of capital (Smith 1979). These trends have been apparent in North Tyneside: the area has suffered a major loss of open green space due to the transformation of both brownfield and greenfield areas to business parks and housing developments. This is also true for Whitehouse Farm and the wider area that has very limited natural and semi-natural green spaces and a very low environmental performance (NTC 2015b).

In 2011, Bellway Homes (North East) Limited submitted an application to develop 366 executive houses at Whitehouse Farm, threatening one of the last open green spaces in the area. The development site was being used daily by residents and was highly valued for its relatively rural character. The proposal was met from the beginning with strong local opposition with residents arguing that executive housing does not meet the needs of the social majority and will only intensify urban sprawl, further destroying the area's geography and threatening the social cohesiveness and well-being of the existing community (Price 2012). The Planning Committee of North Tyneside Council received more than 1000 objections to the proposal[22] and in 2012, among a public protest outside the council chamber, voted unanimously to refuse planning permission.[23] The developers, however, decided to appeal to the Government to overturn the decision and a public inquiry was held in October 2012. This was the first time that Bellway proposed to offer the necessary compensation through offsetting.

In the autumn of 2014, I spent almost a month in Newcastle discussing with several people that have been involved in the public inquiry and with residents and local activists to understand the role that offsetting played in an otherwise classical social-ecological conflict. As they explained to me the developers initially attempted to downsize the impacts of the housing development and what forced them to change their tactics was the emergence of new ecological data toward the appeal showing that Whitehouse Farm was supporting an important species richness for an urban fringe site (Department for Communities and Local Government 2013). The housing development would have direct and permanent impacts on the breeding territories and foraging habitats of farmland bird species due to the complete loss of arable land, affecting up to six species of principal conservation importance (ibid.). The new data made clear that off-site mitigation was required, and this was the time when offsetting was proposed. When this happened there was still no reporting back from the pilots, so Whitehouse Farm was widely perceived as an unannounced pilot.

Many people who have been involved in the process from the beginning but openly opposed the development told me that they were never invited to the first discussions about offsetting and they were only informed that offset sites have been identified when the relevant report was "leaked" to them. However, the developers had commissioned since July 2012 the Environment Bank to calculate ecological losses by using the Defra metrics (Environment Bank 2012a). The Bank had initially determined that 122.5 conservation credits and 836m of hedgerow had to be purchased to offset the development impacts. In a revised report

[22]See http://www.chroniclelive.co.uk/news/north-east-news/go-ahead-given-controversial-plan-build-5855195.

[23]See http://www.nhsn.ncl.ac.uk/news/great-news-for-campaigners/.

published in September 2012, these numbers had changed significantly, and the hedgerow offset requirement was now 105m. The report offered a very brief and quite technical explanation of the rationale behind this new calculation. This came as no surprise since in all the reports produced by the developers' consultants and the conservation brokerage company there was no specific analysis of the underlying logic of offset calculations. Most reports included just a brief description of the application of the Defra metric that was full of jargon and impossible to follow even for someone with ecological knowledge but no background information on the specific work that the consultants and the brokers had undertaken in the area (see e.g., Environment Bank 2012b, p. 8). The same problems also characterized the calculation of environmental gains in the proposed offset site (in Mares Close, Seghill). By using a set of multipliers to include the risk of failure and the uncertainty involved, professional judgment (Environment Bank 2014b) and subjective projections, some of them based on experimental restoration methods and axiomatic assumptions about the future condition of the area, the report concluded that a net gain of biodiversity would be achieved. These calculations were not only reproducing the weaknesses of the Defra metric, which was still under trial, and the absence of a method for scoring condition but also reflected the longstanding problem of downplaying local biodiversity values that occurs when subjective links between value and magnitude are used to assign different significance levels (CIEEM 2016). Despite the impacts of the loss of Whitehouse Farm for the local community and the escalating public debate, the social aspects have never been addressed or acknowledged as relevant. This became obvious in the selection of the offset site that has been primarily guided by land availability and land prices leading to the selection of an area that was 3 miles away from Whitehouse Farm and where public access would be restricted to do not disturb ground-nesting birds (Environment Bank 2014b).

As it became clear from my discussions with people involved in the process, offsetting played an important role in facilitating the approval of the planning application while significantly changing the rules of the debate. The introduction of a new way of calculating environmental compensation limited the ability of those who were not familiar with offset metrics to challenge the assumptions made, including residents who were not only lacking the necessary expertise but were also often excluded from the discussions that were taking place. Shifting the discussion to offsetting was perceived by the local community as an attempt to disorientate the public debate: from discussing the actual housing needs and the impacts of the new development to the community, the focus shifted on the number of biodiversity units lost and gained and on whether no net loss would be achieved. Despite the obvious lack of evidence on how offsetting would be implemented, the vague management proposals, and the largely arbitrary projections for the future condition of the offset site, the inspector argued that there was a "realistic probability" that offsetting would succeed, essentially adopting the argumentation of the brokerage company (see Department for Communities and Local Government 2013). The Whitehouse Farm housing development was, thus, officially identified, as a necessary investment that would address the housing needs of the area. It is important to point out here there were not only conflicting accounts regarding the actual housing shortage (Department for Communities and Local Government 2013), but also housing needs were based on estimations made before the economic recession. As the recent plan of NTC acknowledged, the nadir of the market, particularly during 2008–2011, coincided with the lowest level of housing delivery (NTC Plan 2017): in March 2012, there were 2.986 outstanding units with Planning Permission in North Tyneside and in March 2016 this number had increased to 4765 (ibid.). The use of untrustworthy projections of housing needs[24] and the fetishization of housing numbers that has been encouraged by both the NPPF (2012) and the Housing White paper,[25] along with the absence of an updated local plan, meant in practice that priority was given to the presumption in favor of development.

Offsetting by portraying environmental compensation as an issue of mathematical calculations and credits exchange imposed a scientifically disguised neoliberal conservation agenda to legitimize and neutralize

[24]See http://www.independent.co.uk/environment/should-britain-build-on-its-green-spaces-to-solve-thehousing-crisis-a7568341.html.

[25]See https://www.gov.uk/government/collections/housing-white-paper.

the disassociation of development from its social aspects. This constitutes a paradigmatic case of a techno-managerial solution to a political problem: by favoring technical solutions to socio-ecological conflicts it implies that nature possesses an objective character and that discussions about environmental impacts can be politically neutral. By drawing parallels with the work of critical scholars on the politics of emissions trading mechanisms and climate change (e.g., Felli 2015; Lohmann 2005; Swyngedouw 2010), we see that these characteristics of offsetting can have important depoliticizing effects since they try to encase environmental politics in a narrow defined debate that primarily evolves around ways to measure ecological loss marginalizing and disempowering the less powerful while supporting particular class interests. However, and this is important, depoliticization is not a straightforward process and its success should not be taken for granted. As local opposition in Whitehouse Farm, and many other places across England, showed neoliberal hegemony is not either catholic or a given (for a detailed discussion on the limits of depoliticization and post-politics see Apostolopoulou 2019). Policy-making in the aftermath of the 2008 financial crash has been characterized by austerity politics, the entrenchment of neoliberal, anti-social policies and fast-policy regimes (Peck et al. 2012), showing that consensus is losing ground at least for those whose lives have been significantly worsened. The precarious nature of hegemony and the shifting dialectics between consensus and coercion (Apostolopoulou et al. 2014; see also Sect. 2.4.) have been also evident in both the exclusion of local protesters from decision-making, and in the way that economic coercion has been deployed to discipline policy-making into market logic by building on the decreasing public expenditure for the region, and the potential financial gains from offsetting. Importantly, during the same period with the Whitehouse Farm conflict, local groups in Ponteland, Hexham and Newcastle were also fighting to prevent thousands of new houses being built on green areas. Attempts to manufacture and impose consent had, therefore, a broader audience: they targeted mobilized groups across the country and the wider public opinion with the goal to neutralize opposition to the discernible social impacts of the proliferation of urban development in greenfield sites, the green belt and the open countryside (Apostolopoulou 2009).

6.5.2 The "Feathered" Obstacles to Urban Redevelopment in Lodge Hill[26]

> As far as habitat goes, yes, the Lodge Hill site could be offset because it's not an extremely special habitat, it's an old base of the ministry of defense, it's a successional scrub habitat, it's not a wonderful place as some residents claim.
>
> Conservation broker

> You can't replace the beauty of this place in my humble opinion. You can go to the first roundabout and hear about five nightingales competing against each other. Lodge Hill should be celebrated and put on the map as one of the top treats in Medway, as something great for people.
>
> Local activist

Lodge Hill sits next to the village of Chattenden on the Hoo Peninsula in Medway, one of the largest urban areas in South East England, located about 30 miles east of London. The site is 2.83 km^2 and has been used for various military activities since 1873. It had been gradually decommissioned since the late 1980s. The wider area shares important similarities with North Tyneside: unemployment in some areas of North Kent, including Medway, reached a 25% in the 1980s following a series of closures of the docklands in East London and concurrent deindustrialization. A key part of the recent history of Lodge Hill relates to the Thames Gateway project. The regeneration project along with HS1, which was agreed in 1993 that would pass through North Kent and Medway shortening the distance to London, gradually made Lodge Hill an ideal sleeping/commuting town. The link between new transport geographies and new urbanization patterns became stronger since the 2006 decision to hold the Olympic Games of 2012 in the East London part of the Thames Gateway leading to a significant increase in housing projections in the wider area (see also Allmendinger and Haughton

[26]For a detailed exploration of the Lodge Hill case see Bormpoudakis et al. (2019). the rise and fall of biodiversity offsetting in the Lodge Hill large-scale housing development, South East England. Environment and Planning E: Nature and Space, https://doi.org/10.1177/2514848619884890.

2009). Even though Lodge Hill has been included in all regional planning strategies and plans since the 1995 Thames Gateway Development Plan (Department of the Environment 1995), it was not until September 2007, with house prices in South East England at their peak and house building at a 20-year high, that the development started getting off the ground. The Ministry of Defence contracted Land Securities Group plc., a member of the FTSE 100 index[27] and the biggest property developer in Britain. The development proposal for Lodge Hill included 5000 houses, a retail center, and related amenities, and a promise for thousands new jobs and a new "sustainable" community. Despite the green rhetoric and the promise of bringing prosperity to the wider area, the application was met from the beginning with strong opposition both from local people and several conservation organizations.

From 2008 until 2011, a series of presentations, hearing sessions, and technical meetings were held in which the fact that Lodge Hill was one of the top five sites in Kent for nightingales was discussed and the need to introduce a buffer zone to protect the species was widely acknowledged (Trimedia 2009). It is important to emphasize here that nightingales have declined by 90% in the last 50 years and they appear on the UK's Red List of birds of high conservation concern.[28] In 2012, data from the British Trust of Ornithology (BTO) national nightingale survey indicated that the Lodge Hill population was possibly exceeding the 1% of the national population with around 60% of site population within or right on the edge of the proposed development site. According to these data, impacts on nightingale habitat would be much more significant than it was initially thought, and the required compensation habitat would have to be provided off-site. This is when biodiversity offsetting entered the discussion. Medway Council commissioned the Environment Bank, and in December 2012, a report on offsetting was published (Environment Bank 2012c). In public discourse, offsetting was enthusiastically promoted by its keen supporters, for its potential to enable a development that had the "right" vision for Medway's countryside. Even

---

[27]The Financial Times Stock Exchange (FTSE) 100 is a share index of the 100 companies listed on the London Stock Exchange with the highest market capitalization.

[28]See https://www.rspb.org.uk/get-involved/campaigning/protecting-wildlife-sites-near-you/save-lodge-hill/.

though the Lodge Hill development initially seemed the ideal case for testing offsetting's promises this gradually changed. Several interviewees who participated in consultation meetings explained to me that offsetting was not only used to dismiss the mitigation hierarchy and completely disregard the stages of avoidance and minimization but also, as in the Whitehouse Farm case, to shift the public debate on technical discussions around the appropriate compensation for the loss of nightingale habitat.

In 2013, new data emerged indicating that nightingales on the development site represented 1.3% of the national population and Natural England's Executive Board confirmed that the existing Chattenden Woods SSSI would be extended to include most of the land within the Lodge Hill site. Importantly, the SSSI notification did not exclude offsetting from consideration. Most Medway residents opposed Lodge Hill development with the project becoming a major political issue in the 2013 by-elections. Land Securities, on their side, in February 2014, submitted an updated planning application to Medway Council to address the environmental impacts considering the new SSSI designation. In the new application, the Defra metrics had been abandoned and a new terminology was adopted: instead of offsetting now the term "compensation" was used. However, the core logic of offsetting had not changed: the new report argued that not only no net loss would be achieved but also a net gain and proposed a nightingale compensation site within land owned by the Ministry of Defense in Shoeburyness/Foulness in Essex. As a local conservationist told me: "In theory you could offset the habitat but actually you can't tell the birds to go to the new habitat across the river in Essex which is where they were looking at doing it." Again, as in the Whitehouse Farm case, the offset site was in a different area (across the Thames!) and had no public access.

The fact that the offsetting proposal had been rebranded as "nightingale compensation land" created a lot of confusion to many. Natural England employees argued that Lodge Hill was not an offsetting case anymore since the Defra metrics had been abandoned. Opponents of the development, and particularly conservationists, interpreted this as a communication strategy from governmental officials to distance themselves from a case that was becoming highly controversial. Despite the strong public opposition to the new proposals, Medway Council unanimously approved the updated application in September 2014 subject to its referral to the Secretary of State for Communities and Local Government. The housing needs of the area and the priority given to the presumption in favor of development in the absence of a local plan were again, as in the case of the Whitehouse Farm, among the key reasons for the approval along with the argument that offsetting would sufficiently address the environmental impacts of the development. However, the decision was called-in in February 2015. This meant that ministers, following a public inquiry that was supposed to take place in 2018, would have to judge whether the national significance of the development could override wildlife concerns. In September 2015, Land Securities, citing the increased costs (£11.3 million) incurred by the environmental requirements of the housing development, pulled out of its contract with the Ministry of Defense and the public inquiry that was scheduled for March 2018 had been canceled. A updated Medway Local Plan Development Strategy[29] was published in March 2018 and development within the Lodge Hill SSSI was one of four scenarios that Medway Council set out for consultation, with a rural town centered on Hoo St Werburgh included in all the scenarios. The future of the area remains quite uncertain until now.

The Lodge Hill case, as we have explained elsewhere (see Bormpoudakis et al. 2019) can make an important contribution in theorizing the link between neoliberal natures (Heynen et al. 2007) and the changing economic geographies of capitalism. Transformations in economic and transport geographies have been fueled by the infrastructural and entrepreneurial visions that shaped the Thames Gateway regeneration project and broader economic development in South East England (Sparke 2000). As Marx has explained (see Marx (1973 [1857]) on the annihilation of space with time; see also Harvey 1989) perpetual reductions in the cost and time of movement and the drive for eliminating distance and spatial barriers through advances in technology and transport infrastructure could provide the engine for the circulation of

[29]Available at: https://www.medway.gov.uk/downloads/file/2203/development_strategy_with_cover_2018.

capital and materials, in this case between the emerging mega-region of London and the expanding European market. Changes in economic and transport geographies have, in turn, bound with transformations in urban geographies and played a key role in the production of large-scale housing visions in Medway and the Hoo Peninsula (Sparke 2000). The territorialization of the Thames Gateway was thus strongly influenced and shaped by the changing geographies of the South East (Allmendinger and Haughton 2009). Understanding the evolution of offsetting in Lodge Hill as an outcome of the articulation of transport and urban geographies with neoliberal natures also offers important insights into the kind of alliances that oppose offsetting. As we explain elsewhere (see Bormpoudakis et al. 2019) viewing the Lodge Hill conflict through urbanization lenses, can allow us to understand that rival developers, such as Barratt Strategic, McCulloch Homes, Taylor Wimpey or The Church Commissioners for England, while instrumental in contesting offsetting and the housing development in Lodge Hill, are competitors for housing planning licenses, namely, competitors for land. This shows well both that intra-capitalist competition can influence the dynamics of the neoliberalization of nature and that offsetting can have both positive and negative roles to play in relation to capital accumulation (see also Sects. 4.3 and 5.5 and Apostolopoulou et al. 2018). It is this constellation of factors that explains why Lodge Hill not only has not been the success story that the government was aiming for, but a case of a successfully contested neoliberalism (Leitner et al. 2007), contributing at that time to the decreasing popularity of offsetting in the UK (Bormpoudakis et al. 2019). Finally, the wide and conflicting interests that opposed offsetting in Lodge Hill call us to critically consider the origins, scope, limits, and potentials of particular alliances and better understand how solidarities can be constructed (see Featherstone 2013) and how oppositional politics can be build that will not only transcend place (Katz 2001) to form more universal politics (Featherstone 2005; Harvey 2012) but will also put at their core instead of reactionary or reformist demands the need for radical alternatives to the capitalist production of space and nature.

6.5.3 Green-Washing Urban Development and Large-Scale Infrastructure? The Case of HS2 Railway

> The difficulty for a high-speed railway is that it has to be almost straight so your ability to do anything else is very limited; there will always be a compromise between villages, existing paths, woodlands, and highways. Offsetting does not make HS2 easier because HS2 is a different project, is a Hybrid Bill, something like offsetting is not needed to gain permission – but for other projects there is a risk, there is real danger that developers will see offsetting as an opportunity to say ok, we can do that, we can destroy and offset afterwards. And we can make the numbers to work for us and show a no net loss or even a net gain.
>
> Consultant working in the construction and infrastructure sector

One of the most emblematic examples of the convergence between offsetting, urban development and the post-crisis infrastructure rush in England is the case of the High Speed Two (HS2) trainline. HS2 is a railway network proposed by the UK Government to provide a new link between London, the West Midlands, the East Midlands, South Yorkshire, Leeds, and Manchester.[30] Construction of the proposed scheme started in 2017 and is expected to finish in 2026. HS2 Limited is the company responsible for developing and promoting the new rail network and is funded by grant-in-aid from the government.[31] As we read in the relevant website, the vision of HS2 Limited is "for HS2 to be a catalyst for growth across Britain" and "the backbone of Britain's rail network." Importantly, HS2 is a hybrid Bill. Hybrid Bills are rare, and they are used to secure powers to construct and operate major infrastructure projects of national importance. Hybrid Bills have been also used for the Channel Tunnel Rail Link Act 1999 (HS1) and the Crossrail Act 2008.

[30] High Speed Two (HS2) high-speed north-south railway is being taken forward in phases: Phase One will connect London with Birmingham and the West Midlands. Phase 2a will extend the route to Crewe. Phase 2b will extend the route to Manchester, Leeds and beyond (https://assets.publishing.service.gov.uk/government/uploads/system/uploads/attachment_data/file/627919/B1_Understanding_the_Bill_v1.0.pdf).

[31] See https://www.gov.uk/government/organisations/high-speed-two-limited/about.

HS2 is an Act of Parliament (law) since February 2017, namely, the High-Speed Rail (London-West Midlands) Act that gives major powers to the government and the constructors of the new trainline. For the construction of phase 1 of HS2 these powers[32] include: (i) the power (for the Secretary of State) to acquire land or rights over land compulsorily on a permanent basis within the Act limits necessary for the works to be carried out, (ii) the extinction and exclusion of rights over private land as well as the exclusion of new rights of way over certain pieces of land within the Act limits, (iii) the temporary possession and use of land (for example in the case of construction sites), (iv) planning permission for the carrying out of development authorized by the Bill, (v) deregulation, namely, provisions disapplying or modifying existing regulatory regimes, such as those for listed buildings, ancient monuments, burial grounds and commons and open spaces. Regarding the latter we read in the Act that "no enactment regulating the use of commons, town or village greens, open spaces or allotments, and no enactment specially regulating any land of any of those kinds, prevents or restricts the necessary works for HS2 phase 1," and, finally, (vi) the right to carry out works to reinstate businesses or facilities within the Act limits and the enforcement of environmental covenants. This basically means that HS2 has the power to grab and exploit land in any way that it is necessary for the project. Non-surprisingly the new trainline as well as the Hybrid Bill have been met with massive public opposition across the country. Criticism targeted all aspects of the scheme, including its highly contested national importance. As a representative of 51m[33] and a local activist participating in STOP HS2, respectively, told me:

> The line, HS2, is a linear feature that cuts through the whole county, it's a huge barrier running from North to the South and it's not just the line but also all the related infrastructure. Almost all residents don't want to see it, they don't want to hear it. Is it necessary? It comes down to politics. It's a crazy route for us, it's rhetoric, it's all about speed and then it became all about capacity. But even if we bypass all these, then we will

---

[32] See http://www.legislation.gov.uk/ukpga/2017/7/pdfs/ukpga_20170007_en.pdf.

[33] 51m is a group of 17 local authorities that has joined together in a national campaign to challenge the HS2 rail project. For more information see: https://www.51m.co.uk/about-51m/.

> see that they are not doing this properly. They want to deliver this as quickly and as cheaply as possible.

And:

> I think the main problem with HS2 is that it's a solution looking for a problem. If it is as they say now about capacity, it's a very inefficient way of improving capacity. I mean it's a crazy way of doing it, you're not attacking the problem head-on.

Resistance against HS2 included the formation of over 100 action groups, some more active than others, that have been involved in the STOP HS2 campaign. These included several parish councils and residents' associations, namely, bodies that were not set up specifically to challenge HS2. As an interviewee who has been involved in the formation of STOP HS2 explained to me:

> STOP HS2 is very much a grassroots community campaign. As I say, there are action groups up and down the line. Everywhere that is affected by the line, at least on Phase 1, has got an action group. There's a few gaps on the eastern side of Phase 2 and a few gaps in some urban areas but it's pretty much everywhere that's got some form of community involvement.

Interviewees repeatedly told me during our discussions that local communities were convinced that the new trainline would not address any social needs but, on the contrary, would pose severe threats to their quality of life while bringing real loss for many since, as we saw above, the Hybrid Bill enabled a massive land grabbing by the Government, leading to farmers losing their land, people losing their houses and local communities losing woodlands and parks. As a local activist pointed out:

> There are sites within the limits of the Bill that they will take and it will be a compulsory purchase – the government can take land if they want to develop it for economic reasons of national interest; so essentially the HS2 Bill gives land grab powers to the government.

The severe environmental impacts of the scheme featured a key position among the several reasons that fueled public opposition to HS2. Interestingly, even though the project was expected to directly affect, inter alia, 36 ancient woodlands,[34] HS2 Ltd, as stated in the relevant Environmental Statement (HS2 2013, p. 25), in line with governmental policy to halt overall loss of biodiversity by 2020, claimed that "is seeking to ensure that the Proposed Scheme[35] results in no net loss in biodiversity," "at the route-wide level" (HS2 2013, p. 17). In order to achieve that goal, HS2 Ltd utilized a modified version of the Defra offset pilot methodology to compare the habitats present pre- and post-construction and inform the level of compensation provision required. Even though offsetting has been used to calculate biodiversity losses and gains due to the trainline, there was no formal requirement and, therefore, offsetting has not formed an official part of the project's Environmental Impact Assessment. HS2 Ltd decided to use offsetting voluntarily primarily because at the time that the environmental statement was published offsetting was very popular in some government cycles. As state officials told me:

> HS2 started when offsetting was a new exciting tool and that's why they used this idea to claim that they will have a net gain, it was much more fashionable at the time. But since then it has been criticized and politicized and it has been proven that it is not as transparent as initially thought.

And:

> Offsetting was the 'baby' of the previous Secretary of State and it was explicitly framed as the most promising way to combine large-scale development with environmental protection so it's not surprising that they tried to test it on a scheme like HS2.

[34] See https://www.woodlandtrust.org.uk/get-involved/campaign-with-us/our-campaigns/hs2-rail-link/.

[35] Phase One London and West Midlands route of the proposed High Speed 2 (HS2) Railway is referred to as the Proposed Scheme.

What is important to point out here is the opportunist way in which offsetting has been used. Even though in the environmental statement the Defra offset pilot metric was characterized as the "best available basis for an offsetting methodology," it was argued that key amendments were necessary due to the outcomes from the pilots and the landscape scale character of the project. As a consultant working for HS2 Ltd explained to me:

> The Defra metric does not explicitly take into consideration species. I do not know also why they did not include connectivity as one of the measures but having tried to do this it's very difficult, most species have very different requirements regarding connectivity. We weren't aware of any other method of measuring ecological losses that's why we used this method, there is nothing else out there that would give us figures. But we did not have to do it, we felt we should do it.

HS2 Ltd proceeded to a series of amendments of the Defra metric. Firstly, an additional "very high" score has been added under habitat distinctiveness to consider habitats of principal importance identified in Section 41 of the Natural Environment and Rural Communities (NERC) Act (2006) which cannot be adequately re-created if lost. Secondly, the distinctiveness score attributed to all habitats that form part of an area that qualifies as the habitat of principal importance type open mosaic habitat on previously developed land was increased to ensure that the value of these habitats is fully recognized within the calculation. In particular, the report applied a number "8" distinctiveness for ancient semi-natural woodland, mature lowland heathland, and lowland fen. Thirdly, the application of a variable condition weighting for habitats of low distinctiveness has been removed. This meant that all low distinctiveness habitats would automatically receive a condition weighting of poor recognizing that condition has negligible effect on the overall value of those habitats that are intrinsically of low distinctiveness. Fourthly, greater consideration of the importance of both habitats lost and gained (in relation to the function of ecological networks) was incorporated into the spatial risk multipliers, in order to recognize the landscape scale of the project and its impacts. Fifthly, the blanket one-step restriction on the change in condition was removed and replaced with the condition that for high distinctiveness target habitats a maximum future target condition of moderate can be claimed. Finally, a general rule in the HS2 offsetting methodology was that in the absence of access to conduct a survey a moderate condition (2 points) would be assumed. Most of the above amendments received harsh criticism by conservationists, ecologists and grassroots organizations that, inter alia, expressed major worries for the complete absence of transparency:

> They told us that we'll see the amendments and the calculations, but we have never seen the actual calculation, we've only seen the method they followed. I won't be surprised if the actual calculations are not included in the environmental statement.

One of the most controversial issues has been the claim of HS2 Ltd, and of some governmental officials, that ancient woodlands should not be excluded from offset calculations. This has been communicated by the supporters of the scheme as a pragmatic, rational choice that would ensure the highest possible compensation for a loss that was inevitable. As a consultant told me:

> Can you use the metric for ancient woodlands? If you accept that ancient woodlands are a bit like coral reefs or tropical forests, completely irreplaceable, you can never replace them. But if you leave it out how can you ever demonstrate no net loss? It's better to include it and then debate how you will compensate for their loss instead of leaving this out.

The opportunistic way offsetting had been used was also obvious in the fact that the metric had been applied to compare the number of biodiversity units before construction and the number of units after construction but not to identify the necessary compensation measures. As another consultant explained to me:

Interviewee: The approach for mitigation was very much based on professional judgment, approaches developed over decades of working on projects, so we are not using the offsetting approach in that sense, people did not understand that, even though we repeatedly repeated it—if I can say that.

Interviewer: And why have you decided to use the metric but not the whole approach?

Interviewee: Given the escalating controversy around offsetting we felt that we would have a debate on figures, and we didn't feel as a project that such a discussion would be useful. Professional judgment is a better option for us, it's more difficult to question ecologists that work in this business for 20 or 30 years.

The intention of HS2 Ltd was to deliver the compensatory habitats by using the powers under the Hybrid Bill and not through the establishment of offsetting agreements with third parties. The following excerpt from my discussion with two members of STOP HS2 gives an interesting perspective on these tactics:

Interviewee 1: It's just because they buy the land within the Bill limits and they're using it for construction, so it makes sense from their point of view to do the offsetting in places that they've already messed up.

Interviewee 2: Like we've got an example of a small farm in Kenilworth and because it's such a small farm and HS2 is going through it they've done the cost analysis and think 'You know what, it's not worth building a bridge for him to get from one side of his farm to the other'.

Interviewee 1: So, he loses 61% of his land and basically, he's finished. HS2 Ltd has effectively put him out of business by saying this is where we're putting the offsetting without any consultation. And he is happy that he even had the opportunity to sell some of his land to them.

Some ecologists working for local authorities and participating in 51m explained to me that based on their own testing of the metric for some affected areas there would be a quite substantial deficit in biodiversity units. This meant that it would be highly unlikely that enough units could be found within the construction boundaries and that since HS2 Ltd was planning to compensate only within these boundaries it would make sense for them to use the Defra metric only to measure gains and losses and not to calculate the delivery of the necessary compensation. However, to keep all options open, HS2 Ltd had also pointed out that offsetting agreements with third parties might be required in the future to deliver additional measures.

HS2 even though does not offer a typical case of the implementation of offsetting, it constitutes an indicative example of how the policy and the rhetoric of No Net Loss could be voluntary and purposefully used to green-wash controversial large-scale projects with major social, economic, and environmental impacts. Moreover, even though, as explained above, offsetting was not necessary for granting planning permission, it was perceived as a tool that could be used to downplay both the impacts of the new trainline and the delivery of compensation. This has important implications for understanding how biodiversity governance is being reconfigured through the convergence of offsetting and large-scale infrastructure or more broadly urban development. As a consultant with a key role in the designation of offsetting in the UK put it:

> I suppose HS2 is an example where if it goes through without significant modification then biodiversity offsetting will have been used to fast-track and green-wash large-scale infrastructure projects, or megaprojects if you like, because it will have allowed them to bypass some controls that otherwise would have come into play. And to offset only within a narrow corridor around the railway itself means that ecologically speaking obviously you're not going to have similarly undisturbed habitat and so it's perfectly clear that you cannot claim Not Net Loss.

What offsetting was essentially trying to achieve and legitimize in the case of HS2 was not only the reframing of a highly controversial megaproject as green and clean but also the delivery of environmental compensation in direct accordance with the patters of large-scale infrastructure, making conservation a by-product of socially environmentally destructive economic growth.

6.6 When the Win-Win Rhetoric Meets TINA: The Tyranny of Pragmatism

> New construction and development are essential for economic growth, yet business-as-usual practices have often resulted in a significant loss of biodiversity. The Business and Biodiversity Offsets Programme (BBOP) is working to change that equation, by helping companies to conserve biodiversity in an ecologically effective and economically efficient manner as they pursue their business goals.
>
> BBOP[36]

> Whatever you do, inevitably, there is less habitat left, so you must look at the quality of what you put back, so if you turn an arable field into a woodland or grassland then you have an improvement. So, even though you lose land you can end up with a net balance and make your loss an actual gain.
>
> Environmental Consultant

As discussed in previous chapters, in line with the global discourse on the benefits of market environmentalism, offsetting was launched as a win-win, innovative approach to the calculation of compensation that was going to reconcile longstanding conflicts between economic development and environmental protection. Escalating criticism along with disappointing results from the pilots gradually led even keen supporters of the policy to a U-turn. People who have been characterizing offsetting as a groundbreaking policy were now trying to underestimate the changes that it was bringing about arguing that previous calculations of compensation have been very qualitative, vague, and diverse across the country and that offsetting had the potential to offer an accurate, quantitative measurement mainstreaming the value of biodiversity and standardizing calculations. It was surprising to note during my discussions with various actors how the origins of the policy, most notably the strong influence of the BBOP in shaping offsetting in England and the role of major industries in promoting its widespread implementation, were either ignored or

[36] See http://bbop.forest-trends.org.

sidelined to elide the policy from its ideological underpinnings and from any specific economic or class interests. The unwillingness to admit the policy's pro-development character was accompanied by an attempt to deny any direct link between the increasing adoption of offsetting and the entrenchment of neoliberal policies in the post-2008 era. Proponents of offsetting were basically trying to prove that the criticism against it has been fueled by a hostile ideological position against market-based instruments and the economic valuation of nature and a preference for state regulation and that is was stemming from a misinterpretation of the policy's goals. Considering the gradual demystification of offsetting's promises, even the Environment Bank decided to change its core slogan from offsetting to "accounting." Rebranding offsetting in the light of criticism has been a common tactics especially in the case of controversial projects that have been met with public opposition, as the Lodge Hill case described above.[37]

The emphasis on the pragmatic benefits of offsetting was not of course entirely new. The evolution of the policy in England, from its first inception until its pilot implementation, has been marked by a consistent attempt of its proponents to frame it as a realistic solution to the failure of the planning system to provide adequate compensation for the residual impacts of development. Importantly, as the policy was losing its glory a noticeable shift took place: the win-win rhetoric gradually gave its place to the argument that offsetting was basically the *only* available and feasible solution. This meant that resisting or opposing environmentally destructive projects was utopic and the only way forward was to accept that impacts on biodiversity are unlikely to cease or even abate in the near future (Gordon et al. 2015), and thus find ways to deal with these inevitable impacts. This essentially forecloses any serious ecological critique of current patterns of economic growth. But if conservationists and environmentalists accept this narrative, then they are adopting the agenda of capitalists, and they endorse a dystopian future where biodiversity loss is continuous, and chiefly directed by the interests of different sections of capital. In that scenario, conservation

[37]This changing rhetoric has been described quite explicitly in one of the interviews I had with a conservation broker (see Chapter 7).

is restricted to simply directing, or redirecting, where the destructive footprint of economic development will fall, without any guarantee that what is protected today will not be developed in further cycles of offsetting tomorrow (Apostolopoulou and Adams 2017a). Foreclosing the possibility of challenging the drivers of environmental destruction and implying that there is no alternative to the capitalist production of nature has not been unique to offsetting. The resurgence of Thatcher's TINA slogan has become a hegemonic discourse after the financial crash of 2008 and is now characterizing all spheres of public policy and governmental rhetoric in most places across the globe, reflecting capitalism's inability to offer a positive vision or a convincing new win-win discourse. In England, this was reflected in the fact that despite variations most people accepted that halting unnecessary or environmentally and socially destructive development and directly questioning current patterns of economic growth was highly unlikely in a context of prolonged austerity and recession. Interestingly, after the first applications of offsetting on the ground most people did not have high hopes for the future and could see that long-term problems of the planning system, such as subjective judgments and absence of appropriate data, were still dominant. Nonetheless they were unable to see a viable alternative. As an interviewee argued:

> What happened [he refers to a case where offsetting has been implemented] in not great planning practice I guess and in an ideal world the offset couldn't have been offset because it would have had much stronger protection. But under the previous system the results weren't great basically and the government's own research backs that up. Planning Policy Statement 9 that dealt with biodiversity for the National Planning Policy Framework was inconsistently applied and there was good stuff in there, much better stuff than we've got now. But things have changed and in the new reality that we are facing, and this is a reality of a weakened environmental legislation, offsetting seems to introduce at least some common standards and methods of calculating residual impacts [...] It's not ideal but this is what we have now.

A good example of how offsetting became from "exciting" and "groundbreaking" just a "pragmatic" solution to unresolved problems has been the way in which "in perpetuity" has been translated on the ground.

In the one of the most advanced pilots in England, people participating in offsetting explained to me that, in the absence of explicit national guidance, they have been working in 30-year offsets. The argument was that even though the loss is permanent, there is evidence that from a planning legal perspective in perpetuity is defined as 25 plus 5 years because this is the "logical" time period that you could secure an offset site. Local officers acknowledged that there are different views on the definition but insisted that a robust biodiversity offsetting system would have to confront the private landowners and, thus, it cannot be expected from them to undertake conservation works on their land in perpetuity. As a local planning officer argued:

> After that 30-year period the developer and his successors has to manage that land in the same positive manner but for no income. They just won't sign up for it. And I cannot blame them. So, we must be reasonable if we want offsetting to work. Many landowners like the idea of offsetting because, for example, they can get an income from the biodiversity found in their farm. For offsetting to work they should be able to make their living from it. If they are going to have requirements that they must manage the land in a certain way after the period of the scheme, they will put that cost in their offset and it will be very expensive. That is not viable. I understand the point of view that it should be in perpetuity. But I don't think that it's reasonable. 30 years of offset schemes, because of the inflation, are already expensive. [...] This is a positive way for development and it also works from a biodiversity perspective, but the developers are paying for it. You can't convince them from a theoretical point of view. You have to be pragmatic and practical.

Of course not all interviewees shared this view. Some of the people I discussed with felt that the way "in perpetuity" was being translated was one of the signs of the major divergence between the rhetoric of offsetting's supporters and its actual implementation on the ground, a reflection of the fact that adherence to pragmatism can ultimately lead to endless concessions:

> I feel that we'll get a poor system which is poorly delivered. The big problem was that we started off on the basis that the sites will exist in

perpetuity and that became diluted. Perpetuity came to be 25/30 years. But the thing about biodiversity loss and enhancement is time. You can plan woodland, you can start grassland off, but it needs time to mature. And it needs in some cases a tremendous amount of time. So, when you lose something, everybody agrees that you can't replicate it exactly and that it needs a long time to mature. While it is maturing you have got biodiversity loss. You have lost something important in the process which cannot be picked up in the metrics.

References

Allmendinger, P., & Haughton, G. (2009). Soft spaces, fuzzy boundaries, and metagovernance: The new spatial planning in the Thames Gateway. *Environment and Planning A, 41,* 617–633.

Allmendinger, P., & Haughton, G. (2013). The evolution and trajectories of English spatial governance: 'Neoliberal' episodes in planning. *Planning Practice & Research, 28,* 6–26.

Apostolopoulou, E. (2009). *The social conflicts during the implementation of conservation policy in protected areas: Analysis and appraisal of conservation policies in Greece* (PhD thesis). Aristotle University of Thessaloniki, Greece. Available at: https://www.didaktorika.gr/eadd/handle/10442/22497.

Apostolopoulou, E. (2019). Beyond post-politics: Offsetting, depoliticization and contestation in a community struggle against executive housing. *Transactions of the Institute of British Geographers.* https://doi.org/10.1111/tran.12354.

Apostolopoulou, E., & Adams, W. M. (2015). Neoliberal capitalism and conservation in the post-crisis era: The dialectics of 'green' and 'un-green' grabbing in Greece and the UK. *Antipode, 47,* 15–35.

Apostolopoulou, E., & Adams, W. M. (2017a). Biodiversity offsetting and conservation: Reframing nature to save it. *Oryx, 51,* 23–31.

Apostolopoulou, E., & Adams, W. M. (2017b). Biodiversity offsetting and the reframing of conservation: A reply to ten Kate & von Hase and Dempsey & Collard. *Oryx, 51,* 40–42.

Apostolopoulou, E., & Adams, W. M. (2019). Cutting nature to fit: Urbanization, neoliberalism and biodiversity offsetting in England. *Geoforum, 98,* 214–225.

Apostolopoulou, E., Adams, W. M., & Greco, E. (2018). Biodiversity offsetting and the production of "equivalent natures": A Marxist critique. *ACME: An International Journal for Critical Geographies, 17*(3), 861–892.

Apostolopoulou, E., Bormpoudakis, D., Paloniemi, R., Cent, J., Grodzińska-Jurczak, M., Pietrzyk-Kaszyńska, A., et al. (2014). Governance rescaling and the neoliberalization of nature: The case of biodiversity conservation in four EU countries. *International Journal of Sustainable Development and World Ecology, 21,* 481–494.

Blowers, A. (1987). Transition or transformation? Environmental policy under thatcher. *Public Administration, 65,* 277–294.

Bormpoudakis, D., Tzanopoulos, J., & Apostolopoulou, E. (2019). The rise and fall of biodiversity offsetting in the Lodge Hill large-scale housing development, South East England. *Environment and Planning E: Nature and Space.* https://doi.org/10.1177/2514848619884890.

CIEEM. (2016). *Guidelines for ecological impact assessment in the UK and Ireland: Terrestrial, freshwater and coastal chartered.* Institute of Ecology and Environmental Management, Winchester.

Cloke, P. (Ed.). (1999). *Policy and change in thatcher's Britain.* Oxford: Pergamon.

Conservative Party. (2010). *Open source planning* (Policy Green Paper No. 14). London.

Cowell, R. (2013). The greenest government ever? Planning and sustainability in England after the May 2010 elections. *Planning Practice and Research, 28,* 27–44.

CPRE (Campaign to Protect Rural England). (2015). *Set up to fail: Why housing targets based on flawed numbers threaten our countryside.* London, UK. Available at: http://www.cpre.org.uk/magazine/features/item/4167-set-up-to-fail.

Darling, E. (2005). The city in the country: Wilderness gentrification and the rent gap. *Environment and Planning A: Economy and Space, 37,* 1015–1032.

Defra (Department for Environment, Food and Rural Affairs). (2012). *Red tape challenge—Environment theme implementation plan.* Department for Environment, Food and Rural Affairs. London, UK. Available at: https://assets.publishing.service.gov.uk/government/uploads/system/uploads/attachment_data/file/69598/pb13819-red-tape-environment.pdf.

Defra (Department for Environment, Food and Rural Affairs). (2013). *Biodiversity offsetting in England green paper.* Department for Environment,

Food and Rural Affairs. London, UK. Available at: https://consult.defra.gov.uk/biodiversity/biodiversity_offsetting/supporting_documents/20130903Biodiversity%20offsetting%20green%20paper.pdf.

Department for Communities and Local Government. (2013). *Appeal by Bellway Homes (North East) Ltd.* Land at Whitehouse Farm, West Moor, Newcastle-upon-Tyne. Application ref: 11/02337/FUL. APP/W4515/A/12/2175554. Available at: https://acp.planninginspectorate.gov.uk/ViewCase.aspx?Caseid=2175554&CoID=0.

Department of the Environment. (1985). *Lifting the burden.* London: HMSO.

Department of the Environment. (1995). *The Thames gateway planning framework (RPG9A).* London: HMSO.

Engels, F. (1872). The housing question. Part III, "Supplement on proudhon and the housing question". In K. Marx & E. Engels, *Selected works* (Vol. 2). International Publishers.

Environment Bank. (2012a). *Conservation credit requirement scoping analysis of Whitehouse Farm.* Killingworth: The Environment Bank Ltd.

Environment Bank. (2012b). *Offset search: Offset for Whitehouse Farm.* Killingworth: The Environment Bank Ltd.

Environment Bank. (2012c). Biodiversity offsetting to compensate for nightingale habitat loss at Lodge Hill, Kent. Ripon: The Environment Bank Ltd.

Environment Bank. (2014a). *Biodiversity offsetting information sheet 4 February 2014, guidance for developers.*

Environment Bank. (2014b). *Biodiversity offsetting scheme, Whitehouse Farm.* Killingworth: The Environment Bank Ltd.

Featherstone, D. (2005). Towards the relational construction of militant particularisms: Or why the geographies of past struggles matter for resistance to neoliberal globalisation. *Antipode, 37,* 250–271.

Featherstone, D. (2013). "Gramsci in action": Space, politics, and the making of solidarities. In M. Ekers, G. Hart, S. Kipfer, et al. (Eds.), *Gramsci: Space, nature, politics* (pp. 65–82). Chichester: Wiley.

Featherstone, D., Ince, A., Mackinnon, D., Strauss, K., & Cumbers, A. (2012). Progressive localism and the construction of political alternatives. *Transactions of the Institute of British Geographers, 37,* 177–182.

Felli, R. (2015). Environment, not planning: The neoliberal depoliticisation of environmental policy by means of emissions trading. *Environmental Politics, 24,* 641–660.

Gordon, A., Bull, J. W., Wilcox, C., & Maron, M. (2015). Perverse incentives risk undermining biodiversity offset policies. *Journal of Applied Ecology, 52,* 532–537.

Gray, M., & Barford, A. (2018). The depths of the cuts: The uneven geography of local government austerity. *Cambridge Journal of Regions, Economy and Society, 11,* 541–563.

Hannis, M., & Sullivan, S. (2012). *Offsetting nature? Habitat banking and biodiversity offsets in the English Land Use Planning System.* Green House, UK.

Harvey, D. (1989). From managerialism to entrepreneurialism: The transformation in urban governance in late capitalism. *Geografiska Annaler: Series B, 71,* 3–17.

Harvey, D. (2012). *Rebel cities: From the right to the city to the urban revolution.* London and New York: Verso.

Harvey, D. (2014). *Seventeen contradictions and the end of capitalism.* London, UK: Profile Books.

Heynen, N., McCarthy, J., Prudham, S., & Robbins, R. (Eds.). (2007). *Neoliberal environments: False promises and unnatural consequences.* London and New York: Routledge.

HM Government. (2010). *The coalition: Our programme for government.* London: Cabinet Office.

HM Government. (2011). *The natural choice: Securing the value of nature.* Available at: https://assets.publishing.service.gov.uk/government/uploads/system/uploads/attachment_data/file/228842/8082.pdf.

HM Government. (2013). *Biodiversity offsetting* (p. 1126). Defra: Impact Assessment.

HM Government. (2017). *Cutting red tape: Review of house building.* Available at: https://assets.publishing.service.gov.uk/government/uploads/system/uploads/attachment_data/file/591039/house-building-cutting-red-tape-review.pdf.

HM Treasury. (2013). *Infrastructure plan 2013.* London, UK.

HS2. (2013). *London-West Midlands environmental statement* (Vol. 5). Technical Appendices. Scope and Methodology Report Addendum (Ct-001-000/2). Available at: https://www.gov.uk/government/uploads/system/uploads/attachment_data/file/260153/Vol5_Scope_and_methodology_report_addendum_CT-001-000.2.pdf.

IEEP. (2016). *Study on specific design elements of biodiversity offsets: Biodiversity metrics and mechanisms for securing long term conservation benefits.* DG ENVIRONMENT: ENV.B.2/ETU/2013/0060r.

Katz, C. (2001). On the grounds of globalization: A topography for feminist political engagement. *Signs: Journal of Women in Culture and Society, 2,* 1213–1234.

Lefebvre, H. (1976). Reflections on the politics of space. *Antipode, 8,* 30–37 (M. J. Enders, Trans.).
Leitner, H., Peck, J., & Sheppard, E. S. (2007). *Contesting neoliberalism: Urban frontiers.* New York and London: Guilford Press.
Lohmann, L. (2005). Marketing and making carbon dumps: Commodification, calculation and counterfactuals in climate change mitigation. *Science as Culture, 14,* 203–235.
Maclennan, D., & O'Sullivan, A. (2013). Localism, devolution and housing policies. *Housing Studies, 28,* 599–615.
Madanipour, A. (2009). Newcastle upon Tyne. In J. Punter (Ed.), *Urban design and the british urban renaissance* (pp. 132–147). Abingdon: Routledge.
Marsden, T., & Sonnino, R. (2008). Rural development and the regional state: Denying multifunctional agriculture in the UK. *Journal of Rural Studies, 24,* 422–431.
Marx, K. (1973 [1857]). *Grundrisse.* London: Penguin.
Mathiesen, K. (2015). *How and where did UK lose city-sized area of green space in just six years?* Available at: https://www.theguardian.com/environment/2015/jul/02/how-where-did-uk-lose-green-space-bigger-than-a-city-six-years.
Nayak, A. (2003). Last of the 'real geordies'? White masculinities and the subcultural response to deindustrialisation. *Environment and Planning D: Society and Space, 21,* 7–25.
NPPF. (2012). *National planning policy framework.* London: Department for Communities and Local Government. Available at: https://www.gov.uk/government/uploads/system/uploads/attachment_data/file/6077/2116950.pdf.
NTC (North Tyneside Council). (2015a). *Indices of multiple deprivation.* Available at: https://my.northtyneside.gov.uk/sites/default/files/web-page-related-files/IMD%202015%20Headline%20Analysis%20and%20Mapping%20-%20Data%20Files.pdf.
NTC (North Tyneside Council). (2015b). *Green space strategy.* Available at: http://my.northtyneside.gov.uk/sites/default/files/web-page-related-files/Green%20Space%20Strategy%202015%20%28Sub%20Header%29.pdf.
NTC (North Tyneside Council). (2017). *North Tyneside local plan.* Available at: https://my.northtyneside.gov.uk/sites/default/files/web-page-related-files/North%20Tyneside%20Local%20Plan%202017-2032.pdf. Accessed 7 November 2019.
Peck, J. (2012). Austerity urbanism: American cities under extreme economy. *City, 16,* 626–655.

Peck, J., Theodore, N., & Brenner, N. (2012). Neoliberalism resurgent? Market rule after the great recession. *The South Atlantic Quarterly, 111,* 2.

Price, N. (2012). *Proof of evidence of nick price on behalf of West Moor Residents' association Rule 6 party* (Planning Inspectorate Reference: APP/W4515/A/12/2175554).

Rotherham, I. D. (2015). *The rise and fall of countryside management: A historical account.* Earthscan: Routledge.

Smith, N. (1979). Toward a theory of gentrification a back to the city movement by capital, not people. *Journal of the American Planning Association, 45,* 538–548.

Sparke, M. (2000). "Chunnel visions": Unpacking the anticipatory geographies of an Anglo-European Borderland. *Journal of Borderlands Studies, 15,* 187–219.

Stockdale, A. (2010). The diverse geographies of rural gentrification in Scotland. *Journal of Rural Studies, 26,* 31–40.

Swyngedouw, E. (2010). Impossible sustainability and the post-political condition. In M. Cerreta, M. Concillio, & G. Monno (Eds.), *Making strategies in spatial planning* (pp. 185–205). The Netherlands: Springer.

Trimedia. (2009). *Lodge Hill stage two public consultation—Workshops: Feedback analysis.* Available at: http://www.medway.gov.uk/pdf/Statement_of_Community_Involvement_Final_Evidence_.pdf.

Tucker, G. M., Allen, B., Conway, M., Dickie, I., Hart, K., Rayment, M., et al. (2014). *Policy options for an EU no net loss initiative.* Report to the European Commission (with Annexes), Institute for European Environmental Policy (IEEP), London.

7

Discussing with the Supporters and the Opponents of Biodiversity Offsetting

E. Apostolopoulou, *Nature Swapped and Nature Lost*,
https://doi.org/10.1007/978-3-030-46788-3_7

If offsetting is implemented as it should be within the context of the planning system and the mitigation hierarchy is obeyed to its fullest extent, so at first you avoid and then you try and retain what habitats you can on site and then finally you do compensation for those residual impacts that you absolutely cannot avoid, then it may be fine. But what I see is that people skip straight to offsetting. They say, "Oh we tried to avoid it but it's too hard." And they then try to use offsetting to manipulate the system, numbers—like statistics—can be very easily manipulated and used to disorient public debate.

—Former conservation broker

7.1 Introduction

> Realist writing, of which history offers many widely varying examples, is likewise conditioned by the question of how, when and for what class it is made use of.
>
> Bertolt Brecht

Academic writing is increasingly equated to publications in scientific journals often with quite strict word limits. A book is a chance for an author to expose her/his opinion without the restrictions of space and style that most journals impose. This comes with a burden and a responsibility: the book will remain, and it has to be able to stand-alone and make sense. However, especially if someone is committed to critical social science research and fieldwork, a book can also give voice and space to other people. All these research participants without whom a social scientist would be unable to grasp the ideological and political opinions that shape, support, or contest the current, capitalist, production of nature. In this book, it is my conscious choice to present these conflicting opinions on a controversial issue, as offsetting, without using only my own interpretation. This constitutes both an acknowledgment of the limits that positionality may impose to the description of the socioeconomic and political reality, and an attempt to allow these different and often opposing positions to reach the reader so that she/he would be able to draw her/his own conclusions without my intervention. In what follows, I firstly present two interviews with supporters of offsetting and then two interviews with opponents.

7.2 Discussing with the Supporters of Offsetting

7.2.1 Interview with a Conservation Broker

Interviewer: What do you think about the evolution of biodiversity offsetting in England so far?

Interviewee: Offsetting hasn't evolved in the way we hoped and thought that it would. There are two reasons for this. One is that the political drive for offsetting in the UK has waned slightly, partly because the individual minister involved in it was very keen in offsetting and has subsequently gone and partly because it got a very rough treatment in the British press, in the British media, and it was heavily criticized and was most commonly associated with negative language, like license to trash. The further reason was that the main environmental NGOs supported it quietly and privately, whereas a small number of small NGOs opposed it very publicly and forcefully. And so the British media played out offsetting as an anti-environmental measure, which of course it was never intended to be. We are environmentalists and we are proposing offsetting and other ecosystem services paying schemes as a way of putting a financial value on the natural environment that it doesn't have. We were fed up to trying and improve the British planning system, which now underestimates and ignores the ecological value of most sites that are developed. So, now we have a planning system that simply does not work for the environment and the reason that it does not work is that there is no accounting and, therefore, no accountability.

Interviewer: There is already a system for providing compensation; I guess you believe that it is not enough?

Interviewee: So, there is provision for compensation that's right. Indeed, there is a legal framework for planning authorities to deliver No Net Loss for biodiversity. But until you try to measure what it is that is being lost, you cannot account for it, literally there is no accountability. So, rather than focus on offsetting and compensation, we are now focusing on straightforward accountability. We say to the planning authorities, not to developers or businesses but to the planning authorities, how is it you are accounting biodiversity, how you are meeting your legal duties to conserve biodiversity, to deliver No Net Loss. Because unless you quantify the individual impacts of a development across your district, how do you know you are delivering No Net Loss? And the answer is without a tool to quantify you can't. So, we are planning to get authorities to use the Defra metric in their planning applications. And just to give you one example. In 2008, just before the economic crisis with a lot of building going on, according to government figures residential housing development was a 100-billion-pound industry and they leave it, via section 106 payments, they leave it 5 billion pounds of payment from that industry, 5% was given in payment. And of those payments, how much went to the environment? And the answer is 5 million. So, only 1 thousand of the money went to the environment. So, there is money to build roads, and hospitals and playing fields, and school playgrounds, and 1 thousand of that money is going to the environment. So, we have a system now that is failing to account for the environment. The use of the metric whether or not it ever leads to offsetting, because actually when you are applying the metric 9 times out of 10 you simply strengthen the avoidance of the mitigation part of the hierarchy, because when developers see how much it costs to buy credits to do offsetting, they do more avoidance and they do more mitigation. And so we are focusing now not on offsetting, because it has such a bad name in the press, but on better accounting.

Interviewer: Do you mean that you abandon the term?

Interviewee: Yes.

Interviewer: Is this a recent decision?

Interviewee: Yes, it is, so, if you see at our website it now says better accounting for our life. And well of course we will use the phrase if someone says what you do when we get to the end of the hierarchy. Well, when you get to the end of the hierarchy you need to compensate. Because otherwise it is going to be a net loss. And you can compensate by paying to deliver biodiversity gain at a different site. That is biodiversity offsetting, yes, it is. But that has always been, that is not the point. The point is to have better accountability and better mitigation.

Interviewer: Do you think in this way you will disconnect what you are doing from the criticism against offsetting?

Interviewee: Partly that. I mean we don't want to pretend that we are doing something different but when we talk to a planning authority about offsetting, they just say: well we are not interested because it is a license to trash. And we say no it isn't, it was portrayed in that way in the press but that's not what it is about.

Interviewer: So, you think offsetting is not a helpful term anymore.

Interviewee: It's a very unhelpful term. And I will give you a further example. I was to … [the interviewee mentions a city] last week talking to the EU commission, there they say that the whole European No Net Loss strategy is now uncertain because the environment has fallen down the way the list of the commission priorities and if they are going to do anything for the environment they aren't comfortable talking about offsetting and No Net Loss, because they are getting so much criticism from the European environmental NGOs. Even in Europe they are thinking tackling the problem from this compensatory viewpoint.

Interviewer: Why do you think that this is happening, that the term receives so much criticism? Is it the idea of offsetting that sounds strange because it implies that you can offset nature somewhere else?

Interviewee: Yes.

Interviewer: Because this is my feeling from my fieldwork, this is the part that makes people oppose to offsetting, that we cannot offset nature.

Interviewee: Yes, true, but why they think that? Why can't we offset nature?

Interviewer: Many research participants referred, for example, to cultural values and to local communities' relations to nature. My impression from fieldwork is that it is a quite disturbing idea for people to hear that we can build something here and offset the damage somewhere else.

Interviewee: Sure. Well that's a very accurate analysis of what people feel about offsetting. But that's not what offsetting is. The whole people's perception is based on what's come out from the media about how offsetting is a disaster. So, firstly, two or three key things. Bringing in an offsetting metric into the planning system is not moving from a system where bio-environment is protected, and we are now lessening the protection because we can offset. The issue is we have an environment that is completely unprotected, and destroyed and no one takes any account of it. And so, when you ask people if it is a good idea, they create the counterfactual that we are moving from a protected system to a less protected system. But that is not the counter factual. Can we bring in a system to take better account of the environment? That's the one thing. The counterfactual that the environmental NGOs and the British Press have portrayed is totally incorrect. Second issue is does offsetting deals with cultural values? No, it doesn't. And it has never come to do that. My passion is biodiversity conservation, and habitats and species are being destroyed not because they do or don't have cultural values but because they have no economic value. So, when someone says to me: I don't want you to destroy that wood that I live next to, because you can offset it, the answer is that wood is going to be destroyed. It has nothing to do with whether you can offset it or not. Planning permission has been given to destroy that wood in that field. Do we try and get some compensation for it or not? Because the environment is being destroyed wherever you live. And a second comment on the locality issue is, where you have a large impact locally, you must offset that impact locally. And there is no case ever been recorded of habitat lost in

Essex and offsetting in Cambria. Never ever happened. All the examples of offsetting are being done very locally.

Interviewer: Well there are examples where offsetting probably won't happen very locally, as the Lodge Hill case. The latest proposal is to locate the offset in Essex.

Interviewee: Let me talk you about Lodge Hill. The proposal for compensation for Lodge Hill was to create eight to nine thousand hectares of wet woodland being managed by the local Wildlife Trust in perpetuity either around Kent or Essex. Why? Because those are the best sites for the nightingales. The question was about the impact on the nightingales and whether that impact could be compensated for that species. Not for humans, not for the dog walkers who want to walk their dogs in the local woods, the issue was whether for the nightingales it was better. And my opinion was that it was better for the nightingales. In my opinion the X [to respect anonymity names have been deleted] campaigned dishonestly claimed that the compensation wasn't good for the nightingales. If you wanted to compensate the nightingales the best thing you can do was to take the money from the developers, let them build their houses in that ex-MoD[1] site and create a ribbon of wet woodland alongside the site. Because when nightingales migrate, they are moving a long migration and they are looking for places to occupy. So, there are two issues there. What is best for biodiversity and what's best for people. In that case, Lodge Hill used by dog walkers as a dog toilet, was that good for local people? No, it wasn't. But there is an issue of local cultural use of sites and how they can be replicated. I would say again that in the UK most of the development is not on publicly owned land. The vast majority is on private land being sold by private individuals and there is no public access.

[1]MoD: Ministry of Defense.

So, this hypothetical conflict between people and offsetting is actually very rare. Because most of the land that has been developed has no public access, it is not publicly owned. And therefore, taken as a whole biodiversity offsetting, when you take small environmental impacts from lots of bits of farmland and then when you join them together to create a new habitat scheme, actually you are improving public access. As it happens now there is a piece of farmland and here is someone's house and that piece of farmland is sold for development, and these people don't like it because they don't want a house built next to them and it's true that offsetting, creating a low-flower grassland meadow five miles up the road doesn't help these people because they are having a house built next to them, that's true. But in terms of the environment a large low-flower meadow near there is better than it is to have a small piece of farmland there.

Interviewer: What about the criticism that offsetting can facilitate planning permission? There are cases where permissions were initially rejected but then because of offsetting they were eventually approved.

Interviewee: I know of no examples.

Interviewer: You don't think that there is evidence that this is happening?

Interviewee: No, there isn't. In my opinion there is no evidence. The legal position is clear. Bringing in an offsetting metric makes no difference to the designated status of any site and makes no difference to the planning process. I think that it is true that if a planning committee is faced with an undesignated site that has some wildlife value, they are required to take that into account. And they will have local people campaigning that this is an important site. And for the planning committee that is a challenging balance to make. And when you apply the metric you quantify that and, therefore, you could argue that when you quantify and make it simple and the developer says; yes there is

8.5 biodiversity units here but we're going to create 12.5 units over there because you put numbers on that you make it simpler. And the planning committee then says we understand that, we are taking it under consideration, but the numbers demonstrate that this biodiversity is being offset over there. You could argue that there is a psychological factor that makes easier it for planning committees to say I agree. That should make their decision-making easier. Because arguments are being presented clearer. And I would argue that 9 times out of 10 in the absence of these metrics they just give the planning committee the permission anyway. So, again it is the counterfactual I said earlier. Are we bringing in a system that makes it worse? Or are we bringing in a system that accounts for everything? In my opinion we make it worse 1% at the time and we make it better 99% of the time. So, even if there are one or two examples that it looks that bringing in offsetting enabled the planning decision; I can show nine examples for every one of those where applying the offsetting metric has delivered gains for biodiversity.

Interviewer: Let's go back to the criticism. If the policy is as good as you describe it then how do you explain the criticism?

Interviewee: Off the record I've given up trying to understand. Because this is a complex discussion, you cannot have it in the public domain. Because newspapers never give you more than 30 words to say whether it is good or bad. And indeed, much of the opposition to offsetting is coming from very sensible organizations that are not actually opposed to offsetting. They are nervous about offsetting and most of them have been saying that this needs to be done. But they are not allowed to say that. They are just asked by the newspapers whether they are for it or against it. And the British media always want to have a black or white, is it good or bad. And of course, the answer is if it is done badly it is bad, and if it is done well then it could be good. And even when it is done well it is only going to work for 90% of the cases, there will be some habitats, ancient woodland for example or some designated SSSIs, you can never offset. But the point is we are losing our wildlife. The environmental issue is not ancient woodland or limestone pavements or SSSIs. It is the 94% of the country that is not designated where all the development now is focused on poor quality grassland. And 90% of those developments are going ahead with no accounting.

Interviewer: So, you don't agree with the implementation of offsetting in SSSI or ancient woodland?

Interviewee: I don't think it should be applied but what to do when someone gives permission to damage an SSSI? What to do when there is a government that approved a national plan to build a railway? That railway is going to be built and they are going to destroy ancient woodland. And we should make them pay for it. We should oppose the destruction of ancient woodland and we should try to persuade them to build tunnels and bridges and change the routes, because conserve ancient woodlands in situ is the best you can do. But if at the end of the day the government says they will use this land and destroy ancient woodlands then what would you do? In my opinion you apply offsetting metrics, you have huge multipliers and you make them recreate at very big ratios, 10, 20, 30 to 1. You make them replant woodland. Absolutely you must not destroy ancient woodland. But is the railway going to be built? Yes. Is it going to destroy ancient woodland? Yes. Do we need the government to pay for that and compensate? Yes. And building huge national forests is a very exciting thing to do. If HS2 gives 300 million pounds for environmental compensation, we could have an extraordinary wood planting scheme. Would it replace the ancient woodland that has been lost? No, not in my lifetime and in my grandchild's lifetime. But in 50 years' time we could have a tremendous young wood growing in and for me the counterfactual is that if you don't apply offsetting for HS2 is it going to prevent HS2 from being built? No.

Interviewer: I would argue that there is also the option to say I will stop the development. I mean there is an escalating criticism on whether we need more economic growth or more large-scale infrastructure that is based on valid arguments and makes sense. To support environmentally destructive and socially controversial development and, at the same time, being committed to protect the environment, I am not sure it makes sense.

Interviewee: I know you are right and philosophically I am right there. We are too materialistic, just too much focused on GDP, we don't understand what makes people happy, we don't value the environment. If I was the prime minister, I would stop development, no more development, new houses, brownfield sites, new houses in the cities. But I am not, and I don't argue to be, and there is no D option to tick and the reason why is that we are a tiny minority. And people want new houses, new cars, jobs, that's the society we live in. This is not a case of stop development or do offsetting. This is development is going to happen, how can we make people understand what it needs to be paid for it. Your question is about enabling development; they nibbled at the edges of that. And there are some examples that are uncomfortable for people that want to bring in compensation schemes. The issue is that development is happening, and no compensation is being found for it.

Interviewer: Something else that is relevant to what we are discussing—what do you think about the wording of the green paper?

Interviewee: People who wanted the conservative party to bring in offsetting drafted that wording. And it was written in the language of the conservative party, it was a piece of political rhetoric that was of its time. The reason why offsetting has been stopped is that it wasn't put forward by the government, has nothing to do with the environmental

NGOs not liking it, it has been stopped because the treasury thinks that it will slow development down. So, it is interesting because there is a wide public perception that offsetting has been slowed down because it is environmentally unpopular. And that's what the NGOs and the media like to think but it's not true. Offsetting has been slowed down because the economists don't like it, because the chancellor and three or four people from his team were not convinced that it would be helpful for developers. We tried to convince the government that although bringing in offsetting may make developers to pay more money for conservation in some cases, it would be beneficial to them. Because the bottom line of it is that if you are creating 400 million pounds of habitat creation who is paying for that? And the answer is that developers are paying. And the government would say we are not putting another green tax to developers. So, our argument was that because it simplifies the process, it speeds the process up, it is good for them. And even though their money may go from their pocket to a conservation NGO or a broker, they end up saving a lot of money because their planning application only takes two months instead of six or twelve months. And you know what? We convinced developers of that fact but the economists in London said to the chancellor if we bring this in it is going to slow down the housing program. And that is why offsetting is not happening as much as we were hoping for. It has nothing to do with the "Friends of the Earth".

Interviewer: Let's go back to what you just said, that you managed to convince the developers. Why were they convinced?

Interviewee: One issue is uncertainty. They hate it. So not knowing how long the planning application is going to take or how much environmental impact they will have to deliver is an anathema to them. They hate that. The second issue is that when they offset, they discharge their environmental obligations, so they part their obligations onto someone else.

That was very powerful for big housing developers. Instead of having to go and find your own offsetting scheme and look after it, and manage it and appoint managers, you just give to someone else the money—to a broker I mean—and will do it for you. And the third thing, which I think was the least important, was that they were able to demonstrate environmental sustainability. So, they were able to say we have done the calculation, we know there is an impact, but we offset it, we paid this money and now that local nature is reserved, that was paid by us. So, for the developers it was about reducing cost and time in the planning process, discharging their environmental obligations onto someone else, both of which are happy to pay for. And, thirdly, having environmental credential so they can say we are green builders, we don't damage the environment. So developers were happy, they were convinced. And that's why we were disappointed by the treasury. Because the treasury asked the developers whether this was ok, and the developers said it's fine, we'd rather didn't have to do it, but if you are going to do something for the environment and we must comply we can work with this specific policy.

Interviewer: So, it would have worked better if the government were more determined.

Interviewee: Yes. The other thing is that the minister involved was dealing with badges, and flooding and offsetting, and he was getting so much criticism for offsetting. It's interesting that people now say that because the government said you could restore ancient woodlands, you know they just brought the whole thing into disrepute, but the person who is blamed for saying that, he never did say that. The press misquoted it. So, that was very disappointing.

Interviewer: What about the international experience with offsetting?

Interviewee: Well these arguments are very Anglo-centric. It is not even the UK. It is England. And they had nothing similar in the US or Australia. They are having something similar in South Africa, I think. Where there again they started the process, there is no federal top down approach anymore, so they are starting to do their own thing. That is happening in the UK as well. We are working with two or three big counties where they're bringing in offsetting.

Interviewer: So where are we now? I mean we are waiting for the final report for the pilots. Do you think we will have a law?

Interviewee: No. There won't be a law. They won't regulate. And the report from the pilots is on the minister's desk now and I suspect they will release that just before the elections, very quietly and the media will be interested in the elections, and then there will be a couple of headlines.

Interviewer: However, I have in mind many cases where offsetting is proceeding, so in practice offsetting is happening.

Interviewee: Yes, it is, particularly in Warwickshire. It is just not as widespread as we hoped and not linked so far to big conservation banks as it was initially planned. So, yes, it's happening. But this is not new. Compensation has been happening for the past 30 years. All over the country you can see wildlife sites that are the results of compensation from planning.

Interviewer: Yes, but the metric is something new, the Environment Bank is something new.

Interviewee: Exactly. So, offsetting has been happening for 30 years. It is just not been done consistently, or rigorously or by using the metric. And I think it will continue to go on happening. And hopefully it will continue happening with the metric.

Interviewer: Do you fear land speculation?

Interviewee: Well, I think doing conservation should be valued. I think they should profit from it. I think it is a perfectly legitimate land use. And if a farmer instead of growing corn to make money, he wants to grow orchids to make money, why not? And people should be paid to do conservation management and grow those orchids. We subsidized wheat for taxpayers, we subsidized wheat enormously, why shouldn't we pay the farmer to grow orchids? I'd rather grow orchids than grow wheat. In terms of whether there is speculation, from my experience so far, it could conceivably happen if the demand is more than the supply.

By the way I don't know if you know, but where it is expected to get more progress is through habitat banking. And the reason is that what developers don't like is uncertainty on timing and price. So, under normal offsetting scheme you have a development here and you try and find an offset locally, but you don't know how much it's going to cost. So, the developer says OK I will buy some credits to compensate my impact, how much are they and where can I buy them? And you don't know how much they are or where you can find them. The way round that is to set up habitat banks and this is what is now happening. There are 10 habitat banks spread across the county, each of them is 40 hectares, each of them is grassland restoration and costs about 1,3 million pounds. So, the cost for the 10 banks is 13 million pounds. And each bank yields 240 credits. A hectare of grassland restoration yields about six credits and each bank yield 240 credits. The cost is 13 million so at the start of the scheme the county council buys all the credits: 2400 credits for 13 million pounds. And that money then goes to a secure bank account to pay the farmers to do the work. Then every time a development has a planning permission, the metric is applied, and the developer says I need to buy 2.7 credits, and the planning authority says yes, we will sell these to you, they are £5000 each and you buy them now. Because the habitats banks are big economies of scale it is very cost effective to conservation management.

Interviewer: And how the land became available for habitat banking?

Interviewee: It is all private land. And that is an issue because there is no public access to the banks. But what you are offsetting is not major impact from large schemes, it's lots and lots of small developers building two houses on farmland and having to buy three credits, £15,000.

Interviewer: This was the idea from the beginning with the Environment Bank, there was this map online where landowners could register their sites, but it hasn't worked very well so far, I think.

Interviewee: No. You are right. But the key difference is when you pay them. Because for habitat banking you pay the farmer and the land is managed and that thing is being set up now. For the original idea all those dots on the map were mainly wildlife trust farmers, and they hadn't started the management. So, those were people registering an interest saying if we were to be paid, we would do conservation management. But because there was no development next door they weren't being paid. And because everybody thought that there was going to be a government scheme people were coming forward saying when it starts up I've got a site that can be used. Those 10 habitat banks I am taking about, it hasn't started yet, will be paid by the planning authority and money will go to the farmers.

Interviewer: How will the price of the credit be determined?

Interviewee: The price has been worked out from the cost of doing the land management. And in fact, the farmer is paid 600 pounds per hectare per annum and it rises very slow with the inflation. That price was selected because it will be economically viable for mixed farmers to take their land out of production for 30 years. So, the price was driven by the cost of doing the land management. And it also pays for the monitoring and reporting fees for the lifetime of the thing.

Interviewer: And the ownership remains the same?

Interviewee: The ownership stays with the farmer, and it is a commercial contract and if he does what he is saying he is going to do he gets the payment every year and if he doesn't the payment stops, and if he destroys the land he has created he has to pay the money back.

Interviewer: This is a clear example of what we were saying before, making conservation profitable.

Interviewee: Exactly. This is attractive for those farmers because they have a very uncertain business model. The price of corn, the price of phosphate goes up, they got no idea from one year to the next. Whereas this is guaranteed constant income. So, they take 10% of their farm, put it into a conservation bank, they know at least they've got guarantee. Financially it works for them. The issue is where does the money come from? There are private investors, but the simplest model is for the planning authority to borrow the money. The legal problem is that there is a potential conflict of interest because the planning authorities are telling the developers you must buy credit and then selling them the credits. So, they will be challenged by the developers. But I think it can be demonstrated that authorities aren't making any profit, they just financing environmental schemes, and then the developers are just paying for these and brokers are calculating given the housing demand in the region. If the metric was appropriately applied to all of that residential demand, those 10 banks, 13 million pounds will sell out in five years. So, in five years 13 million pounds are going to the environment. And then after five years you do another 10 banks.

Interviewer: So, you think that this is going to work.

Interviewee: If this works and the scheme has been under designation for a year now, if this works 1000 hectares of low-flower meadow restoration will be created which is more than the environmental movement has ever done in any decade ever. And hopefully there is going to be another one in the next five years, and the money will pay for the management and the monitoring for the next 30 years. So, you asked me about the debate. We've decided that there is no point talking or arguing to defend offsetting in the media because you just get criticized. There is a very good, new launched genuine debate about where offsetting can work and what you need to make it work. But we cannot have it in public. Because all the Sunday Times wants to know is are you for it or against it, please tell me in 50 words why this is good. We've decided that offsetting does have a very powerful environmental role but mainly away from protected sites, mainly in the wider country site where lots and lots of low ecological value land is being built all the time.

Interviewer: Will you leave it all to the private sector? I have in mind species banking in the US where the price of credits is not exactly calculated as you said. It's similar to carbon offsetting where prices may change according to demand and supply for example or are influenced by economic recession.

Interviewee: So, I think that there is no parallel. Biodiversity offsetting is completely different to that. Because of the spatial literacy. Many of the offsetting multipliers have to do with where you are and what habitat is. So, unless you standardize it, so if you are a developer and you are building on woodland you can't buy the credit, you must go and compensate for woodland. So, I don't think it can turn into that market driven system. And absolutely you must depend on the public sector as well. It depends what you mean by that. There is no reason at all why wildlife trusts cannot be brought forward to act as habitat bankers. And there is no reason at all why they shouldn't be paid for that land management. Our experience is that it is much cheaper to do it with farmers because farmers own the land, have the livestock and know how to manage it. And the Wildlife Trust must go and buy land and employ staff and bring in. So, simply in terms of cost, this is the most cost effective.

7.2.2 Interview with a Consultant Working for the Housing Industry

Interviewer: Let's start our discussion from the confusion that has been created on whether what you are doing is offsetting or compensation.

Interviewee: Yes, we did start using offsetting and the Defra metric. The problem with the Defra metric is that it was only ever intended for habitat loss. When first attempts were made to implement the Defra approach, figures started to emerge in the region, 600 hectares for compensation that would be needed which was—we felt, and our client felt—significantly out of proportion to the amount of habitat for the species[2] that it was actually lost. Well there are multipliers in the metric to deal with multiple risk factors—the time it will take to establish the habitat, the degree of technical difficulty of establishing the habitat etc. So we thought that in terms of the habitat of this particular species is probably one of the easiest and most straightforward habitats to create, so in terms of the degree of technical difficulty is very low, in terms of the guarantee of it growing again very low requirement for multiplying in that respect because this habitat grows very quickly and at fairly predictable rates and so the end of the exercise that we went through and the figures we were getting were higher and higher and higher, everybody looked at them suggested that there needed more habitat compensation—our clients just said well this is just getting ridiculous and out of hand, we need a more realistic, common sense approach to the offsetting of this. And so, we did an exercise where we work out the maximum area of land that a species couple would need based on studies that have been done in other parts of the country as well. We worked on the precautionary basis, we worked on the top edge of

[2]The species is not identified here to ensure the anonymity of the interviewee.

those estimates, so at the maximum extent of what was likely to be required. And we came up with a figure that broadly every pair of the species would require about two hectares of suitable habitat to be able to survive and raise successfully a brood of chicks. And we then multiplied that by the number of breeding pairs that have been identified by the X [the NGO mentioned is not identified to ensure anonymity] to give us a basic figure. We then added some additional sort of precautionary approach and we produced a report that I have here. You see that we looked at all the risks of successfully providing a new habitat. We haven't completely thrown out the idea that there are risks inherent in any compensation or offsetting project so we identified nine risks to the delivery of compensation land as we have called it. And we have addressed each one of those risks. And in addressing some of them we provided multiplying for additional habitat, so for example that a risk would be in the implementation that if we plant say a 100 hectares of compensation land some of the plants might fail. What percentage might fail we can base that on experience of massive, large scale planting schemes so let's say that 5% fails therefore we need to put 5% more in that we would have done in order to cover for that risk. So, we have built the risk factors and the multipliers in that way, by looking at risks in providing suitable habitat and the risks inherent in doing that and then we addressed those risks in our compensation scheme.

Interviewer: Are there other similar cases of compensation for this species?

Interviewee: No, no one has ever tried it before. Which is one of the criticisms that have been leveled against us. There is a huge risk, a huge question mark because nobody has ever done it so how you know whether it is going to work. Our response is that this species is very adaptable; studies have shown that if the suitable habitat exists, this species will move in and use it for breeding. We have in our evidence several case studies. So, the species will move around, and they must do because of the very nature of the type of habitat they need. By that very precise habitat requirements they have once suitable habitat is developed to a stage where is no longer suitable those species would have to move on and find suitable alternatives. Therefore what we are saying is that if we can create the ideal habitat and manage it to keep it at that stage in its succession so that it remains ideal, in optimum condition for breeding, then we will be providing something that you probably won't find anywhere else in the UK, a piece of habitat for this particular species that it's being managed to keep it at exactly the right stage in the succession so that the species would be able to come back generation by generation after generation. In the normal course of events they would use it for 10, 15, 20 years at the most and then it would be at the sub-optimal condition and they will move on.

Interviewer: So, you are not using the Defra metrics anymore?

Interviewee: We abandon the Defra metrics but not the core logic of offsetting. We did that because we found that those metrics weren't working in terms of giving a practical solution in this case and because of the difficulties we found in trying to adapt them for compensation for a species rather than simply habitat.

Interviewer: Are you following the mitigation hierarchy?

Interviewee: Exactly. As you know within the National Planning Policy Framework it says that if you are going through the mitigation hierarchy if you can't avoid you mitigate, and if you can't mitigate, you compensate but there are no prescriptions as to how that compensation should be calculated or provided but obviously we need to make the compensation suitable to replace exactly what is being lost. What we have come up with is that there is a requirement for 304 hectares of habitat to be provided to compensate for the loss of the species and this is based on sustaining the UK population of this species so the habitat that we are providing has to be large enough to support the UK population. That's the target we have set.

Interviewer: There have been several reports from the X [I mention a conservation brokerage company here] about implementing offsetting in the area. Are you using them?

Interviewee: Well at the end of the day they didn't prove to be very helpful. The figures for offsetting were getting larger and larger and as you can imagine if you are making a planning application two things will be going through your mind. Number 1, every time that figure for offsetting goes up then that costs me more money and it gets to a point where you think to yourself so if it carries on going up any further then there is no point because we will get to the point where any profit I might have made up from this development it's just going to go in terms of offsetting. So, either we have to decide that this isn't a site that can be developed, and we walk away and we leave it, or we look at a practical alternative that gives us what we consider a more commonsense and realistic approach.

Interviewer: But you still think that biodiversity offsetting is a good idea?

Interviewee: I am definitely a supporter of the principle of biodiversity offsetting and I believe that there would be a lot more planning cases and development cases in the future if offsetting is introduced where it will be required because there will be residual effects that previously haven't been dealt with, because they weren't regarded as being significant enough to require compensation. If we are going to achieve the target that is NNL—which is what we are all working towards not just in the UK but across Europe and the world—then it seems to me that probably offsetting is the only way we can go. In the future, all the effects must be dealt through avoidance, mitigation or offsetting as the last resort if mitigation is not enough so yes, I am a keen supporter of the principle of the scheme and I have done some work. But as you know the devil is in the details and now of course the other problem is that we haven't got a metric for species and it's difficult to see how possibly we could have a standard metric that would deal with all species, because all species are different. But there are approaches that look at what the risks are of providing the offsetting for a species and then deal with those risks individually. And this is what we have done.

Interviewer: So, following your approach will you achieve No Net Loss?

Interviewee: Well, we are hoping to do better than that. Because we have worked on the precautionary principle, the worst-case scenario. So, we think, well we are certain that we will actually overcompensating because along with the compensation land, a proportion of that population will still carry on breeding at the development site. So, we would hope for a Net Gain out of this process.

Interviewer: In general, do you think that No Net Loss as a conservation goal is achievable?

Interviewee: It depends on how you assess the No Net Loss—doesn't it? Whether you measure it as a loss of habitat so therefore we have lost 5% of broadleaved woodland but if we replace that 5% of broadleaved woodland with 5% neutral grassland we have got a net loss of broadleaved woodland; overall in terms of semi-natural habitat we might have an equivalence—No Net Loss but we have a net loss of that particular habitat. Going further if you assessed it from the ecosystem services point of view, we might have lost important services and we might have been providing offsetting that offsets the habitat loss but does not offset the services that those habitats that have been lost were providing. So yes, you are right, it's complicated and the problem with complicated systems in the political climate, in the development climate, the more complicated they are the less likely they are to be acceptable. So, I think what Defra has tried to come up with, is a system that tries to address the issue without being over-complicated because they knew if it was over-complicated it would never be accepted. And I guess we are starting as a country a long way behind others, like Australia or the USA, and I know from speaking to colleagues, for example in Australia, that they are constantly reviewing and revising their offsetting system based on experience. And I am pretty sure that that's would happen in the UK and other parts of Europe. Offsetting will start off as a basic, simple framework and then we will refine it and improve it. Because we started by striking two pieces of wood together to warm ourselves and cook, and now we have microwaves. And that's how human society works. We start with a simple idea, we get this simple idea up and running and then we perfect it and refine it. And I suspect that's what would happen with offsetting because I am sure that the principle is good, and this is the way things should proceed.

Interviewer: What about the consultation meetings that took place in the area? Are there similar meetings for this new compensation/offsetting proposal?

Interviewee: No.

Interviewer: So why there were meetings before and not now?

Interviewee: I think basically because it became obvious that it wasn't going to be acceptable, it wasn't going anywhere.

Interviewer: In the new report that you produced have you explored sufficiently the avoidance stage of the mitigation hierarchy?

Interviewee: That is not a question I can answer, because that is a political question. Whether we need this number of homes and industry in this area is not an issue for me as an ecologist; that's an issue for central and local government. All I can say is that the council had approved the application despite knowing potential effects on ecology. So that's gives you I think some understanding on how important they think this development is for their economy; they approved it under these circumstances. So, it's their judgment and not mine. And in terms of the needs of course there are in the NPPF tests for the need and in terms of the loss of an SSSI that this area represents the need must be demonstrated to be greater than the impact of the loss of the SSSI. And again, that judgment was made by the local council and they decided that the need was greater than the protection of the SSSI and it was their judgment to make. Now of course the secretary of state has called-in the application to look at and therefore there will be scrutiny around those issues, and we will be waiting to see what the secretary of state decides. But obviously I can't answer. That is for others to determine.

Interviewer: What about the SSSI? This came up after the process had started.

Interviewee: To be fair to the developers when they first looked at developing this site there was an SSSI and it was their intention to completely avoid any impacts on that in the original development scheme. It was only subsequently when the SSSI was extended that these issues began to arise. And so, the first effort was to avoid and in fact within the outline planning application 70 hectares of habitat, of that SSI, is to be retained within the development site. We followed the mitigation hierarchy, avoid losses within the development boundary, where that wasn't possible, we looked at mitigation. Some of the mitigation will be provided on-site but given the scale of mitigation required it can't all be provided on-site. I mean we are talking about the loss of this species because it is an SSSI but you've got to remember that there are also several other EU protected species on this site. So, all of that must be either avoided or mitigated as well. It's a complex case. And when we added up what was being lost and what could be retained on site it was clear we wouldn't be able to mitigate sufficiently within the development site boundaries. So, what our clients did was to secure a piece of land immediately adjacent to the development site to use as an off-site mitigation area. This is a piece of farmland.

Interviewer: Which they own?

Interviewee: Yes, it's owned by the applicant. 86 hectares of additional land will be provided for mitigation—for each one of the habitats or species groups—this is again part of the planning submission—we went through and we worked out the amount of suitable habitat that was being lost, the amount of suitable habitat that could be created on-site and the amount of suitable habitat that could be created off-site and we came to a net figure then of how much it was being lost and how much it was being created. As you can see in the case of terrestrial invertebrates is a net loss of 25 hectares, in the case of others there will be a gain from the mitigation. We tried the best we could to balance the losses with mitigation and compensation to try to achieve—as we said before—a No Net Loss in biodiversity.

Interviewer: So, you are essentially doing offsetting.

Interviewee: In principle yes, we just don't use the Defra metrics. In this case the off-site mitigation area is immediately adjacent to the development site that means that the soils, and the topography are very similar. So, it should be very feasible, in some cases we are not only creating new habitat from scratch, we are translocating the existing habitat and putting it in the off-site mitigation area. So, for example, a large proportion of grassland will be carefully lifted and transported and relayed to create new grassland within the off-site mitigation site.

Interviewer: Is this standard practice?

Interviewee: I wouldn't say standard practice but there are many successful examples now within the UK of habitat translocation particularly with grassland. So, I think we are on safe ground with that.

Interviewer: And what about the offset/compensation site?

Interviewee: The species compensation site, well the indicative site would be in a different area.

Interviewer: There are also Natura 2000 sites in the proposed offset site, is that right?

Interviewee: Yes, there are three, I think. The compensation land does not directly affect any of them. The issue was whether it indirectly affects them because a lot of the bird species from the SPAs use farmland within the offset site for feeding during the winter. So, the question was if we put the compensation site on that land would it have an adverse effect on the integrity of that populations by removing important feeding areas? We did surveys to show that there is far more farmland available in the offset site than the birds are using, we identified which types of farmland the birds were using. They don't use all of it, they tend to use the wet grassland and the arable and that winter cereal crops within it and they don't use the rest. We worked—the exact figures are in the report—between 600 and 700 hectares of farmland which is unlikely to be used by the winter bird species from the SPAs and this is arable land and pasture grassland, improved grassland. So, habitat that currently has no wildlife value at all.

Interviewer: How was the offset site selected? The logic is not clear especially because it is located many miles away from the development site.

Interviewee: That all goes back to the report that we were discussing before, the compensation feasibility study. We looked at habitat suitable for this species—it has got to be within the core range of the species population so that narrows it down to certain counties in South and East of England. We used all necessary factors and we put those onto a map of England, and we identified land that was suitable for providing compensation for this species.

Interviewer: In the proposed site the land is owned by the same owner, so I guess this made it an easier option?

Interviewee: What we did, having identified the land that was suitable, then was to look at land that was both suitable and on the ownership of the owner of the land at the development site. And that's how we came up with the offset site. It meets all the criteria.

Interviewer: In the report it says that 5–9 years are needed for the creation of the compensation area in the offset site. This is an important time lag.

Interviewee: The time lag risk is one of the risks we identified and one of the ways we have dealt with this is through creating additional habitat, which means that to help compensate for the time lag risk we will create more habitat than we need to do.

Interviewer: In ecological terms can we be certain that this will work?

Interviewee: We think that the risk is quite low, so we will start providing the compensation land before we start to clear the habitat at the development site. And the removal of the suitable habitat will be faced so that only the amount of habitat that is required to be taken out at that time for that development will be removed. I know a lot of people are very worried that we will just go there and clear the whole site, that's not going to happen. It will be cleared only as much as is needed for a particular phase of development. And the development is scheduled to take over 21 years. So small amounts of the habitat will be taken and at the same time larger areas of habitat will be created as compensation land for the species. So—but there is the risk of course—is not only the time lag between the clearance and the planting in the ground but also the time lag between the clearance and the replacement habitat reaching optimal, suitable conditions to support breeding species. To deal with that we have added between 20%, well its 40% of additional habitat.

Interviewer: What about local community's opposition? Compensation is being offered in a different area, this is a quite controversial choice.

Interviewee: I don't think it's controversial. I think it's a very straightforward issue. We are not providing compensation for people; we are providing it for a species. That's the legal requirement. We are required to compensate for the loss of the

SSSI. Now the local population I know they have legitimate concerns about the development, but we are not, this process is not about providing them with compensation for loss of habitat because we have got to remember this is private land and local people don't currently have the right to use that land. There are some public rights of way through it but…

Interviewer: Yes, but we are talking about birds, you can hear them singing—you do not need the right to access the site to hear birds singing.

Interviewee: No, true.

Interviewer: The development will profoundly change the character of the area.

Interviewee: It will. But we must never loose site of the fact that the original SSSI—as was originally designated—will not be affected by the development. And about 70 hectares of the newly designated SSSI will also be protected within the development site. So, a very very large area of the existing SSSI, a large proportion would be actually protected. And so, local people will have greater access to that area once the development occurs, if the development occurs, because the plans for the site actually include creating new parks, and new rights of way, footpaths that will enable the local community to have greater access to local nature.

Interviewer: What about the offset site? Is it accessible to local people?

Interviewee: No, and they won't be able to visit that either. One of the beauties of that site from a nature conservation point of view is, number 1 is an island, number 2 is an island where there is no right of public access at all which means that any compensation that we provide on that site would be completely secured. It won't be subject to any disturbance.

Interviewer: However, even some supporters of offsetting acknowledge that offsets should provide compensation for the cultural values of biodiversity.

Interviewee: Yes, I understand. However, what I am sure off is that two things should be kept separately. The scientific aspects and the social aspects. I will give you an example. We have been reviewing sites of importance for conservation for a local authority that is renewing its local plan and some of those sites were designated on the past based on not only their ecological value but their social value as well, local, cultural value. And I made it clear from the start that we will be doing the evaluation purely on ecological terms, that we would make a judgment and rank the importance of all these sites based on strictly ecological criteria. It's then up to the local authority to decide how much importance they attach to the social criteria, they are at the best place to make that judgment because they are the local council and responsible for the local community. This is not our job. And this is one of the best aspects of offsetting, that it clearly separates the ecological from the social.

7.3 Discussing with the Opponents of Biodiversity Offsetting

7.3.1 Interview with an Environmentalist

Interviewer: I understand from what you have been writing that you are not a supporter of offsetting. Do you oppose it in principle or do you think that it can be improved?

Interviewee: Sometimes some things are so enshrined and so fossilised that if you try and change them for the better all that happens is that either you get broken or a little chunk breaks off and then it turns into something even worse, I think that's what's happened with the latest round of capital. I have the same feeling about offsetting. And of course, about other popular ideas, like natural capital. Have you talked to the Natural Capital initiative?

Interviewer: No, I haven't talked to them yet.

Interviewee: Well that seems to be on a parallel track with offsetting about, or underneath, there's a kind of underpinning framework for justifying it. And it's been very interesting seeing how much more acceptable that approach appears to be to conservation organisations who would be happy to criticise offsetting but seem more comfortable with the language of natural capital which I find extraordinary because, you know, it's the same thing. I find that quite worrying, I mean I've got a lot of colleagues and friends in the wildlife trust movement and there seems to be quite a big split within the movement as to people who are supportive of offsetting, on the basis that they think it's going to deliver some gain, as opposed to others who are more like me, are opposed to it in principle regardless of whether it delivers gain or not, on the basis that it's profoundly wrong, we might win a small battle but we're going to strategically lose…

Interviewer: The war.

Interviewee: Exactly. Take a few hundred metres of mud but lose five divisions in doing it. See the example of Lodge Hill, you said it's one of your case studies, right? Conservationists, as far as I know, what they're focusing on at the moment is the Nightingale Compensation Land. They have critiqued the metrics that the consultants of the developers have used to come up with the appropriate compensation. But what they've said is that 'according to our calculation it should be two or three times that much'. Now to my mind they're falling into the trap of playing the same game. By the way, another issue there is the grasslands, they are really nice and really rich, special, of a type that pretty much disappeared from Britain. And there was no proposal to offset them; they were just going to move them, physically pick them up and move them onto an adjacent area of land that they thought would be suitable. 30 hectares of unimproved wildlife-rich grassland, they were going to move them. Well they have been arguing over how much is actually valuable and they have been trying to get the numbers down as low as possible, but the whole area is very rich. So, in some ways I wonder, would it be more honest if they were to say 'we know we can't move them, we know that it won't work, we know we're going to destroy them but we're going to create, five times as much somewhere else and hand it over to a sympathetic owner'. Would that be more honest? That's offsetting isn't it? Well it's compensation. Is it more honest than saying 'it's okay, we'll be able to move it'. So that's for me is quite problematic. Offsetting has a positive message, it's like you want to shift the discussion from the issue that actually I'm destroying nature here and say 'No, no, I have No Net Loss or even a Net Gain'. It's a PR, it's a propaganda.

Interviewer: In the Lodge Hill case why do you think that they decided to use offsetting?

Interviewee: I think it was a very pragmatic decision on the part of the developers and probably the council as well to find a new justification for the damage that they knew they were going to cause. I was looking at the people who were advising the developers, they have got a communications and PR consultancy working for them, and I can only assume that as part of their communication strategy for getting development through planning permission they were looking around and they saw biodiversity offsetting, I mean really I think it might have been at this sort of level. Looking around, saw this new thing called offsetting, because somebody in a workshop somewhere said 'let's use offsetting, this will be the new thing that we'll be able to label our development as being critical and kosher and actually there was going to be no problem with the loss of biodiversity because look at all these clever things, and look at this light up here and...' Do you know what I mean? But I am only speculating. Offsetting is clearly on the agenda of the new government and they have been pushing for it. It's the new win-win idea. On the other hand, the Lodge Hill development has been on the cards for a very long time. And it would be interesting to look back through the paperwork which of course is almost impossible because the Medway council website is so hard, it's hell. But it would be very interesting to see when offsetting was first mentioned.

Interviewer: I think it was in 2012.

Interviewee: I also think it came in in 2012, and that was at the time when the planning inspector was looking at the core strategy. So probably offsetting came in as a way of trying to persuade the planning inspector. Well of course the developers commissioned the X [mentions a conservation brokerage firm] to do a sort of hypothetical offsetting exercise for them, and I think they were hoping that this would influence the planning inspector or Medway Council. And it was very funny because initially they were saying "Oh no, this is all fine" and then afterwards, after it had all gone completely pear shaped they were backing away furiously saying "Oh no, it has nothing to do with us". I thought well no, you are being completely hypocritical because you are on record as saying that this would be a great way of helping the development and No Net Loss and so on. And then when it all went wrong they couldn't run away fast enough.

Interviewer: Now there are many people from the government that don't accept that Lodge Hill is a case of offsetting.

Interviewee: That's the new message. Whatever goes wrong is not offsetting, it's compensation. That's what they are calling it, Nightingale Compensation Land, it's not Nightingale Offsetting.

Interviewer: There is also another case that is quite complicated, the Thaxted case in Essex. I was trying to find information about this but it's difficult because almost everybody who was involved is not there anymore. In the end, I found someone in Essex Council but didn't seem to know a lot or didn't want to share a lot of information.

Interviewee: I think that's a huge issue around offsetting: transparency and democratic accountability. As far as I can see it's zero.

Most of this stuff goes on behind closed doors. Deals done between, Lodge Hill is a classic example of that, the planning officer who puts the case to the committee, basically just repeating what he was being told by the developers on offsetting. No attempt to critique it, and the critique that they had received from some conservation organizations was very downplayed, or even from X [the interviewee mentions a state organization here]. I mean this is quite an interesting one with Lodge Hill because when X was notified, which is about a year ago, they put out a press release saying 'well the SSSI notification doesn't mean the application and the development can't go ahead'.

Interviewer: Because they said that they don't interfere in the planning process.

Interviewee: Exactly, they don't interfere in the planning process and offsetting is kind of on the agenda. Well something happened, I don't know what. Because they have changed their position quite substantially since then and the latest letters, particularly in relation to Nightingale Compensation Land, are much stronger. Probably they have realised that it is such a cause célèbre. On the other hand, maybe this relates to a loss of enthusiasm about offsetting from the government. I was quite surprised how quickly offsetting became adopted when this government came in, but that was few years ago and we are still looking at pilots. There's been a lot of talk about it and it hasn't really taken off. I mean bearing in mind how much development is going on at the moment, you would expect offsetting happening all over the place. On the other hand, given the major debate around offsetting, maybe they changed their tactics and they are implementing this at the local level.

Interviewer: I think this is what they are doing because they do not want a national debate on offsetting, especially before the elections.

Interviewee: Yes, that makes sense. So another kind of aspect to it is the relationship between offsetting and the greenfield/brownfield debate, because there seems to be quite a strong message coming also from politicians that we need to refocus our efforts on developing brownfield sites and this is all mixed up with confusion about what the greenbelt is as well. Brownfield has been a target for development and the notion that brownfield kind of have a biodiversity value is being discounted. I'm getting into the technical details of the metrics now, but are the metrics deliberately downplaying the value of brownfield land for biodiversity in order to make it easier for brownfield developments to flow through the planning process? And if yes, because if that is the case then it kind of undermines the argument that the offsetting process is objective that is of course what seems to be the central tenet really isn't it? It's 'well, we've created an objective way of assessing the value of this site'. But if it's being quite picky about which values, like you were saying earlier, you know, why just focus on biodiversity? There are all these other values, like community use, landscape, there's been talk around ecosystem services offsetting, but that hasn't come to anything yet as far as I can see.

Interviewer: What do you think about conservation/habitat banking?

Interviewee: Well, I think the initial idea was to combine banking with offsetting. And this was the idea of the Environment Bank when they established their registry[3] but so far it hasn't been very successful. What troubles me is how could you put a line around something and say "well that's now a conservation bank", how is that different from what it was before? I've been in Queensland and I went to the port where all the coal goes out of Queensland and it was absolutely extraordinary, the scale of operations there is mind boggling. The size of the coal mines. And the argument

[3]See http://www.environmentbank.com/the-registry.php.

is "it was going to be destroyed anyway so these conservation credits are an improvement". No, you're just giving yourself an excuse.

Interviewer: Yes, this is exactly the argument. I was reading last week on the website of one of the biggest conservation NGOs globally that mining areas 'can be an exciting opportunity for conservation'. This is a strange wording.

Interviewee: Essentially it's buying into the neoliberal dream. And this is what we have to challenge. Because, and this is also the case for ecosystem services, if we're talking about trying to allocate monetary value to natural things, natural features, whether it be biodiversity, water, air, soil, whatever, then we are never, ever going to succeed in doing that because monetary value is a social construct and it's created by capitalism. Money is never going to express the value of nature. It's such a fundamental principle of money that it does not reflect the real, social value of anything, isn't it? And yet people don't seem to be thinking about it in those terms, the natural capital type thing. They have kind of accepted the notion that the concept of money can be applied.

Interviewer: This is a fundamental issue; money cannot express the value of anything, not only of nature.

Interviewee: Yes, exactly. Think about friendship, health. It's like trying to buy democracy isn't it? The idea that you could buy democracy immediately automatically undermines the idea that you could buy democracy. It's a complete paradox. And in the same way the idea that you could somehow protect nature by placing a monetary value on it, it is a paradox isn't it?

Interviewer: Yes, it certainly is.

Interviewee: I think this is something that we have to challenge if we want to challenge offsetting. Nature or places cannot be represented by credits or bought and sold, and it they could, then it's meaningless anyway.

7.3.2 Interview with an Activist, Member of a Local Committee Opposing a Large-Scale Infrastructure Project

Interviewer: Could you please tell me the key reasons that you are against the development?

Interviewee: I was a member of a group of opposition for the recent Gateway planning around Coventry airport. The site contains certain species of flora and fauna, it is a quite extensive site and it was used as one of the county council's pilot schemes in the biodiversity offsetting, the Defra scheme. The offsetting reports identified a net loss in biodiversity on-site but not withstanding that. Biodiversity loss due to the development was one of the key arguments that we had against the proposal but not the only one. The application swaps green belt, it swallows our rural village, lots and lots of reasons of why we are against and remain against the planning application, but offsetting became one of those points, arguments.

Having done a little bit of research on it, I found that the council's chief ecologist [the council is not identified to ensure anonymity] had written a paper, a consultation document in 2012 possibly. And he said that he was looking for its future development, that enhancement in ecology and biodiversity and that sort of thing will be sought across development if we come into section 106 agreements but mostly through biodiversity offsetting. Now, I found that quite worrying, in that although it was only a pilot scheme and although the pilot scheme hadn't been run, the county ecologists said that in the future most of our biodiversity enhancement is going to come through offsetting. I thought this was a little bit presumptuous. It also effectively opened the gate to potential developers, saying that "ah, right, if that is the view of the county council, then that's the direction we will go with our application. If we do a biodiversity offsetting exercise we can tick the box and be good boys with the council and all those will support our planning application". Which again, seems to me, it was leaving the door open to development.

So when we got looking at the detail of the application we found exactly what they have done. The council was having discussions with the applicants and the applicants' consultants. All perfectly normal but what they were doing is that they have been running a sort of offsetting calculation and obviously there has been some "toing and froing" about it. There is a lot of water on one side, major bird populations, and they wanted to get rid of those birds because of their proximity to the airport. So one of the exercises was to go and dig a hole somewhere else further away from the airport to fill it with water and rids. And so the birds would go there and make them feel they are in a similar pond with the pond that exists there at the moment. And all this was run through by the offsetting calculation. But then, they found that actually the cost for digging a hole somewhere else and filling it with water and making the birds to move to is rather excessive. They didn't want to go to that route. But they wouldn't also change the design of the scheme. The highest level of biodiversity, the most important paths of the site, were the paths that were planning on building the distribution centre of big massive sheds. And this wasn't negotiable. They haven't considered of changing the design in order to offset the impacts. Instead the buildings would go in there and that was that. So in effect what they said is that "since it's not negotiable where our development will take place, what we are going to do now is to calculate the compensation for it".

And that was one of the main arguments that we had, that they didn't follow the mitigation hierarchy but they jumped straight in to compensation like that was the only solution. And that is a position we still maintain. The offsetting reports, when you go into the details of it, you then realize that they were actually coming up with a net loss. What the applicant was saying is that he was paying to plant some trees somewhere else if he could find a piece of land from somebody. So, if you could find somebody that has got some land he could give him then the money to plant some trees or something that then that would offset the loss on site.

The council apparently owns some land and they might be prepared to accept the cash from the applicant to do whatever they want to do on that. That is the one option. And they still claim that they will find more land somewhere else to put the compensation on, but that land hasn't been found. And yet throughout the planning application, the developers and their consultants, along with the county ecologists, were saying that the scheme would have a net benefit in terms of biodiversity. With no evidence to back up that claim. So, this went through two planning applications, and then it went into a public inquiry. We won't see the final report until the Secretary of State makes his decision that we anticipate around Christmas.

Interviewer: And you also object the new local plan, is that right?

Interviewee: So yes, this is the other thing that is going on in Warwick at the moment. There is a lot of housing development that is being proposed in the new local plan and a lot of people in the area are against. Generally speaking the overall feeling is that obviously the Conservatives, the government is pushing for growth.

Interviewer: Yes, this has been obvious also from the Defra Green paper on offsetting, and especially its introduction.

Interviewee: Yes, exactly. So the council has taken this on board and their data says the population in this district, based on births and deaths, and current levels of migration, needs something like 4–5 thousands more houses. And, just in case, Warwick wanted to do 12–13 thousands, to be sure we are covered. So, there is a lot of upset about that. Going back to the Gateway situation, to give you an idea of how pushy the council was for growth. The Gateway site became easily a greenbelt. At that point, the council decided to approve the Gateway, or minded to approve referred to the secretary of state. One of the conditions that they were proposing was that the site must remain a greenbelt. It would only come out of the greenbelt if it didn't get fully developed in accordance to the approved plan. The reason for that was that, at that time, they wanted to protect the site against any ad hoc development. They wanted to ensure that if this particular scheme didn't go ahead, because it wasn't financially viable or whatever, then nobody could just build bits here and there ad hoc, without any sort of real control. Any new plan or application will still have to satisfy these very special circumstances. If it is removed from the greenbelt, then the protection of these very special circumstances is lost and in its place you have the presumption in favour of sustainable economic development. So you go from a state where it is preserved or protected to an open road. In the final iteration of the submission draft document that was under consultation last month, we saw that they are now proposing to take the site out of the greenbelt. Although there has been no consultation with the local community or anybody else on greenbelt release. Without asking anybody they changed the local plan saying that they are going to get it out of the greenbelt. So, needless to say, we objected to that as well. And having just completed one public inquiry we are gearing up for what will be called an examination in public. If the council minded to proceed with the newly proposed local plan and submit it to the inspector then we will be seeking to be represented in this process, because, despite the fact that we have already had one public inquiry, due to the obvious amount of opposition to their plans to take the site out of the greenbelt we will go to another inquiry.

Interviewer: Who is involved in the local committee against the Gateway?

Interviewee: Principally it is three parish councils, all three villages around Coventry area. Three parishes. Throughout the original planning application process the three parish councils that have been objected represented the population. They collaborated and they worked together but it was a lot of hard work. So, we managed to put everything together, and we communicated constantly through the process, and, when it actually came to the point that they were minded to approve it, we then wrote to the secretary of state and requested to call the application in. So, what the parish councils agreed to do was to set up what was told 'the community group' with representatives of the three parish councils but also non-parish council members. We also had the local counsellor. He was also Rule 6 party for the inquiry. And several community groups, and they all met on a weekly basis.

Interviewer: And which are the key objections?

Interviewee: So the ecological concerns are one key aspect. Within the community group we have analysed many other aspects, including the economics of the case. We had people from the University helping us and they said that this is madness. It doesn't work. It's crazy. You wouldn't do this. There are also many other concerns, about traffic for example.

Interviewer: Are there issues about access to the site?

Interviewee: There are major issues about access to the site. There is also a major feeling that this was a bit of a Trojan Horse to the expansion of the airport. If you look at the layout of the scheme you can see that. The airport itself was ahead two previous inquiries. It is tempted to effectively compete with Birmingham airport. They want to put some new passenger terminals to increase the number of flights. And twice it has been turned down. Knowing well that this was going to be resisted strongly, they put this application together, which is an office park on one side of the airport with a hotel and various other things, and a large distribution centre on the other side of the airport which has really bad access, the only access is by a country lane by the end of the airport. If you are importing and exporting then you might want some storage next to the airport. Otherwise, why would you put that distribution centre on the wrong side of the airport? You got to get pass the wrong way or over the wrong way to get there.

Interviewer: And which is the current use of the area?

Interviewee: It is farmland and it is also a sewage treatment plant. And that was another of the things they say, that one of the causes for this application was to redevelop a brownfield site because it was previously been used as sewage works. There was a major sewage work down in Coventry, not that far away down the road, and an area that has been used for settlement for a heavily industrialized area. For years there have been heavy metals concentrated there, acid, and all sorts of chemicals what were somewhere out in Coventry's industrial park. And they were too toxic to be used where normally sludge dried, and is used for sub fertilizing to those fields. But material was coming out from Coventry so contaminated that it wasn't suitable for that sort of use, and so it was transported to another area and put into settlement beds there. So the sludge settled there. Those settlement beds, nature has established itself a hence of the major water bodies. But from the council's point of view and from the applicant's point of view it is a brownfield site and they are the heroes, they come along and de-contaminate the field, and cover it with concrete and make it safe [laughs].

Interviewer: So which do you think that is the role of offsetting in this particular case?

Interviewee: What is happening, and I am very much against it, is that offsetting is being used as an excuse to get away with everything. Rather than saying is this development absolutely essential, does it really have to be there in that field, or can it be over here or there, instead of that site, and only use offsetting as an absolute last resolve. What has happened, I mean our experience at the Gateways was: we do what we want and then we use offsetting as a tool to compensate afterwards.

Interviewer: So you think that in theory the metrics could work? They could recreate the place that is being lost?

Interviewee: Well, it will never be the same. And particularly something like ancient woodland. There are people that are better qualified than I to comment on whether digging a hole over there and planting is actually going to make the birds to want to go there. What I can say is that this idea of offsetting can be dangerous. A little story I say to the people about the Gateway. If you take offsetting and you take the wheat field. A field full of wheat has very little ecological value. So, if you concrete over half of it and build a flour mill and a bakery, then in order to compensate for that you take the other half of the wheat field, you dig a hole in it and fill it with water, plant some trees around it put a wall of plant meadow in there and you look at the metrics and you have provided more compensation than you needed for the loss of the wheat-field. But then what are you going to do with the flour mill and your bakery when you have no wheat? So, that's my take on offsetting. You can use the metrics to complete madness. And of course, I can very much see the danger of putting a price to nature. Small example from the Gateway. They did the metrics and then they came out with a net loss, so the applicant says: "tell me what the value of net loss is and I'll just sign you a check".

Interviewer: This is how it works, that's true. Offset credits have a price and developers can buy and sell these credits.

Interviewee: Let me comment on this by referring to our case. The greenbelt is where it is for several reasons. The greenbelt is there to stop cities expanding constantly, and cities become like doughnuts. They grow outwards and outwards and the middle dies. And you can see that all over the world. You have nice and vibrant city centres where all the cafes and bars are, and that surround a circle of unused industrial land. And it is, you know, falling apart before you have to go to nice suburbs where all the residential areas are but it costs a lot to build more houses there. So instead of building them in the middle it is easier to keep growing outwards and building on the fields. I mean that is why we have the greenbelt, to stop that happening. And the local communities really want their villages surrounded by a greenbelt. And if that greenbelt wasn't there, Coventry would just grow and grow and grow. And just swallow up all these villages and they would just become its further suburbs of the city. That has to be prevented in our view. Otherwise this village here in Warwickshire, won't be what it is. It will just be concrete and the landscape will change forever. So offsetting is part of a bigger argument. It is a key element in nature's destruction for urban development and economic growth without any limits. We are not making land, so every time we damage a piece, then that's it. It's gone. Anyway, to go back to your question, I cannot see how all this damage and all this loss can be represented by money.

Interviewer: What you said before is important, whether this development is essential or not.

Interviewee: Exactly! And that is the more fundamental argument in my view. As I said, if you have a debate and your conclusion is that this development is absolutely essential, then you can start discussing mitigation and compensation measures somewhere nearby. Anything else does not make sense to me. Because then you can remove a part of the greenbelt in Birmingham or Coventry for example and then declare the Isle of Skye as greenbelt to compensate [laughs]. It doesn't help much. That is the danger. You can destroy an ecosystem here and in order to compensate for it you recreate one in Cambridge.

Interviewer: So what the local committee thinks about the offsetting proposal?

Interviewee: Well we strongly disagree. What we said is that the greenbelt is greenbelt. And what the applicant is saying is that they are going to build a county park round it where local people can have access to, so that will be our gain, we'll get a County Park out of it. But we cannot have birds in the county park because it is going to be around the airport, we are going to put nets over the water bodies to stop birds going there, they are going to electrocute the fish in a regular base so there is no food for the birds but you can walk around and look at the flowers. But we live in the greenbelt, I mean we are surrounded by fields full of crops and hedgerows and trees. We don't have all access to that field in front of our houses. The farmer wouldn't like it to tremble all over his wheat or whatever. That doesn't mean you can't appreciate it. You can say to the developer: thank you for your offer for the county park, but we don't want that county park. We want the countryside that surrounds us as it is. There are public footpaths through the countryside, numerous ways by which people can gain access to nature.

Interviewer: I understand that all those who object this project think that it will change completely the character of the area.

Interviewee: Absolutely. The site is very large and you can break it up into different parcels. There is the area where there is the old sewage work and farm, but other parts are still farmed and there are sheep all over there. And there is a test track that they used and they were testing their tanks. That hasn't been touched for years, and years. Decades. And that apparently has got some phenomenal flora and fauna. Because nobody goes there, because it is redundant, apparently it is still a sort of heaven for wildlife. And there is another part of the side where there are other water bodies that were supposed to be used to expand the sewage works, but they have never been used. Then you got the airport which uses a big flat open expands of grass and strip of concrete and beyond that you have got another piece of land that is rolling farmland. Part of it is crops; part of it is cattle grazing. And there is a site that is important due to a Roman fort that was re-constructed. It was obviously built in that location because it overlooks a valley. On that section by the fort that rolls down to the A45 is where they want to build the hotels and the office buildings. That will give a lot due to the historic setting. Then on the other side of the airport, where you have got the tank truck and sewage works and all the real biodiversity hot spots, they want to put the warehouses or aircraft hangars. So, what they are going to do is to build big high buildings around the development and screen the development from all over the countryside. They are going to build 20 meter high buildings and then they are going to put some 6 meter high so you cannot see the 20 meter high buildings and then they are going to put a footpath round so you can have access to this county park. Gateway is a big issue for the area but also the local plan for the area as well. Within the local plan they are now proposing to get the Gateway area out of the greenbelt, removing the greenbelt protection and put in place a presumption in favour of sustainable development. All these along with major new housing developments. I am not sure how can anyone seriously claim that offsetting is the solution to all these or can compensate for this profound loss of nature.

Interviewer: Are you in contact with any other local groups? Because there are similar cases in other places around the country where offsetting has been proposed as a solution.

Interviewee: No, we haven't but we would like to do so. Unfortunately the inquiry demanded a phenomenal amount of work and talk since 2012.

Interviewer: Biodiversity offsetting was included in the proposal from the beginning?

Interviewee: All the way through. They suggested getting into this when we were doing the pre-planning application consultations. My view is that the council let them argue about biodiversity offsetting.

Interviewer: Do you have any documents that you could give me?

Interviewee: Sure, there are many documents. We have got an economics proof of evidence, written by a lecturer of the University. We have got the social case, a proof on ecology and landscape. We have a document on water contamination and the relation to the historic use of the site. This is something I forgot to mention, the level of contamination of the sewage banks. They can't move them because of the level of contamination. The plan includes the usage of this contaminated material in order to build/re-create the site. And of course there are serious fears about the exposure of this contaminated material to the weather and the rain. This is absolutely crazy.

Interviewer: And the council has accepted this?

Interviewee: All in favor. They actually hired somebody whose job is to get the Gateway through. There is supposed to be an independent assessment as part of the planning application. In 2012, we were saying that there are millions of square feet and offices and warehouses that were empty. So, why on earth do you want to build more? There is no demand for them and there are several other sites that already exist that could be redeveloped and improved first. So, that was within the debts of recession. In 2012, the applicant was saying: oh yes, but this site is in the wrong place, the wrong size, shape, they don't have the right planning permission, and ultimately all these sites are rubbish and no good, and so they have to have the new Gateway. By the time we get to 2014 and we go to the public inquiry the applicant changes his tune completely. Now, all these sites are fantastic. They are all going to be built up by the end of next year and we are going to run out of land so we got to have the new Gateway now. So, economic and business arguments are at least problematic.

Interviewer: This reminds me of the Olympic Games in Greece. The key idea seemed to be that because it's the Olympics, mega-projects are imperative and we would approve them without any in-depth consideration of their impacts.

Interviewee: Yes. That is the Olympics. They will put us on the map. It is a very similar situation. Suddenly, common sense is coming out of the window. The Gateway between Warwick and Coventry is like: Peterborough has got a Gateway; we should also have a Gateway, because everyone should have one of these massive developments. Why should we miss out?

Interviewer: A last question about offsetting. In offsetting the compensation refers only to selected ecological aspects, what about other values? Like cultural aspects or people's relationship to place?

Interviewer: Somebody else mentioned the same thing. He said that we know how much it costs to replace hedgerow but what about, how you compensate the person that lives next to that hedgerow one day, and lives next to a massive development the next day and has to go many miles down the road to get and visit a hedgerow. Now, what's the value of the social impact? So the applicants proof on ecology [the interviewee is showing me a map of the area], so the development proposal is to build massive sheds here. Bring a road round here across the end of the wrong way and back through here, and then cross this field, a new fly over junction the A45, then links the new road back to this one here. Coventry city centre and the railway station are here; there are amazing road works at the moment. That would be A46 towards the A45 and it becomes again A46. At the moment, Coventry airport has a terminal over here. It is not part of the planning application, because the planning application is for office development, and technology park and a logistics distribution centre. When you look at it you start to think, hang on a minute. What would suggest that this is not going to be a bunch of aircraft hangers, at the end up here, and that over where that they applied to build all these offices and a hotel, why will they build a hotel there? We now have a new access road with a direct link into Coventry city centre and railway station. I wonder when they are going to come back with: "oh dear, we have not thought of it yet, let's just stick the passengers terminal on that side of the airport next to the hotel and the new access road". But as I said, that's where all the biodiversity is. In that water bodies and all around here. It is a little heaven for nature. That's where they want to build the big sheds. Then, what they are going to do? They are going to give us a county park, where we can all take the dog for a walk [laughs]. This is the logic of offsetting. We really should be grateful, shouldn't we?

And this is the biodiversity offsetting report [the interviewee shows me the report]. This is the calculation of the metrics, this is the end result. Deficit. And yet everyone stood in front of the planning committee and said: "but it is going to have a positive impact on biodiversity". Well according to that it is not. So that's the applicants' side of it. Many people ignore the fact that the community here is really against this, there are 800 and plus letters of objection. It is quite obvious that the local community doesn't want this, and you are an elected council completely ignoring this. This is actually registered active parkland [the interviewee is showing me a picture]. This is Stoney Deer park which is part of the Stoney estate. Stoney Abbey and the estate are great to stare listed landscapes. And, at the moment, this here sits on the edge of the plateau. The airport is actually on the plateau so this right here prevents it screens the view. The other side of that is Coventry City centre. So, your view is screened by that ridge and the woods.

[Showing another picture] That is not going to be affected by the development; they are going to take that bridge out. This is actually a bridge to nowhere; it is just a bridge over a river in the middle of the countryside. It only has access from one side of the river to the other side of the river for lords and ladies [laughs]. We are not claiming that this is, you know, pristine, it is a manmade landscape but nonetheless it is a valuable landscape. And no biodiversity offsetting can ever replicate this or make better this loss for us.

8

Afterword

E. Apostolopoulou, *Nature Swapped and Nature Lost*,
https://doi.org/10.1007/978-3-030-46788-3_8

How beauteous mankind is! O brave new world,
  That has such people in't!

—William Shakespeare (1611)[1]

One never notices what has been done; one can only see what remains to be done.

—Marie Curie, Letter to her brother (March 18, 1894)

[1]The term "Brave new world" is from Shakespeare's The Tempest. Aldous Huxley took the term and offered a dystopian version of the brave new world in opposition to Shakespeare's optimism expressed in the quote above.

8.1 Toward a Radical Retheorization of Offsetting: Uneven Geographies, Inequality, and Social-Environmental Justice

Urbanization is today's defining global development trend (United Nations 2018). Even though it has been traditionally linked with the concentration of people in cities, urbanization is now extending beyond city borders, perforating the urban-rural divide (EEA 2017). From highways crossing indigenous land and erstwhile hinterlands, like the Amazon, to pipelines crossing the Carpathian Mountains, for some the wildest part of Europe, a planetary form of urbanization (Brenner and Schmid 2015; Merrifield 2013) is being projected into novel geographies, connecting erstwhile rural places with an expanding urban fabric via material, energetic, informational, and infrastructural links (Brenner and Schmid 2011, 2015). Urban expansion is a major driver of global social-environmental change and directly affects human development by influencing access to land and resources and transforming livelihoods. Urban land cover is expected to grow by 200% until 2030 in comparison to 2000, with the highest rates of land conversion projected to occur in so-called biodiversity hotspots (Seto et al. 2012).

Urban expansion is inextricably linked with large investment in infrastructure and has been followed by a widespread construction rush[2] in the aftermath of the 2008 financial crash. This has been evident not

[2]Gerritsen (2009) in a report published in March 2009 about the expected global infrastructure boom writes: "The global infrastructure boom will intensively unfold between 2009–2015 and will transform how the world looks, gets educated, moves goods and services, creates wealth, treats the sick, cares for the poor, powers its homes and businesses, and wages war. The amounts of infrastructure money about to slosh into the world economy defy imagination: the Obama administration will spend $150billion of its $787 billion stimulus plan on infrastructure and is expected to add to that; China has pledged $585 billion and stands ready to do more; India is expected to spend $500 billion on infrastructure over from now till 2015; the EU $252 billion; Japan $129 billion; Canada $12 billion; Australia $4.7 billion, Singapore $13.8 billion; Germany $42 billion; and so on. CIBC World Markets estimates total infrastructure spending over the next 20 years at $35 trillion. Some $3 trillion of fiscal adrenalin will be injected into the global economy in the next 24 months alone. The only other time in human history when this much money has moved this quickly into the global economy was during WWII and that event of course reset the world order for the next 60 years, with America at the helm".

only in European countries, including the UK, but also internationally, with indicative examples Australia, China[3] and the US. It is in this context that biodiversity offsets and compensation mechanisms have been gaining increasing popularity as evidenced in their institutionalization in several countries across the globe. This is largely based on their promise to facilitate economic development while also keeping the "stock" of natural capital constant. Importantly, as we have seen in previous chapters, offsetting and compensation mechanisms are both a product and a trigger of the broader shift toward the economic valuation of nature (OECD 2018) and a key element of the neoliberalization of nature. Conceptually, they resemble the polluter pays principle, reflecting similar ideas to carbon offsetting but with a focus on biodiversity (Moilanen and Kotiaho 2018). Even though the varieties of offsetting and compensation mechanisms are numerous, from restoring and recreating habitats, establishing protected areas and regulating human activities, to buying the necessary amount of credits "off-the-shelf" from a habitat bank, there are four important common characteristics in all these mechanisms that significantly change the way environmental destruction caused by economic development is governed. Firstly, by measuring losses and gains in terms of priced credits, offsetting and compensation mechanisms reconstruct conservation (and environmental protection more generally) around the measurement of a putatively quantified economic value of nature assuming that elements of non-human nature can be isolated from their spatial, evolutionary, historical, social, and cultural context (see also Moreno-Mateos et al. 2015). Secondly, by introducing the idea and practice of trading of ecological losses and gains across space and time, they disconnect compensation from the places where the impacts of development occur creating socially and geographically uneven outcomes. Thirdly, by claiming that nature recreated, now or in the future, can be of higher quality than nature lost, they imply that environmental destruction can be the driver of

[3]A reference should be made here to China's Belt and Road Initiative (BRI), the single largest infrastructure project since the Marshall Plan that is expected to cost at least US$1 trillion. Called the "21st century Silk Road", the initiative involves the development of trade routes across the globe, crossing sensitive ecosystems while enabling a massive number of large-scale infrastructure investments profoundly changing places and livelihoods.

environmental improvement. The latter has been vividly expressed in emerging strategic alliances between environmental NGOs and donors and industries that figure among the biggest polluters globally. Fourthly, by being placed at the last stage of the mitigation hierarchy they enable the approval of projects that would otherwise have been rejected facilitating economic growth. Offsetting mechanisms initiate not only a radical reconceptualization of nature as a virtual ledger (Igoe et al. 2010) but also an active *remaking* of socionatures according to this image manifesting the deepening of the real and formal subsumption of nature. As Smith argued (2006, p. 23) if one takes a wider geographical perspective on such policies it is tempting to paraphrase Engels' assessment of the housing question: "the bourgeoisie has no solution to the environmental problem, they simply move it around."

The above perceive particular importance if we consider that offsetting is now being proposed as the most appropriate solution for providing compensation in almost every large-scale project across the globe in both urban and rural areas. As mentioned in Chapter 2, as of 2018 there were 12,983 offset projects worldwide extending over 153,679 km^2 (Bull and Strange 2018) and channeling more than \$4.8 billion yearly (Bennett et al. 2017a). In 93% of cases, the implementation of offsetting has been driven by compliance with regulatory obligations; with the energy, transportation, and mining/minerals sectors dominating demand (Bennett et al. 2017a; Madsen et al. 2011; IUCN et al. 2018) whereas most offsets have been located on state-owned and managed land.[4] These data underline the tight linkage between offsetting and urbanization as discussed in Chapter 4 and the pro-development character of offsetting and compensation mechanisms. Indeed, the popularization of offsetting in the aftermath of the 2008 financial crash has been related to the entrenchment of neoliberal policies, the further entrepreneurialization and urbanization of rural areas and the imperative to deregulate and simplify the environmental and planning legislation not only in post-crisis England (Apostolopoulou 2019; Apostolopoulou and Adams 2015, 2017a; Apostolopoulou et al. 2014) but also in many places globally (Bennett et al. 2017a, b). It has, therefore, been a manifestation of the

[4]This refers to the global scale and variations are observed in different countries.

emergence of an increasingly *symbiotic* relationship between neoliberal of conservation, urban development and economic growth. The reframing of nature as a movable amenity is intertwined with the new territorialities that the changes in urban, transport, and more broadly economic geographies bring about and aims to legitimize and render common sense the idea that nature, either a protected area, a forest, an endangered species or an urban green space, can be simply (re)located where the interests of particular sections of capital dictate. This implies that thanks to recent scientific advancements, primarily in natural capital accounting and restoration ecology, non-human nature should not be considered a barrier to infrastructure expansion, large-scale housing developments or extractive activities since its destruction can be fully compensated for (as the No Net Loss goal implies).

Drawing on the above observations, I theorized in this book *offsetting* as a mechanism that transforms non-human nature to a set of replaceable and exchangeable parts and conservation to a system of exchange of ecological losses and gains across space and time facilitating the reordering of the landscape according to economic development and urban patterns, enacting and often legitimizing a profound land use change driven by the interests of specific sections of capital. I also theorize offset *sites* as places where land is "saved" through various forms of conservation activity to enable economic development, in most cases urban development, in a development site where nature is allowed to be destroyed. Offset sites are the products of the complex dialectics of green and un-green grabbing and their establishment has often led to a double grabbing of land (in both the development and the offset site) with dramatic consequences for local livelihoods, social reproduction and communities' quality of life. Conservation through offsets becomes, therefore, implicated in wider patterns of spatial planning, urban expansion and capital accumulation, where the prices attached to commensurable ecological values that have taken the form of offset credits, are used to deliver economically efficient outcomes that serve specific capitalist interests. The, so far largely ignored in the relevant literature, theory of rent is of crucial important here because it provides a sharp analytical tool for understanding new forms of ecological compensation and new social-ecological conflicts around them as struggles over value appropriation demonstrating that offsetting may not create value but it certainly generates opportunities for rent (Apostolopoulou et al. 2018; Greco and Apostolopoulou 2019). An important observation here, that reiterates the importance of analyzing offsetting through the lenses of the theory of rent, is the relocation of ecological values from urban to rural areas that offsetting has quite often generated. Evidence from England showed that offsetting has mostly favored the location of offset sites close to already existing protected areas, areas of high nature value, or any place away from urbanized areas. This movement or relocation of resources across the landscape is not random but a natural byproduct of offsetting. As Womble and Doyle (2012) point out, research in various locations in the US, including Florida, Illinois, Oregon, and North Carolina, has shown that off-site compensation through mitigation banks caused the migration of wetlands and streams from higher population density, urban or suburban areas to low population density, rural areas. As the authors highlight this systematic spatial redistribution is quite intuitive since it is largely driven by land prices: urban land and associated wetlands are high-value locations for development whereas mitigation bankers can purchase rural compensation sites at lower land prices. Acquisition of lower-value, rural land reduces the production costs per compensation credit for third-party mitigation sponsors giving to banks in rural areas a comparative advantage over credit providers in expensive, high-density areas (Womble and Doyle 2012). As we have seen in previous chapters, the fact that the location of offset sites has been driven by land prices in England, generated valid concerns that offsetting would intensify uneven geographical development and social-environmental inequality, for example by potentially offsetting in the North environmental destruction caused in the South where land is more expensive.

The retheorization of offsetting proposed here can contribute at shifting the analytical focus of the critical literature toward unraveling the role of offsets and other market-based compensation mechanisms in driving a land use change that serves specific class interests, reinforcing a socially and geographically uneven production of nature, and transforming places and livelihoods. By drawing on the labour theory of value and the theory of rent it can also show that offsetting is not a mechanism that brings about the commodification of nature but primarily plays a central role in processes of rent extraction and, therefore, it has both positive and negative roles to play in relation to capital accumulation. Of course this does not negate the fact that offsetting is a form of for-profit conservation that aims to facilitate economic growth and urban development benefiting particular sections of capital. Importantly, a radical retheorization of offsetting as a process that *remakes socionatures* in accordance to the patterns of an increasingly social-environmentally disruptive urban development along with a broader shift in critical research and praxis toward unraveling people's consequential entanglements with ecosystems and places that experience loss and transformation (Apostolopoulou 2019), could give to urban political ecology and the political ecology of conservation the unity they so far miss by extending the study of urbanization beyond the confines of the city and the study of neoliberal conservation beyond the confines of protected natures and the countryside. This can also radically transform critical conservation studies and bring them in a more direct dialogue with critical urban studies. The need for the approach described above is further reinforced by the fact that offsetting and compensation mechanisms target places and ecosystems within and beyond the traditional city, and thus, transcend the traditional focus of the political ecology of conservation and neoliberal conservation literatures to rural areas. Moreover, and relatedly, the creation of an offset in a rural area is often linked with the provision of compensation for the loss of urban nature involving, inter alia, cases where urban communities face displacement, exploitation or/and dispossession from green spaces due to privatizations, housing developments, infrastructure projects, land extraction, gentrification, regeneration and urban redevelopment. Postcolonial comparative urbanism (Lees 2012) and countertopography (Katz 2004, 2011) have a key role to play in tracing those links as both methodological and conceptual tools that can invoke connections among places as different as Iñapari city in the Peruvian Amazon, where offsetting is being proposed to compensate for the loss of nature due to a new highway that is opposed by Indigenous groups, and Harmondsworth Moor on the outskirts of London, where offsetting is being proposed to compensate for the loss of nature due to airport expansion. Such connections are necessary to produce the spatialized abstractions that are required for building a framework capable of unraveling the uneven and contested geographies of offsetting and grappling with the various ways offsetting and urbanization remake socionatures both in highly urbanized regions and erstwhile rural areas of the Global South and North.

A final note is important here. As mentioned above offsets, as most of the instruments that bring about the economic valuation of nature, are not vehicles for the production of value but for the extraction of rent (Apostolopoulou et al. 2018; Greco 2014, 2015; Greco and Apostolopoulou 2019) which in the broader context of the financialization of nature (Purcell et al. 2019) has extended far beyond land, involving abstract nature (Apostolopoulou and Adams 2019; Apostolopoulou et al. 2018; Felli 2014). Even though this means that no abstract value, as understood in capitalism, is embodied in offsets, it does not change the fact that the process of abstraction from the irreducible and incommensurable heterogeneity of use values toward the equivalence of exchange value, as expressed in representing ecological losses in terms of biodiversity units and/or credits or carbon emissions in terms of carbon credits, is a concrete, real process of abstraction which violently erases all qualitative differences toward the ghastly homogeneity of equivalence (Greco 2015; Sayer 1987). This violent abstraction from use values is the foundational violence on which capitalism rests (Greco and Apostolopoulou 2019) and ignoring it lies at the core of many recent—hopeless—attempts (see e.g., Costanza et al. 2014; De Groot et al. 2012) to address the increasing environmental contradictions of capitalism and reduce its destructive ecological effects by giving a price to nature (Greco and Apostolopoulou 2019). It is this violent abstraction that lies at the core of offsetting's unevenness inevitably linking such policies with the deepening of environmental injustice and social inequality.

8.2 The Contradictions of the Capitalist Production of Nature and the Centrality of Class

If you're not angry, what kind of person are you?

—Ken Loach (2015)[5]

Even though the increasing adoption of offsetting policies, and more broadly the shift toward neoliberal forms of conservation globally, constitute a reactionary response to the challenge to govern nature–society relationships, they should not be seen as a disruption to the history of nature conservation and environmental protection under capitalism but rather as its continuation. Offsetting and other initiatives that aim to put a price to environmental destruction and environmental protection portray non-human nature as an external, ahistorical "natural" entity, consisting of a stock of resources (or natural capital). This not only deepens nature–society separation but also significantly dismantles the hopes that some environmentalists and ecologists had after the mainstreaming of the need to incorporate people into environmental governance (Apostolopoulou et al. 2012a, b) and integrate environmental concerns into policy-making (Paloniemi et al. 2012). Ironically enough, an "innovative" market-based tool like offsetting is based on a resurgence of a reverse protectionism: it enables the destruction of nature in a development site based on economic criteria, and (at least in theory) it drives the designation of an offset site based on ecological criteria. This implies that nature can be destroyed for the sake of economic development, and thus for "humans'" sake, but it has to be saved for "nature's" sake as if humans magically do not matter anymore. Using the argument of environmental protection to exclude people from nature while simultaneously supporting the interests of specific industries lies at the core of conservation since the emergence of the first national parks. Indeed, despite the idealized perception of "virgin" nature, which

[5]See https://www.theguardian.com/film/2016/oct/15/ken-laoch-film-i-daniel-blake-kes-cathy-come-home-interview-simon-hattenstone.

was particularly attractive to the ultra-nationalist and fascist ideologies of the twentieth century (Blamires and Jackson 2006), the dominant example of protected areas emerged in a certain historical context, that of the consolidation of capitalism and border expansion in the late nineteenth century in the US and was based on a conservative and reactionary ideology of a negative relationship between society and non-human nature (Nash 1989). The establishment of protected areas included the violent relocation of local populations from their land, with enormous social, economic, and cultural consequences, to serve specific economic, investment, and tourist interests. National parks have thus largely served from the very beginning of their creation the interests of the ruling classes (Ghimire and Pimbert 1997) and have been strongly influenced by both imperialism and colonialism. Yellowstone is an emblematic example of a park whose establishment was accompanied by the deprivation of the rights of Indigenous people and their violent relocation[6] whereas its legislation was largely determined by the interests of the Northern Pacific Railroad Company, which sought to build a commercial road that would cross the West and retain the monopoly of its use (Abakerli 2001).

To understand the contradictory history of conservation a class analysis is necessary and this is why environmentalism without class has historically failed to understand the environmental contradictions of capitalism. The hegemonic discourse of the bourgeoisie identifies its class interests as the interests of humans in general (or of an abstract humanity) while, simultaneously, likes to portray as technical what in effect constitutes an instance of class. In the case of offsetting, this is clearly shown in the way that ecological arguments are being used to legitimate the disentanglement of compensation from its social aspects but are not being considered capable of blocking environmentally destructive development. A class analysis can explain this paradox by showing that the main reason that the social aspects of ecological degradation and loss are framed as irrelevant is because they are antithetical to the interests of particular industries. Offsetting attempts to mask

[6]Indigenous resistance struggles led to the death of hundreds of Indians in Yellowstone (Kemf 1993; Keller and Turek 1998).

the unequal power relations that drive the production of new social-ecological configurations by portraying the remaking of socionatures as a matter of expert knowledge whose goal is to offer compensation to an unspecified subject and its supporters in most cases remain silent on who gains and who loses from the trading of ecological losses and gains across space and time (Apostolopoulou 2009). Moreover, offsetting, and market-based environmentalism more broadly, is both a product of the win-win discourse of sustainable development (Adams 2009) which consists on an orchestrated attempt to dissolve the environmental contradictions of capitalism and a manifestation of the increasing dominance of corporate interests in the agenda of mainstream environmentalism. As with all policies, even though there is a key driver behind the emergence of these mechanisms, namely, to find a way to reconcile economic growth, urban development and environmental protection in favor of the former, there is also a conjunction of factors that further explains their specific characteristics and the hegemonic discourses that surround them. I summarize these in the following five key aspects with a particular focus on the aftermath of the 2008 financial crash.

Firstly, it is important to remember that historically, the introduction and popularization of the concept of biodiversity to describe the variation of non-human nature has gradually created a standardized and abstracted language to encompass living diversity at all scales from genes to ecosystems, and through this leveling of difference it facilitated the management and control of nature (Wilson and Peter 1988; Takacs 1996). Biodiversity offsetting builds on this managerialism and extends it: by portraying habitats, species and ecosystems through priced credits, it attempts to fundamentally change the way non-human nature is represented, mapped, managed, and experienced. It implies that by establishing the "proper" metrics, different socionatures and different places can be rendered equivalent. The underlying assumption of nature as movable and interchangeable brings about a deeply problematic reframing of nature–society relationships, enabling the social and eco-spatial rearrangement of socionatures. Offsetting's reductionism results from capitalism's treatment of non-human nature as the provider of the material conditions that are necessary for capital accumulation (Burkett 1997), or as a provider of ecosystem services according to the current mainstream terminology, and reflects how capitalist commodity production subjugates use value to exchange value. It also reflects and reinforces the logic of market environmentalism on which conservation policies are increasingly based (Apostolopoulou and Adams 2015; Brockington and Duffy 2010; McAfee 2012) and according to which the solution to the ecological crisis is the proper accounting of natural capital and the internalization of environmental externalities into economic systems and decision-making processes (Bayon's et al. 2008 analysis offers an indicative example of this logic). Its underlying ideology is neoliberalism, a variant of economic thinking that extends the econometric principles of measurability and efficiency-maximizing behaviors to previously untouched domains as reflected by the colonization of social (Fine and Milonakis 2009) and increasingly natural sciences by economics (Apostolopoulou et al. 2018). The emphasis on the economic valuation of nature even though emerges from the historically specific relations created by capitalism as a value system, based on exploitative capitalist class relations, perform an ideological function, which is that of naturalizing the process of concrete abstraction (Sayer 1987) on which value in capitalism is based. Its ultimate goal is to change society's shared, common sense (Gramsci 1971) of what socionatural interactions are, how they work historically and what the alternative options to the capitalist production of nature could potentially be.

Secondly, in the midst of increasing public awareness for the impacts of climate change and biodiversity loss, including the recent climate strikes as well as communities struggles against the oil and gas industry across the globe, it is not easy for all capitalists to disregard environmental concerns, at least without any pretension. This acknowledgment may of course take several forms: from attempts to achieve some reformist reforms to efforts to greenwash socially environmentally destructive projects. Whatever the exact motivation, the need to address sustainability concerns and tick the box of corporate social responsibility is quite popular, at least in some sectors that seem willing to support pro-development conservation policies like offsetting in an effort to gain their social license to trash.

Thirdly, there is again, as in the 1970s, a resurgent discussion around the limits to growth mostly fueled by increasing concerns for the effects of climate change. This partly reflects the pragmatic concerns of sections of capital about the depletion of natural resources (as reflected in the key objectives of governmental policies or/and international agreements) but nonetheless it is a deeply contradictory process. These concerns, no matter how real, are in most cases translated in a reductionist understanding of the climate and biodiversity crises and in technocratic policies whose primary objective is to find ways to limit the despoliation of natural capital in the hands of particular sections of capital by expanding the use of market-based instruments. Offsetting's No Net Loss slogan and its emphasis on the exchange of priced credits reflect precisely this attempt and the inability in achieving this, including the documented weaknesses of offset metrics, is a manifestation of capitalism's attempt to resolve in idealistic terms real contradictions. The most important of these is the attempt to keep the availability of natural resources to support future capital accumulation (by compensating the loss of nature due to development), while, at the same time, facilitating development to support current capital accumulation. This contradiction lies at the core of offsetting and constitutes so far an insurmountable barrier to any meaningful resolution of social-ecological conflicts.

Fourthly, capitalism's growth agenda under planetary urbanization and within the context of a new gold rush requires stalking the land and deep sea or even space with "asteroid" mining (sic) being for some a new ecological frontier raising extraordinary questions like "who owns the moon"[7]! Developers and governments are currently initiating a giant land grab against any remaining green space, including protected areas and open green spaces in big cities. The latter along with the increasing inability to find new avenues for profit and capital accumulation has led particular sections of capital to admit that the key reason they now look at conservation banking is land acquisition and rent seeking. Indeed, big landowners and major industries increasingly see biodiversity and carbon credits as valuable assets. Finally, we cannot leave unnoticed the convergence of conservation and capitalism, most notably the increasing alignment of BINGOs (big NGOs) with corporations and multinational

[7]See https://www.theguardian.com/business/2016/feb/06/asteroid-mining-space-minerals-legal-issues.

donors and their collaboration with industries that figure among major polluters. This signals the historical defeat for mainstream environmentalism (and conservation) to form an independent agenda and shows that most BINGOs have no intention to be involved in oppositional politics but consciously choose to ally with capital forces.

The case of England that has been explored in detail in this book provides ample evidence of the above and of the role that offsetting was expected to play in a highly urbanized, advanced economy of the Global North. The way the policy has been implemented so far adds the following observations to what has been already mentioned above. Firstly, offsetting in England not only facilitated the granting of planning permission in the majority of the case studies I explored but has been also actively used to attract developers and unblock urban development (see Chapter 6). This is important because of it significantly weakness the argument that offsetting is only a pragmatic response to the inevitability of ecological loss due to infrastructure development and economic growth. It shows that offsetting can play a much more substantial role at least in areas where local authorities promise to provide a pro-development context by using offsetting and compensation mechanisms in favor of developers. Secondly, and relatedly, there were also cases where although there was no formal responsibility to implement offsetting, specific industries used it to prove the environmental friendly character of their interventions and to green-wash large-scale urban development projects. Thirdly, when the implementation of offsetting proved to be less simple and straightforward from what it was initially expected, developers either abandoned it or made major amendments to the guidelines offered by the Defra Green paper. Fourthly, offsetting has been extensively used as a vehicle to discipline conservation into pragmatic market logic by arguing that within the context of the austerity-driven shifts that have occurred in British social, economic, and political life since 2010 (Gray and Barford 2018), the only way to save nature or open space is through allying with developers. Conservation NGOs have been invited to be part of offsetting's new consensus since their involvement would give them access to more funding and more conservation land to manage. Fifthly, offsetting often succeeded to disorientate and depoliticize the public debate by shifting the discussion to the narrow issue of calculations of ecological losses and gains, marginalizing many local communities. This was part of the attempt to conceal the class aspects of the policy and the multiple ways through which, under the surface of equivalence, it intensified inequality by deepening uneven geographical development and social, environmental and spatial injustice. Overall, the case of England shows well the importance of paying attention to the interplay between offsetting, urbanization and the neoliberalization of nature and space, and, thus, of shifting the focus of the literature to the way offsetting is used to support the production of space(s), place(s) and nature(s). Offsetting exposes the environmental contradictions of capitalism by offering a concrete manifestation of the uneven and increasingly contradictory processes that drive the production of the ecology of both urban and rural environments under capitalism (see also Heynen et al. 2006). The uneven outcomes of the spatial and temporal relocation of compensation are particularly intense since the policy disregards social aspects and local particularities by framing as irrelevant to its goals the social consequences of land use change, the ties between communities and places, or issues of control and access to land and nature (Apostolopoulou 2019; Apostolopoulou and Adams 2019, 2017a, b; Bormpoudakis et al. 2019).

A final note is important here on the importance of understanding environmental politics as class politics. As I have discussed elsewhere (Apostolopoulou 2019) a materialist theory of ideology (Marx 1887) allows us to see "the political" as the historically produced bourgeois political and understand how politics dialectically intersect with economics. In that sense, a non-metaphysical, material, and class understanding of politics by means of a theory of hegemony (Fisken 2012; Gramsci 1992; Loftus 2014; Mitchell et al. 2015; Thomas 2009) is crucial for understanding the origins and politics of offsetting, market-based environmentalism and more broadly of neoliberal environmental and planning policies, the limits of their depoliticing effects, and also the limits of capitalism's hegemony and, therefore, it is of major importance for oppositional politics.

8.3 The Dystopia of Offsetting's Ahistorical and Asocial Non-places: It's not a Flat World, After All

> There is a fine line between acknowledging the extent and seriousness of the troubles and succumbing to abstract futurism and its affects of sublime despair and its politics of sublime indifference.
>
> —Donna Haraway (2016)

Thomas Friedman (2005) in the introduction of his well-known book "The World Is Flat" writes after his trip in Bangalore, India's Silicon Valley: "Columbus reported to his king and queen that the world was round, and he went down in history as the man who first made this discovery. I returned home and shared my discovery only with my wife, and only in a whisper. 'Honey', I confided, 'I think the world is flat'" (Friedman 2005, p. 5). The idea to portray ecological losses and gains, in the form of replaceable, equivalent, and ultimately exchangeable biodiversity, carbon, wetland, or water credits may surely draw inspiration from Friedman's assertion that our world is flat, constituted of non-places for the loss of which nobody will care deeply. This is what those who conceived the policy in England most probably had in their minds: a strategic alliance between developers and conservationists based on the apportionment of development and conservation land as if people do not exist. In this ideal technocratic scenario, developers would earn their right to pollute and destroy by paying conservationists to materialize their noble vision: the establishment of large, joined conservation banks to manage nature for the sake of humanity and future generations with no disturbance from fellow humans that public access to nature inevitably brings about. If people react to this, then some footpaths and cycle routes could also be created to give local communities the chance to walk around the banks and enjoy preserved nature in a proper manner. Offsetting's cynicism is not of course a British creation; it is rather an acute reflection of capitalism's relation to nature in the twenty-first century, in the so-called Anthropo-obScene as Swyngedouw and Ernston (2018) call it to expose the depoliticizing effects of the (widely adopted) term Anthropocene. In an era where the sublime of money ruins, nature is non-surprisingly largely conceived as being constituted from disconnected and historically void pieces of land and collections of species and habitats that in their totality constitute the precious for capitalism's survival (or to be precise mostly for some specific sections of capital) "natural capital." This is the image of a non-human nature that has no local referents, no real sense of place, and no social relevance. And it is rendering this image as commonsensical (in Gramscian terms) that allows portraying as rational, pragmatic and feasible to offset the destruction of existing and real nature (that is occurring *now*) based on projections about the creation of new nature, currently non-existent, which will be equivalent or even better (in the *future*). Moreover, and relatedly, offsetting and conservation banking by restricting access to and control of non-human nature lead to the creation of a peculiar form of fetishized nature (Katz 1998). This nature sometimes is—or used to be—publicly owned since access to offset sites located in state-owned land may be restricted to enable development in private land (see Hackett 2015, 2016). In other cases, as often happened in England, land may be private and thus the loss of public access may be seen as "legitimate" and "legal" since you can feel sorry for the loss of nature you do not own but you cannot protest for losing something that was not yours in the first place. To move this further offsetting also has the potential, already identified in England, to expand the neoliberal dream of special economic zones (SEZs) to the realm of environmental and planning policies. SEZs are areas where business and trade laws are different from the rest of the country, including loosened labour regulations and lower taxations (or even the so-called "tax holidays"), to encourage investments. According to the World Bank,[8] special economic zones typically include a "geographically limited area, usually physically secured (fenced-in); single management or administration; eligibility for benefits based upon physical location within the zone; separate customs area (duty-free benefits) and streamlined procedures." Offsetting by being driven by land prices and by often relocating ecosystem values from urban to rural areas could potentially lead to the creation of a form of "special

[8] See http://documents.worldbank.org/curated/en/343901468330977533/Special-economic-zone-performance-lessons-learned-and-implication-for-zone-development.

nature zones" offering a beneficial for developers planning process. In the offsetting scenario conservation and the preservation of ecosystems and open green spaces becomes, therefore, a byproduct of economic growth and urban development patterns creating special zones of development and special zones of nature, lifeless non-places located where it makes economic sense to place them.

Offsetting's narrative and vision is clearly not only cynical and revanchist but also deeply dystopian. It takes its inspiration from neoliberal capitalism's ignorance for social reproduction and for the needs of the social majority and implements this same ignorance that has been disproportionally felt by the working class particularly since the 2008 financial crash, inter alia, in the collapse of public health, pensions and education, to the realm of nature–society relationships. The intensification of social conflicts over environmental issues in the era following the 2008 financial crash has risen in parallel to rising food insecurity, urban poverty and the escalating housing crisis, but also to the further privatization and economic exploitation or abandonment of public spaces, accompanied by an unequal access to and use of open and safe green spaces along lines of class, race, ethnicity, and gender, among others (Anguelovski 2013; Madanipour 2010). But the fact that under prolonged austerity and a never-ending crisis of capitalism social needs do not matter has major political implications and reflects the decreasing legitimatization of the neoliberal project. As David Harvey argued in a recent interview to Monthly Review[9]: "neoliberalism no longer commands the consent of the mass of the population. It has lost its legitimacy. I already pointed out in The Brief History of Neoliberalism (2005) that neoliberalism could not survive without entering into an alliance with state authoritarianism. It now is moving toward an alliance with neo-fascism, because as we see from all the protest movements around the world, everyone now sees neoliberalism is about lining the pockets of the rich at the expense of the people (this was not so evident in the 1980s and early 1990s)." The political developments in the post-2008 era denote even to the most skeptical observer capitalism's inability to offer a convincing

[9]Available at: https://mronline.org/2019/02/16/the-neoliberal-project-is-alive-but-has-lost-its-legitimacy-david-harvey/.

win-win vision for nature–society relationships as expressed in the resurgence of the revanchist rhetoric of the 1990s about the end of history along with a dystopian TINA dogma. The recent developments in the US, including, inter alia, Trump's unprecedented attack to environmental policy,[10] the election of Jair Bolsanaro in Brazil who is determined to continue what Pinochet started in Chile in the 1970s, the intensification of social-environmental injustices across the globe and the often violent repression of environmental struggles and social movements corroborate this observation as also do the empty promises of the corporate sector. In July 2019, eight of the world's largest asset managers have pledged to account for climate risk in their investments. With a combined $15 trillion of assets under management, the global investment companies, including Blackrock, Goldman Sachs, BNP Paribas, HSBC, Natixis, Amundi, State Street, and Northern Trust, said they would support the implementation of an initiative backed by the French president Emmanuel Macron to pressure companies to become more "climate-friendly".[11] The same day that the press and the French government was celebrating these "inspirational" commitments, leaders across the globe, including French officials, were condemning climate strikes and the radicalization of youth. According to their logic, climate emergency or the sixth extinction primarily require corporate action since according to the neoliberal rhetoric the same agents that drive environmental destruction are also the most relevant to choose which is the appropriate response to the destruction they are causing. However, vague corporate consensus on the need to tackle climate change, centered on reducing and stabilizing CO_2, or to halt biodiversity loss, centered on keeping the stock of natural capital constant or finding ways to substitute it through new technologies, has brought no real change on the ground as has become evident in the inability to reach even the most modest agreed targets on both climate change and biodiversity loss. Recent data corroborate this by showing a record level of increase in atmospheric CO_2 for 2018.[12]

[10]See https://www.jacobinmag.com/2017/02/trump-epa-environment-conservation-reagan-bush/.

[11]See https://www.climatechangenews.com/2019/07/12/asset-managers-worth-15-trillion-make-climate-risk-promise-macron/.

[12]See https://www.theguardian.com/environment/2019/nov/25/climate-heating-greenhouse-gases-hit-new-high-un-reports.

Similarly, according to the 1500-page report that IPBES published in May 2019[13] one million plant and animal species are now at risk of extinction, posing a dire threat to ecosystems and livelihoods globally by, inter alia, negatively affecting food security and clear water provision. Increasing mainstream calls for adopting a vaguely defined sustainability agenda, accelerating the monetary valuation of nature and making "radical" lifestyle changes are not only unconvincing but also fundamentally ignore that there is no equal responsibility to "humans" in general. Indeed, most emissions come from a relatively small number of companies and a very small percentage of the overall population: 100 major corporations are responsible for the 71% of the emissions created since 1970 whereas the richest 10% of people across the globe are responsible for 49% of lifestyle consumption emissions. Similarly, the key driver of global biodiversity loss and species extinction is land use change and habitat loss due to the extraction of mineral, metal and energy resources, urbanization, construction and infrastructure, urban regeneration, large-scale agriculture, land-take and deforestation, land degradation, and habitat fragmentation.[14] As Naomi Klein has recently argued: "so many environmental responses have just been minor tweaks to an economy based on endless consumption; take your electric car to the drive-thru for an Impossible Burger and a Coke with a paper straw. Look, of course it's better than the alternative. But it's nowhere close to the depth of change required if we hope to actually pull our planet back from the brink".[15]

Neil Smith argued in 2008, by borrowing a quote from Jurgen Habermas, that neoliberalism as an economic doctrine has been killed by social movements and economists but remains in a "zombie" state because no global challenge has been erected against it. Neoliberal natures are in a similar state: still dominant but dead (Apostolopoulou et al. 2014; Smith 2008). Even though neoliberal hegemony as consent

[13]Available at: https://ipbes.net/news/ipbes-global-assessment-summary-policymakers-pdf.

[14]For more information see GEO-6, Part A: State of the Global Environment, Chapter 7: Biodiversity. Available at: https://wedocs.unep.org/bitstream/handle/20.500.11822/27659/GEO6_CH6.pdf?sequence=1&isAllowed=y.

[15]Available at: https://www.democracynow.org/2019/9/17/naomi_klein_the_intercept_trump_straw.

is still evident, it is increasingly dialectically intertwined with domination by force, including economic coercion, especially in the context of prolonged austerity and crisis. But as long as conservation, and environmental politics, remain trapped in the tyranny of pragmatism things will only get worse. Framing environmental problems as threatening the survival of humankind is sustained by what Mike Davis called "ecologies of fear" and shifts the focus on techno-managerial and behavioral transformations, that are not only organized within the horizons of the liberal-capitalist order (Swyngedouw 2010) but also, they rather easily take a dystopian turn. Offsetting is an emblematic example of the latter. It brings forward a utilitarian and fetishistic disposition toward nature and whether offset sites are commodities or commodity forms they nonetheless manifest a treatment of nature as a sum of alienable objects (Harvey 2006). As Marx, following Aristotle, argued, money is a radical lever, the great leveler of all things. Representing biodiversity units as exchange values means representing them as merely different quantities and consequently as if they do not contain an atom of use value (Marx 1887). The contradiction between use-value and exchange-value holds a central position in Marx's (1887) value analysis (Harvey 2014) and is a key component of a critical stance against approaches arguing that simply correcting exchange values, namely, putting a monetary value to nature can resolve the environmental contradictions of capitalism. Essentially, these approaches by rooting value to consumption/exchange shift the discussion away from the exploitative character of capitalist production that subsumes both nature and labor toward secondary and disorienting issues like subjective consumers preferences. But rooting value in production yields far more radical political questions allowing us to imagine alternative forms of production, oriented toward social and ecological needs (Huber 2017).

So, what is to be done? What is the alternative if we reject not only offsetting but altogether recent attempts to put an economic price to nature? Against the pessimistic acceptance that the only solution we are left with is to protect some biodiversity or carbon "units," against the pragmatism of oversimplistic neo-Malthusian explanations or the cynical neoliberal admittance that nature will not be saved until capitalists can gain more from its conservation rather than its destruction, but also against the naïve faith to capitalism's ability to infinitely reproduce itself which sees as a remedy for the loss of "Eden" the production of virtual natures in the countryside or elitist green enclaves in cities, and by consciously acknowledging the possibility that futuristic dystopian allegories are not produced by mere fiction but by our own capitalistic world and its contradictions, I argue that it is urgent to shape a radical anti-capitalist environmental agenda. Such an agenda should be capable to confront post-political apocalyptic framings of the environmental crisis (Swyngedouw 2010) and catastrophic projections fueled by the Anthropocene narrative, but also apologetic techno-managerial solutions, like the ones offered by the various eco-modernist approaches (see e.g., Asafu-Adjaye et al. 2015) which, at the end of the day, celebrate capitalism as an eternal production system. This would require placing at the core of our political agenda issues of social-environmental inequality and justice by explicitly asking how nature is produced, who controls this production and who gains and who loses from it. By drawing on the legacy of the right to the city (see Harvey 2008, 2012; Lefebvre 1968, 1996), we should start talking openly about the right to nature (Apostolopoulou and Adams 2019; Apostolopoulou and Cortes-Vazquez 2018; Cortes-Vazquez and Apostolopoulou 2019), namely, the right to influence and command the processes by which nature–society relationships are made, remade and disrupted by urban development and economic growth. Struggles against urban expansion and large-scale infrastructure increasingly involve environmental aspects and the loss of biodiversity and green spaces due to planetary urbanization is increasingly threatening livelihoods and socionatures with major implications for the quality of life and the social reproduction of several local communities within and beyond the city. Indeed, as the 1970s economic crisis opened up the city as a laboratory for neoliberal experiments (Harvey 2005), the same happened, and is still happening, with the 2008 crisis not only for the city but also for the urbanizing countryside, the city-region and 'urban galaxies' (Soja and Kanai 2006, p. 58). Environmental struggles are becoming much more widespread, more radical, and more determined and are often actively blocking socially environmental destructive projects.[16] They should, thus, be clearly acknowledged by Marxists and ecosocialists as class struggles (see also Harvey 2012; Lefebvre 1970) that have the potential to disrupt urbanization, economic growth and capital accumulation. For scholars to actively embrace such struggles and contribute to their unification and radicalization, it would be crucial to consciously put "in command" within radical theory a philosophy of praxis (Loftus 2014) by formulating "adequate theoretical 'translations' of the concrete social and political relations and practices of resistance" that can give rise to a "political of a completely different type" (Thomas 2009, p. 35). A critique of the historically and geographically specific bourgeois political that constitutes the hegemonic neoliberal understanding of nature–society relationships can, therefore, only be achieved through direct involvement with the messy reality of everyday politics to unravel the political processes through which particular socio-environmental configurations are produced, configured, and contested (or not), and identify the political strategies through which the production of socially-environmentally just socio-natures would be possible.

Environmental politics have always been a terrain of class struggle and intra-capital competition as the contradictions around the implementation of offsetting clearly show. But offsetting is not an isolated policy and if it was it would not probably worth a book dedicated to it. It is part of a much bigger argument, a key element in nature's destruction for urban development and economic growth without any limits. Land is not reproducible so every time that land is damaged, it is simply gone (unless we accept the futurism of sci-fi movies which describes a future for the working class in a deserted, ruined earth and a future for the bourgeoisie in a new "green" outerspace). The 'right to nature' puts forward the question of who commands the inner connection between the economic exploitation of nature, urbanization and capital accumulation refocusing the discussion on who drives the production of nature and space and whose class interests this production serves. It is a political and activist position that is future orientated, seeking

[16] At the time of writing the Environmental Justice Atlas has already reported 3.062 social conflicts around environmental issues across the globe (see https://ejatlas.org/).

the transformation of both social relations and nature-society relations. It is a collective social right (Apostolopoulou 2020) that highlights the necessity to deepen and extend existing rights, including abolishing property rights, not as liberal-democratic rights guaranteed by the state, but as political claims of mobilized groups that have the potential to initiate a generalizing political awakening among citizens (Purcell and Tyman 2015; Lefebvre 2003). The proliferation of social-environmental conflicts and movements configure that, against hegemonic narratives about the end of history, the production of nature becomes one of the key political issues of the Anthropocene. What is now imperative is to realize that policy-making under a prolonged global crisis of capitalism will not stop producing revanchist and dystopian policies, like biodiversity offsetting, as long as it builds on the inability of progressive politics to propose radical alternatives to the existing capitalist order (Apostolopoulou 2019). Ultimately, defending the Right to Nature involves a struggle for gaining real social control over the production of nature and space that can unite social-environmental movements across the Global South and North, from remote protected areas to the heartlands of megacities. This struggle will be a struggle for communism, and, as Harvey (2014), argues it will entail a struggle for social control to determine what is and is not socially necessary according not to the logic of the market but to the needs of the social majority. In this struggle, nature conservation should be seen as a social movement built upon collaboration and solidarity, fighting for fair and just nature–society relationships but also for the beauty of everyday socionatures and places that the calculus of capitalism would never capture—as it does not capture so many other things that matter for people. It will be a struggle that has the potential to stand against the increasing conceptual and material alienation[17] between society and nature and bring forward the dialectical unity between humans and non-human nature, theorizing

[17]The concept of alienation (Entfremdung) is dominant in the classic German philosophy, including the important work of Hegel. For Marx, human alienation from non-human nature is intrinsic to value's formal abstraction from use value (Burkett 1999) and is related to the alienation of labor (Marx 1964). As we read in the Economic and Philosophical Manuscripts of (Marx 1844: 1964, p. 29) "the more the worker by his labor appropriates the external world, sensuous nature, the more he deprives himself of means of life in two respects: first, in that the sensuous external world more and more ceases to be an object belonging to his labor – to

environmental struggles as struggles to de-alienate non-human nature and reintegrate it into the web of social connections. This is where we must go in search of our better selves. Or in Margaret Atwood's (2019) words: "*You take the first step, and to save yourself from the consequences, you take the next one. In times like ours, there are only two directions: up or plummet.*"

References

Abakerli, S. (2001). A critique of development and conservation policies in environmentally sensitive regions in Brazil. *Geoforum, 32,* 551–565.

Adams, W. M. (2009). *Green development: Environment and sustainability in a developing world*. London: Routledge.

Anguelovski, I. (2013). New directions in urban environmental justice: Rebuilding community, addressing trauma, and remaking place. *Journal of Planning Education and Research, 33,* 160–175.

Apostolopoulou, E. (2009). The social conflicts during the implementation of conservation policy in protected areas: Analysis and appraisal of conservation policies in Greece. PhD thesis. Aristotle University of Thessaloniki, Greece. Available at: https://www.didaktorika.gr/eadd/handle/10442/22497.

Apostolopoulou, E. (2019). Beyond post-politics: Offsetting, depoliticization and contestation in a community struggle against executive housing. *Transactions of the Institute of British Geographers*. https://doi.org/10.1111/tran.12354.

Apostolopoulou, E. (2020). The right to nature as a social right. *Social Policy* (in press; in Greek).

Apostolopoulou, E., & Adams, W. M. (2015). Neoliberal capitalism and conservation in the post-crisis era: The dialectics of 'green' and 'un-green' grabbing in Greece and the UK. *Antipode, 47,* 15–35.

Apostolopoulou, E., & Adams, W. M. (2017a). Biodiversity offsetting and conservation: Reframing nature to save it. *Oryx, 51,* 23–31.

be his labor's means of life; and, second, in that it more and more ceases to be means of life in the immediate sense, means for the physical subsistence of the worker".

Apostolopoulou, E., & Adams, W. M. (2017b). Biodiversity offsetting and the reframing of conservation: A reply to ten Kate & von Hase and Dempsey & Collard. *Oryx, 51,* 40–42.

Apostolopoulou, E., & Adams, W. M. (2019). Cutting nature to fit: Urbanization, neoliberalism and biodiversity offsetting in England. *Geoforum, 98,* 214–225.

Apostolopoulou, E., Adams, W. M., & Greco, E. (2018). Biodiversity offsetting and the production of "equivalent natures": A Marxist critique. *ACME: An International Journal for Critical Geographies, 17*(3), 861–892.

Apostolopoulou, E., Bormpoudakis, D., Paloniemi, R., Cent, J., Grodzińska-Jurczak, M., Pietrzyk-Kaszyńska, A., et al. (2014). Governance rescaling and the neoliberalization of nature: the case of biodiversity conservation in four EU countries. *International Journal of Sustainable Development and World Ecology, 21,* 481–494.

Apostolopoulou, E., & Cortes-Vazquez, J. (Eds.). (2018). *The right to nature: An academic-activist dialogue on environmental justice, social movements and neoliberal natures.* Abingdon: Routledge-Earthscan.

Apostolopoulou, E., Drakou, E. G., & Pediaditi, K. (2012a). Participation in the management of Greek Natura 2000 sites: Evidence from a cross-level analysis. *Journal of Environmental Management, 113,* 308–318.

Apostolopoulou, E., Drakou, E., Santoro, F., & Pantis, J. D. (2012b). Investigating the barriers to adopting a "human-in-nature" view in Greek biodiversity conservation. *International Journal of Sustainable Development and World Ecology, 19,* 515–525.

Asafu-Adjaye, J., Blomqvist, L., Brand, S., Brook, B., De Fries, R., Ellis, E., et al. (2015). *An ecomodernist manifesto.* Available at: http://www.ecomodernism.org/manifesto.

Atwood, M. (2019). *The testaments.* London: Chatto & Windus.

Bayon, R., Fox, J., & Carroll, N. (2008). *Conservation and biodiversity banking: A guide to setting up and running biodiversity credit trading systems.* London: Earthscan.

Bennett, G., Chavarria, A., Ruef, F., & Leonardi, A. (2017a). *State of European markets 2017: Biodiversity offsets and compensation.* Ecosystem Marketplace, Forest Trends.

Bennett, G., Gallant, M., & ten Kate, K. (2017b). *State of biodiversity mitigation 2017.* Markets and Compensation for Global Infrastructure Development. Ecosystem Marketplace, Forest Trends, Washington DC.

Blamires, C., & Jackson, P. (2006). *World fascism: A historical encyclopedia* (Vol. 1). Santa Barbara, CA: ABC-CLIO.

Bormpoudakis, D., Tzanopoulos, J., & Apostolopoulou, E. (2019). The rise and fall of biodiversity offsetting in the Lodge Hill large-scale housing development, South East England. *Environment and Planning E: Nature and Space*. https://doi.org/10.1177/2514848619884890.

Brenner, N., & Schmid, C. (2011). Planetary Urbanization. In M. Gandy (Ed.), *Urban Constellations* (pp. 10–13). Berlin: Jovis.

Brenner, N., & Schmid, C. (2015). Towards a new epistemology of the urban? *City, 19,* 151–182.

Brockington, D., & Duffy, R. (2010). Capitalism and conservation: The production and reproduction of biodiversity conservation. *Antipode, 42,* 469–484.

Bull, J. W., & Strange, N. (2018). The global extent of biodiversity offset implementation under no net loss policies. *Nature Sustainability, 1,* 790.

Burkett, P. (1997). Nature in marx reconsidered a silver anniversary assessment of alfred schmidt's concept of nature in Marx. *Organization & Environment, 10,* 164–183.

Burkett, P. (1999). *Marx and nature: A red and green perspective*. New York: St. Martin's Press.

Cortes-Vazquez, J. A., & Apostolopoulou, E. (2019). Against Neoliberal Natures: Environmental Movements, Radical Practice and "the Right to Nature". *Geoforum, 98,* 202–205.

Costanza, R., de Groot, R., Sutton, P., et al. (2014). Changes in the global value of ecosystem services. *Global Environmental Change, 26,* 152–158.

De Groot, R., Brander, L., Van Der Ploeg, S., Costanza, R., Bernard, F., Braat, L., et al. (2012). Global estimates of the value of ecosystems and their services in monetary units. *Ecosystem Services, 1,* 50–61.

European Environment Agency (EEA). (2017). *About urban environment*. Available at: https://www.eea.europa.eu/themes/sustainability-transitions/urban-environment.

Felli, R. (2014). On climate rent. *Historical Materialism, 22,* 251–280.

Fine, B., & Milonakis, D. (2009). *From economics imperialism to freakonomics: The shifting boundaries between economics and other social sciences*. London: Routledge.

Fisken, T. D. (2012). *The turn to the political: Post-Marxism and Marx's critique of politics* (Doctoral dissertation). UC Berkeley.

Friedman, T. L. (2005). *The world is flat: A brief history of the twenty-first century*.

Gerritsen, E. J. (2009). *The global infrastructure boom of 2009–2015: Strategic economic consequences for America, China and the global economy* (CRGP

Working Paper 48). Stanford, CA: Collaboratory for Research on Global Projects (CRGP).
Ghimire, K., & Pimbert, M. P. (1997). *Social change and conservation.* London: Earthscan.
Gramsci, A. (1971). *Selections from the prison notebooks* (Q. Hoare & G. Nowell-Smith, Eds. & Trans.). New York: International Publishers.
Gramsci, A. (1992). *Prison notebooks* (J. A. Buttigieg, Trans.) New York, NY: Columbia University Press.
Gray, M., & Barford, A. (2018). The depths of the cuts: The uneven geography of local government austerity. *Cambridge Journal of Regions, Economy and Society, 11,* 541–563.
Greco, E. (2014). Land, property and rent. In A. Fredriksen, S. Bracking, E. Greco, J. Igoe, R. Morgan, & S. Sullivan (Eds.), *A conceptual map of the study of value* (Working Paper No. 2, pp. 41–44). Leverhulme Centre for the Study of Value, University of Manchester, Manchester, UK.
Greco, E. (2015). *Value or rent? Leverhulme centre for the study of value* (Working Paper No. 12). University of Manchester.
Greco, E., & Apostolopoulou, E. (2019). Value, rent and nature: The centrality of class. *Dialogues in Human Geography, 10*(1).
Hackett, R. (2015). Offsetting dispossession? Terrestrial conservation offsets and First Nation treaty rights in Alberta, Canada. *Geoforum, 60,* 62–71.
Hackett, R. (2016). 'Shell games', displacement and the reordering of boreal landscapes in Alberta, Canada. *Area, 48,* 153–160.
Haraway, D. J. (2016). *Staying with the trouble: Making kin in the Chthulucene.* Durham: Duke University Press.
Harvey, D. (2005). *A brief history of neoliberalism.* Oxford: Oxford University Press.
Harvey, D. (2006). *The limits to capital.* London and New York: Verso.
Harvey, D. (2008). The right to the city. *New Left Review, 53,* 23–40.
Harvey, D. (2012). *Rebel cities: From the right to the city to the urban revolution.* London and New York: Verso.
Harvey, D. (2014). *Seventeen contradictions and the end of capitalism.* London, UK: Profile Books.
Heynen, N., Kaika, M., & Swyngedouw, E. (Eds.). (2006). *In the nature of cities: Urban political ecology and the politics of urban metabolism.* London and New York: Routledge.
Huber, M. (2017). Value, nature and labor: A defense of Marx. *Capitalism Nature Socialism, 28,* 39–52.

Igoe, J., Neves, K., & Brockington, D. (2010). A spectacular eco-tour around the historic bloc: Theorising the convergence of biodiversity conservation and capitalist expansion. *Antipode, 42,* 486–512.
IUCN, The Biodiversity Consultancy, & IGF. (2018). *Understanding government biodiversity offset policies in the mining sector*. Available at: https://www.iucn.org/sites/dev/files/content/documents/understanding_government_biodiversity_offset_policies_in_the_mining_sector_november_2017.pdf.
Katz, C. (1998). Whose nature, whose culture?: Private productions of space and the 'preservation' of nature. In B. Braun & N. Castree (Eds.), *Remaking reality: Nature at the millennium* (pp. 45–62). London and New York: Routledge.
Katz, C. (2004). *Growing up global: Economic restructuring and children's everyday lives*. Minneapolis: University of Minnesota Press.
Katz, C. (2011). Accumulation, excess, childhood: Toward a countertopography of risk and waste. *Documents D'anàlisi Geogràfica, 57,* 47–60.
Keller, R., & Turek, M. (1998). *American Indians and National Parks*. Tucson: University of Arizona Press.
Kemf, E. (1993). *Indigenous peoples and protected areas: The law of mother Earth*. London: Earthscan.
Lees, L. (2012). The geography of gentrification: Thinking through comparative urbanism. *Progress in Human Geography, 36,* 155–171.
Lefebvre, H. (1968). *Le Droit à la Ville*. Paris: Anthropos.
Lefebvre, H. (1970). *La révolution urbaine* (Vol. 216). Paris: Gallimard.
Lefebvre, H. (1996). *Writings on cities* (Vol. 63[2]). Oxford: Blackwell.
Lefebvre, H. (2003 [1990]). From the social pact to the contract of citizenship. In S. Elden, E. Lebas, & E. Kofman (Eds.), *Henri Lefebvre: Key writings* (pp. 238–254). New York: Continuum.
Loftus, A. (2014). Against a speculative leftism. In J. Wilson & E. Swyngedouw (Eds.), *The post-political and its discontents: Spaces of depoliticization, spectres of radical politics* (pp. 229–243). Edinburgh, UK: Edinburgh University Press.
Madanipour, A. (Ed.). (2010). *Whose public space international case studies in urban design and development*. London, UK: Routledge.
Madsen, B., Carroll, N., Kandy, D., & Bennett, G. (2011). *State of biodiversity markets report: Offset and compensation programs worldwide*. Washington, DC: Forest Trends.
Marx, K. (1887). *Capital: A critique of political economy* (Vol. I). Moscow: Progress Publishers. Available at: https://www.marxists.org/archive/marx/works/1867-c1/.

Marx, K. (1964). *The economic and philosophic manuscripts of 1844*. New York: International Publishers.

McAfee, K. (2012). The contradictory logic of global ecosystem services markets. *Development and Change, 43,* 105–131.

Merrifield, A. (2013). The urban question under planetary urbanization. *International Journal of Urban and Regional Research, 37,* 909–922.

Mitchell, D., Attoh, K., & Staeheli, L. (2015). Whose city? What politics? Contentious and non-contentious spaces on Colorado's Front Range. *Urban Studies, 52,* 2633–2648.

Moilanen, A., & Kotiaho, J. S. (2018). *Planning biodiversity offsets*. Nordic Council of Ministers.

Moreno-Mateos, D., Maris, V., Béchet, A., et al. (2015). The true loss caused by biodiversity offsets. *Biological Conservation, 192,* 552–559.

Nash, R. (1989). *The rights of nature: A history of environmental ethics*. Wisconsin: University of Wisconsin Press.

Organisation for Economic Co-operation and Development (OECD). (2018). *Cost-benefit analysis and the environment: Further developments and policy use*. Paris: OECD Publishing. Available at: https://doi.org/10.1787/9789264085169-en.

Paloniemi, R., Apostolopoulou, E., Primmer, E., Grodzinska-Jurczak, M., Henle, K., Ring, I., et al. (2012). Biodiversity conservation across scales: Lessons from a science–policy dialogue. *Nature Conservation, 2,* 7–19.

Purcell, M., & Tyman, S. K. (2015). Cultivating food as a right to the city. *Local Environment, 20,* 1132–1147.

Purcell, T. F., Loftus, A., & March, H. (2019). Value–rent–finance. *Progress in Human Geography*. https://doi.org/10.1177/0309132519838064.

Sayer, D. (1987). *The violence of abstraction: The analytic foundations of historical materialism*. Basil: Blackwell.

Seto, K. C., Güneralp, B., & Hutyra, L. R. (2012). Global forecasts of urban expansion to 2030 and direct impacts on biodiversity and carbon pools. *Proceedings of the National Academy of Sciences, 109,* 16083–16088.

Smith, N. (2006). Nature as accumulation strategy. In L. Panitch & C. Leys (Eds.), *Socialist register 2007: Coming to terms with nature* (pp. 16–36). London: Merlin.

Smith, N. (2008). Comment: Neo-liberalism: Dominant but dead. *Focaal, 2008,* 155–157.

Soja, E., & Kanai, J. (2006). The urbanization of the world. In N. Brenner (Ed.), *Implosions/explosions: Towards a study of planetary urbanization* (pp. 142–159). Berlin: Jovis.

Swyngedouw, E. (2010). Impossible sustainability and the post-political condition. In M. Cerreta, M. Concillio, & G. Monno (Eds.), *Making strategies in spatial planning* (pp. 185–205). The Netherlands: Springer.

Swyngedouw, E., & Ernstson, H. (2018). Interrupting the anthropo-obscene: Immuno-biopolitics and depoliticizing ontologies in the anthropocene. *Theory, Culture & Society, 35*, 3–30.

Takacs, D. (1996). *The Idea of biodiversity: Philosophies of paradise*. Baltimore: Johns Hopkins University Press.

Thomas, P. (2009). *The Gramscian moment: Philosophy, hegemony and Marxism*. Leiden: Brill.

United Nations (UN). (2018). *World urbanization prospects* (The 2018 Revision). New York: UN.

Wilson, E. O., & Peter, F. M. (1988). *Biodiversity*. Washington, DC: National Academy Press.

Womble, P., & Doyle, M. (2012). The geography of trading ecosystem services: A case study of wetland and stream compensatory mitigation markets. *Harvard Environmental Law Review, 36*(1), 229–296.

References

Abakerli, S. (2001). A critique of development and conservation policies in environmentally sensitive regions in Brazil. *Geoforum, 32*, 551–565.

Adams, W. M. (2009). *Green development: Environment and sustainability in a developing world*. London: Routledge.

Adams, W. M. (2017a). Sleeping with the enemy? Biodiversity conservation, corporations and the green economy. *Journal of Political Ecology, 24*, 243–257.

Adams, W. M. (2017b). Geographies of conservation I: De-extinction and precision conservation. *Progress in Human Geography, 41*, 534–545.

Adams, W. M., Hodge, I. D., & Sandbrook, L. (2014). New spaces for nature: The re-territorialisation of biodiversity conservation under neoliberalism in the UK. *Transactions of the Institute of British Geographers, 39*, 574–588.

Adams, W. M., & Hutton, J. (2007). People, parks and poverty: Political ecology and biodiversity conservation. *Conservation and Society, 5*(2), 147–183.

Agrawal, A., & Redford, K. (2009). Conservation and displacement: An overview. *Conservation and Society, 7*, 1–10.

Airport Expansion Consultation. (2018). *The case for heathrow expansion*. London: Airport Expansion Consultation.

E. Apostolopoulou, *Nature Swapped and Nature Lost*,
https://doi.org/10.1007/978-3-030-46788-3

Allmendinger, P., & Haughton, G. (2009). Soft spaces, fuzzy boundaries, and metagovernance: The new spatial planning in the Thames Gateway. *Environment and Planning A, 41,* 617–633.

Allmendinger, P., & Haughton, G. (2012). Post-political spatial planning in England: A crisis of consensus? *Transactions of the Institute of British Geographers, 37,* 89–103.

Allmendinger, P., & Haughton, G. (2013). The evolution and trajectories of English spatial governance: 'Neoliberal' episodes in planning. *Planning Practice & Research, 28,* 6–26.

Altvater, E. (1990). The foundations of life (nature) and the maintenance of life (work). *International Journal of Political Economy, 20,* 10–34.

Altvater, E. (1993). *The future of the market: An essay on the regulation of money and nature after the collapse of "actually existing socialism"*. London: Verso.

Andreucci, D., García-Lamarca, M., Wedekind, J., & Swyngedouw, E. (2017). 'Value Grabbing': A political ecology of rent. *Capitalism Nature Socialism, 28,* 28–47.

Angelo, H., & Wachsmuth, D. (2015). Urbanizing urban political ecology: A critique of methodological cityism. *International Journal of Urban and Regional Research, 39,* 16–27.

Anguelovski, I. (2013). New directions in urban environmental justice: Rebuilding community, addressing trauma, and remaking place. *Journal of Planning Education and Research, 33,* 160–175.

Apostolopoulou, E. (2009). *The social conflicts during the implementation of conservation policy in protected areas: Analysis and appraisal of conservation policies in Greece* (PhD thesis). Aristotle University of Thessaloniki, Greece. Available at: https://www.didaktorika.gr/eadd/handle/10442/22497.

Apostolopoulou, E. (2010). A critique of the dominant development ideology for nature-society relationship. *Utopia (Ουτοπία), 91,* 87–106. (in Greek).

Apostolopoulou, E. (2016). Biodiversity offsetting in England: Governance rescaling, socio-spatial injustices, and the neoliberalization of nature. *Web Ecology, 16,* 67–71.

Apostolopoulou, E. (2019). Beyond post-politics: Offsetting, depoliticization and contestation in a community struggle against executive housing. *Transactions of the Institute of British Geographers.* https://doi.org/10.1111/tran.12354.

Apostolopoulou, E. (2020). The right to nature as a social right. *Social Policy* (in press; in Greek).

Apostolopoulou, E., & Adams, W. M. (2015). Neoliberal capitalism and conservation in the post-crisis era: The dialectics of 'green' and 'un-green' grabbing in Greece and the UK. *Antipode, 47,* 15–35.

Apostolopoulou, E., & Adams, W. M. (2017a). Biodiversity offsetting and conservation: Reframing nature to save it. *Oryx, 51,* 23–31.

Apostolopoulou, E., & Adams, W. M. (2017b). Biodiversity offsetting and the reframing of conservation: A reply to ten Kate & von Hase and Dempsey & Collard. *Oryx, 51,* 40–42.

Apostolopoulou, E., & Adams, W. M. (2019). Cutting nature to fit: Urbanization, neoliberalism and biodiversity offsetting in England. *Geoforum, 98,* 214–225.

Apostolopoulou, E., Adams, W. M., & Greco, E. (2018). Biodiversity offsetting and the production of "equivalent natures": A Marxist critique. *ACME: An International Journal for Critical Geographies, 17*(3), 861–892.

Apostolopoulou, E., Bormpoudakis, D., Paloniemi, R., Cent, J., Grodzińska-Jurczak, M., Pietrzyk-Kaszyńska, A., et al. (2014). Governance rescaling and the neoliberalization of nature: The case of biodiversity conservation in four EU countries. *International Journal of Sustainable Development and World Ecology, 21,* 481–494.

Apostolopoulou, E., & Cortes-Vazquez, J. (Eds.). (2018). *The right to nature: An academic-activist dialogue on environmental justice, social movements and neoliberal natures.* Abingdon: Routledge-Earthscan.

Apostolopoulou, E., Drakou, E. G., & Pediaditi, K. (2012). Participation in the management of Greek Natura 2000 sites: Evidence from a cross-level analysis. *Journal of environmental management, 113,* 308–318.

Apostolopoulou, E., Drakou, E., Santoro, F., & Pantis, J. D. (2012). Investigating the barriers to adopting a "human-in-nature" view in Greek biodiversity conservation. *International Journal of Sustainable Development and World Ecology, 19,* 515–525.

Apostolopoulou, E., & Paloniemi, R. (2012). Frames of scale challenges in Finnish and Greek biodiversity conservation. *Ecology and Society, 17*(4), 9.

Apostolopoulou, E., & Pantis, J. D. (2009). Conceptual gaps in the national strategy for the implementation of the European Natura 2000 conservation policy in Greece. *Biological Conservation, 142,* 221–237.

Apostolopoulou, E., & Pantis, J. D. (2010). Development plans versus conservation: Explanation of emergent conflicts and governmental political handling. *Environment and Planning A, 42,* 982–1000.

Arbelaez, D. M. L., & Sagre, J. D. Q. (2015). Compensaciones de biodiversidad: experiencias en Latinoamérica y aplicación en el context colombiano. *Gestión y Ambiente, 18,* 159–177.

Arboleda, M. (2016a). In the nature of the non-city: Expanded infrastructural networks and the political ecology of planetary urbanisation. *Antipode, 48,* 233–251.

Arboleda, M. (2016b). Spaces of extraction, metropolitan explosions. *International Journal of Urban and Regional Research, 40,* 96–112.

Arler, F. (2006). *Ethics and cost-benefit analysis* (Research Report 4). Aalborg, Denmark: Department of Development and Planning, Aalborg University.

Arsel, M., & Büscher, B. (2012). Nature™ Inc.: Changes and continuities in neoliberal conservation and market-based environmental policy. *Development and Change, 43,* 53–78.

Arsel, M., Hogenboom, B., & Pellegrini, L. (2016). The extractive imperative in Latin America. *The Extractive Industries and Society, 3,* 880–887.

Asafu-Adjaye, J., Blomqvist, L., Brand, S., Brook, B., De Fries, R., Ellis, E., et al. (2015). *An ecomodernist manifesto.* Available at: http://www.ecomodernism.org/manifesto.

Atwood, M. (2019). *The testaments.* London: Chatto & Windus.

Aviation Environment Federation. (2018). *Campaigning against unsustainable expansion.* Available at: https://www.aef.org.uk/campaigning-against-unsustainable-expansion/.

Bachram, H. (2004). Climate fraud and carbon colonialism: The new trade in greenhouse gases. *Capitalism Nature Socialism, 15,* 5–20.

Backhaus, N. (2003). 'Non-place jungle': The construction of authenticity in National Parks of Malaysia. *Indonesia and the Malay world, 31*(89), 151–160.

Badura, T., Ferrini, S., Agarwala, M., et al. (2017). *Valuation for natural capital and ecosystem accounting.* Synthesis Report for the European Commission. Norwich: Centre for Social and Economic Research on the Global Environment, University of East Anglia. Available at: http://ec.europa.eu/environment/nature/capital_accounting/pdf/Valuation_for_natural_capital_and_ecosystem_acounting.pdf.

Bakker, K. (2009). Neoliberal nature, ecological fixes, and the pitfalls of comparative research. *Environment and Planning A, 41,* 1781–1787.

Bakker, K. (2010). The limits of 'neoliberal natures': Debating green neoliberalism. *Progress in Human Geography, 34,* 715–735.

Bartram, R., & Shobrook, S. (2000). Endless/end-less natures: Environmental futures at the Fin de Millennium. *Annals of the Association of American Geographers, 90,* 370–380.

Bayon, R., Fox, J., & Carroll, N. (2008). *Conservation and biodiversity banking: A guide to setting up and running biodiversity credit trading systems*. London: Earthscan.

BBOP (Business and Biodiversity Offsets Programme). (2009a). *Biodiversity offset design handbook* (Appendices). Washington, DC: BBOP.

BBOP (Business and Biodiversity Offsets Programme). (2009b). *BBOP pilot project case study. Akyem Gold Mining project, Eastern Region, Ghana.* Washington, DC: BBOP.

BBOP (Business and Biodiversity Offsets Programme). (2012a). *Standard on biodiversity offsets*. Washington, DC: BBOP. Available at: http://bbop.forest-trends.org/guidelines/Standard.pdf.

BBOP (Business and Biodiversity Offsets Programme). (2012b). *Resource paper: No net loss and loss-gain calculations in biodiversity offsets*. Washington, DC: BBOP.

BBOP (Business and Biodiversity Offsets Programme). (2012c). *Resource paper: Limits to what can be offset*. Washington, DC: BBOP. Available at: http://bbop.forest-trends.org/guidelines/Resource_Paper_Limits.pdf.

Bebbington, A., & Humphreys-Bebbington, D. (2018). Mining, movements and sustainable development: Concepts for a framework. *Sustainable Development, 26,* 441–449.

Becker, B. K. (1988). Significado atual da fronteira: Uma interpretação geopolítica a partir da Amazônia brasileira. In C. Aubertin (Ed.), *Fronteiras* (pp. 60–89). Brasilia: Universidade de Brasilia (UNB)/ORSTOM.

Bennett, G., Chavarria, A., Ruef, F., & Leonardi, A. (2017). *State of European markets 2017: Biodiversity offsets and compensation.* Ecosystem Marketplace, Forest Trends.

Bennett, G., Gallant, M., & ten Kate, K. (2017). *State of biodiversity mitigation 2017*. Markets and Compensation for Global Infrastructure Development. Ecosystem Marketplace, Forest Trends, Washington DC.

Bernhardt, E. S., & Palmer, M. A. (2011). The environmental costs of mountaintop mining valley fill operations for aquatic ecosystems of the Central Appalachians. In R. S. Ostfeld & W. H. Schlesinger (Eds.), *Year in ecology and conservation biology* (pp. 39–57). New York: New York Academy of Sciences.

Betancur, A. C., & Villa, W. (2016). *Mining and indigenous peoples in Colombia.* IWGIA.

Bezombes, L., Gaucherand, S., Kerbiriou, C., Reinert, M. E., & Spiegelberger, T. (2017). Ecological equivalence assessment methods: What trade-offs between operationality, scientific basis and comprehensiveness? *Environmental Management, 60,* 216–230.

Bidaud, C., Schreckenberg, K., Rabeharison, M., et al. (2017). The sweet and the bitter: Intertwined positive and negative social impacts of a biodiversity offset. *Conservation and Society, 15,* 1–13.

Birch, K., Levidow, L., & Papaioannou, T. (2010). Sustainable capital? The neoliberalization of nature and knowledge in the European "knowledge-based bio-economy". *Sustainability, 2,* 2898–2918.

Bitsakis, E. (Μπιτσάκης, Ε. Ι.). (1997). Οικολογία: μια μαρξιστική προσέγγιση. *Ουτοπία, 26*, 23–52 (in Greek).

Blaikie, P., & Brookfield, H. (1987). *Land degradation and society*. London: Methuen.

Blamires, C., & Jackson, P. (2006). *World fascism: A historical encyclopedia* (Vol. 1). Santa Barbara, CA: ABC-CLIO.

Blowers, A. (1987). Transition or transformation? Environmental policy under thatcher. *Public Administration, 65,* 277–294.

Böhm, S., Misoczky, M. C., & Moog, S. (2012). Greening capitalism? A Marxist critique of carbon markets. *Organization Studies, 33,* 1617–1638.

Bond, P. (2012). Emissions trading, new enclosures and eco-social contestation. *Antipode, 44,* 684–701.

Bonneuil, C. (2015). Tell me where you come from, I will tell you who you are: A genealogy of biodiversity offsetting mechanisms in historical context. *Biological Conservation, 192,* 485–491.

Bormpoudakis, D., Tzanopoulos, J., & Apostolopoulou, E. (2019). The rise and fall of biodiversity offsetting in the Lodge Hill large-scale housing development, South East England. *Environment and Planning E: Nature and Space.* https://doi.org/10.1177/2514848619884890.

Borras, S. M., Jr., Franco, J. C., Gómez, S., Kay, C., & Spoor, M. (2012). Land grabbing in Latin America and the Caribbean. *The Journal of Peasant Studies, 39,* 845–872.

Bracking, S., Bond, P., Brockington, D., Büscher, B., Igoe, J. J., Sullivan, S., & Woodhouse, P. (2014). *Human, non-human and environmental value systems: An impossible frontier?* (LSCV Working Paper Series No. 1). Leverhulme Centre for the Study of Value, School of Environment, Education and Development, The University of Manchester.

Brenner, N. (Ed.). (2014). *Implosions/explosions: Towards a study of planetary urbanization.* Berlin: Jovis.

Brenner, N., & Katsikis, N. (2014). Is the mediterranean urban? In N. Brenner (Ed.), *Implosions/explosions: Towards a study of planetary urbanization* (pp. 428–459). Berlin: Jovis.

Brenner, N., Peck, J., & Theodore, N. (2010). Variegated neoliberalization: Geographies, modalities, pathways. *Global Networks, 10,* 182–222.

Brenner, N., & Schmid, C. (2011). Planetary urbanization. In M. Gandy (Ed.), *Urban constellations* (pp. 10–13). Berlin: Jovis.

Brenner, N., & Schmid, C. (2014). Planetary urbanization. In N. Brenner (Ed.), *Implosions/explosions: Towards a study of planetary urbanization* (pp. 160–163). Berlin: Jovis.

Brenner, N., & Schmid, C. (2015). Towards a new epistemology of the urban? *City, 19,* 151–182.

Brenner, N., & Theodore, N. (2002). Cities and the geographies of "actually existing neoliberalism". *Antipode, 34,* 349–379.

Bridge, G. (2011). Resource geographies 1: Making carbon economies, old and new. *Progress in Human Geography, 35,* 820–834.

Brockington, D. (2014). Celebrity spectacle, post-democratic politics, and Nature™ Inc. In B. Büscher, W. Dressler, & R. Fletcher (Eds.), *Nature Inc.: Environmental conservation in the neoliberal age* (pp. 108–126). Tucson: University of Arizona Press.

Brockington, D., & Duffy, R. (2010). Capitalism and conservation: The production and reproduction of biodiversity conservation. *Antipode, 42,* 469–484.

Brockington, D., & Igoe, J. (2006). Eviction for conservation: A global overview. *Conservation and society, 4*(3), 424–470.

Brown, K., & Corbera, E. (2003). Exploring equity and sustainable development in the new carbon economy. *Climate Policy, 3,* S41–S56.

Bryant, G. (2016). The politics of carbon market design: Rethinking the techno-politics and post-politics of climate change. *Antipode, 48,* 877–898.

Bull, J. W., Lloyd, S. P., & Strange, N. (2017). Implementation gap between the theory and practice of biodiversity offset multipliers. *Conservation Letters, 10,* 656–669.

Bull, J. W., & Strange, N. (2018). The global extent of biodiversity offset implementation under no net loss policies. *Nature Sustainability, 1,* 790.

Bull, J. W., Suttle, K. B., Gordon, A., Singh, N. J., & Milner-Gulland, E. J. (2013). Biodiversity offsets in theory and practice. *Oryx, 47,* 369–380.

Bumpus, A. G., & Liverman, D. M. (2008). Accumulation by decarbonization and the governance of carbon offsets. *Economic Geography, 84,* 127–155.

Bumpus, A. G. (2011). The matter of carbon: Understanding the materiality of tCO_2e in carbon offsets. *Antipode, 43,* 612–638.

Burchardt, H. J., & Dietz, K. (2014). (Neo-)extractivism, a new challenge for development theory from Latin America. *Third World Quarterly, 35,* 468–486.

Burgin, S. (2008). BioBanking: An environmental scientist's view of the role of biodiversity banking offsets in conservation. *Biodiversity and Conservation, 17,* 807–816.

Burkett, P. (1997). Nature in marx reconsidered a silver anniversary assessment of alfred schmidt's concept of nature in Marx. *Organization & Environment, 10,* 164–183.

Burkett, P. (1999). *Marx and nature: A red and green perspective.* New York: St. Martin's Press.

Burkett, P. (2003). The value problem in ecological economics lessons from the physiocrats and Marx. *Organization & Environment, 16,* 137–167.

Burkett, P. (2006). *Marxism and ecological economics: Toward a red and green political economy* (Vol. 11). Historical Materialism Book Series. Boston: Brill.

Büscher, B. (2010). Anti-politics as political strategy: Neoliberalism and trans-frontier conservation in Southern Africa. *Development and Change, 41,* 29–51.

Büscher, B., & Arsel, M. (2012). Neoliberal conservation, uneven geographical development and the dynamics of contemporary capitalism. *Tijdschrift voor Economische en Sociale Geografie, 103,* 129–135.

Büscher, B., & Dressler, W. (2012). Commodity conservation: The restructuring of community conservation in South Africa and the Philippines. *Geoforum, 43*(3), 367–376.

Büscher, B., Dressler, W., & Fletcher, R. (Eds.). (2014). *Nature Inc.: Environmental conservation in the neoliberal age.* Tucson: University of Arizona Press.

Büscher, B., Sullivan, S., Neves, K., Igoe, J., & Brockington, D. (2012). Towards a synthesized critique of neoliberal biodiversity conservation. *Capitalism Nature Socialism, 23,* 4–29.

CAF. (2013). *Carretera Interoceánica Sur del Perú: Retos e Innovación.* Bogota: CAF.

Çaglar, A., & Schiller, N. G. (2018). *Migrants and city-making: Dispossession, displacement, and urban regeneration.* Durham: Duke University Press.

Cahill, D. (2011). Beyond neoliberalism? Crisis and the prospects for progressive alternatives. *New Political Science, 33,* 479–492.

Callon, M. (2007). What does it mean to say that economics is performative? In D. MacKenzie, F. Muneisa, & L. Sui (Eds.), *Do economists make markets? On the performativity of economics* (pp. 311–357). Princeton: Princeton University Press.

Callon, M., & Muniesa, F. (2003). Les marchés économiques comme dispositifs collectifs de calcul. *Réseaux, 21,* 189–233.

Calvário, R., Velegrakis, G., & Kaika, M. (2017). The political ecology of austerity: An analysis of socio-environmental conflict under crisis in Greece. *Capitalism Nature Socialism, 28*(3), 69–87.

Carbones del Cerrejón Limited y Conservación Internacional. (2012). *Propuesta modelo de compensación ambiental aplicado a la intervención por los permisos de aprovechamiento forestal*. Departamento de Gestión Ambiental. Carbones del Cerrejón Limited.

Carver, L., & Sullivan, S. (2017). How economic contexts shape calculations of yield in biodiversity offsetting. *Conservation Biology, 31,* 1053–1065.

Castree, N. (2003). Commodifying what nature? *Progress in Human Geography, 27,* 273–297.

Castree, N. (2008a). Neoliberalising nature: The logics of deregulation and reregulation. *Environment and Planning A, 40,* 131–152.

Castree, N. (2008b). Neoliberalising nature: Processes, effects, and evaluations. *Environment and Planning A, 40,* 153–173.

Castree, N. (2009). Researching neoliberal environmental governance: A reply to Karen Bakker. *Environment and Planning A, 41,* 1788–1794.

Castree, N. (2010). Neoliberalism and the biophysical environment: A synthesis and evaluation of the research. *Environment and Society: Advances in Research, 1,* 5–45.

CEE Bankwatch Network. (2017). *Land lost but not forgotten: Impacts of the trans-Adriatic Pipeline on the land and livelihoods of farmers*. Available at: https://bankwatch.org/publication/land-lost-but-not-forgotten.

Christians, C., Mears, L., Whitmore, A., & Rhoades, H. (2018). Self-determination as resistance: Re-asserting control over natural resources in Colombia. In E. Apostolopoulou & J. Cortes-Vazquez (Eds.), *The right to nature: Environmental justice, social movements and neoliberal natures* (pp. 17–28). Abingdon: Routledge-Earthscan.

CIEEM. (2016). *Guidelines for ecological impact assessment in the UK and Ireland: Terrestrial, freshwater and coastal chartered.* Institute of Ecology and Environmental Management, Winchester.

Clark, B., & York, R. (2005). Dialectical materialism and nature: An alternative to economism and deep ecology. *Organization & Environment, 18,* 318–337.

Cloke, P. (Ed.). (1999). *Policy and change in thatcher's Britain.* Oxford: Pergamon.

Collingwood Environmental Planning Limited. (2013). *Evaluation of the biodiversity offsetting pilot phase, WC 1051.* Summary of Interim Report Collingwood Environmental Planning Limited in partnership with The Institute for European Environmental Policy (IEEP).

Commonwealth of Australia. (2012). *Environment protection and Biodiversity Conservation Act 1999 environmental offsets policy.* Department of the Environment, Canberra, ACT, Australia. Available at: http://www.environment.gov.au/system/files/resources/12630bb4-2c10-4c8e-815f-2d7862bf87e7/files/offsets-policy_2.pdf.

Conservative Party. (2010). *Open source planning* (Policy Green Paper No. 14). London.

Convention on Biological Diversity. (2010). *The convention on biological diversity and the 10th meeting of the conference of parties (COP 10).* Decision X/2 on Strategic Plan for Biodiversity 2011–2020.

Conway, M., Rayment, M., White, A., & Berman, S. (2013). *Exploring potential demand for and supply of habitat banking in the EU and appropriate design elements for a habitat banking scheme* (Final Report Submitted to DG Environment). UK: ICF GHK and BIO Intelligence Service.

Coplan, K. S. (2017). The missing element of environmental cost-benefit analysis: Compensation for the loss of regulatory benefits. *Georgetown International Environmental Law Review, 30,* 281.

Corbera, E., & Brown, K. (2010). Offsetting benefits? Analysing access to forest carbon. *Environment and Planning A, 42,* 1739–1761.

Corson, C. (2010). Shifting environmental governance in a neoliberal world: US aid for conservation. *Antipode, 42,* 576–602.

Corson, C., MacDonald, K. I., & Neimark, B. (2013). Grabbing green: Markets, environmental governance and the materialization of natural capital. *Human Geography, 6,* 1–15.

Cortes-Vazquez, J. A., & Apostolopoulou, E. (2019). Against neoliberal natures: Environmental movements, radical practice and "the right to nature". *Geoforum, 98,* 202–205.

Costanza, R., de Groot, R., Sutton, P., et al. (2014). Changes in the global value of ecosystem services. *Global Environmental Change, 26,* 152–158.

Cowell, R. (2013). The greenest government ever? Planning and sustainability in England after the May 2010 elections. *Planning Practice and Research, 28,* 27–44.

CPRE (Campaign to Protect Rural England). (2015). *Set up to fail: Why housing targets based on flawed numbers threaten our countryside*. London, UK. Available at: http://www.cpre.org.uk/magazine/features/item/4167-set-up-to-fail.

Crezee, B. (2017). *Preocupantes efectos de una carretera: la interoceánica y la deforestación que causa.* Mongabay. Available at: https://es.mongabay.com/2017/10/preocupantes-efectos-una-carretera-la-interoceanica-la-deforesta cion-causa/.

Cruz, J., Cenzano, A., & Supa, H., et al. (2008). *Impactos ambientales y sociales de la carretera Interoceánica* (Parliamentary Report).

Curran, M., Hellweg, S., & Beck, J. (2013). Is there any empirical support for biodiversity offset policy? *Ecological Applications, 24,* 617–632.

Czarnezki, J. J., & Fiedler, K. (2016). *The neoliberal turn in environmental regulation* (Utah Law Review No. 1, Article 1). Available at: http://dc.law.utah.edu/ulr/vol2016/iss1/1.

d'Amour, C. B., Reitsma, F., Baiocchi, G., et al. (2017). Future urban land expansion and implications for global croplands. *Proceedings of the National Academy of Sciences, 114,* 8939–8944.

Daily, G. C., & Matson, P. A. (2008). Ecosystem services: From theory to implementation. *PNAS, 105,* 9455–9456.

Danchev, S., Paratsiokas, N., & Tsakanikas, A. (2013). *Economic impact from the construction and operation of the trans Adriatic pipeline on Greek territory.* Athens: Foundation for Economic and Industrial Research.

Darling, E. (2005). The city in the country: Wilderness gentrification and the rent gap. *Environment and Planning A: Economy and Space, 37,* 1015–1032.

Dauguet, B. (2015). Biodiversity offsetting as a commodification process: A French case study as a concrete example. *Biological Conservation, 192,* 533–540.

Davies, J. (2013). Rethinking urban power and the local state: Hegemony, domination and resistance in neoliberal cities. *Urban Studies, 51,* 3215–3232.

Dawson, A. (2016). *Extinction.* New York: OR Books.

De Groot, R., Brander, L., Van Der Ploeg, S., Costanza, R., Bernard, F., Braat, L., et al. (2012). Global estimates of the value of ecosystems and their services in monetary units. *Ecosystem Services, 1,* 50–61.

DEC. (2006). *BioBanking: A biodiversity offsets and banking scheme: Conserving and restoring biodiversity in NSW* (Working Paper). Sydney: Department of Environment and Conservation NSW.

DEC. (2007). *BioBanking: Biodiversity banking and offsets scheme: Scheme overview*. Sydney: Department of Environment and Climate Change NSW.

Defra (Department for Environment, Food and Rural Affairs). (2007a). *An introductory guide to valuing ecosystem services*. Department for Environment, Food and Rural Affairs. London, UK. Available at: https://assets.publishing.service.gov.uk/government/uploads/system/uploads/attachment_data/file/69192/pb12852-eco-valuing-071205.pdf.

Defra (Department for Environment, Food and Rural Affairs). (2007b). *Securing a healthy natural environment: An action plan for embedding an ecosystems approach*. Department for Environment, Food and Rural Affairs. London, UK. Available at: https://ecosystemsknowledge.net/sites/default/files/wp-content/uploads/Defra%20eco-actionplan.pdf.

Defra (Department for Environment, Food and Rural Affairs). (2010). *Payments for ecosystem services: A short introduction*. Department for Environment, Food and Rural Affairs. London, UK. Available at: http://www.fwr.org/WQreg/Appendices/payments-ecosystem.pdf.

Defra (Department for Environment, Food and Rural Affairs). (2012a). *An overview of the local nature partnership role*. Department for Environment, Food and Rural Affairs. London, UK. Available at: https://assets.publishing.service.gov.uk/government/uploads/system/uploads/attachment_data/file/192580/local-nature-partnerships-overview120402.pdf.

Defra (Department for Environment, Food and Rural Affairs). (2012b). *Red tape challenge—Environment theme implementation plan*. Department for Environment, Food and Rural Affairs. London, UK. Available at: https://assets.publishing.service.gov.uk/government/uploads/system/uploads/attachment_data/file/69598/pb13819-red-tape-environment.pdf.

Defra (Department for Environment, Food and Rural Affairs). (2013). *Biodiversity offsetting in England green paper*. Department for Environment, Food and Rural Affairs. London, UK. Available at: https://consult.defra.gov.uk/biodiversity/biodiversity_offsetting/supporting_documents/20130903Biodiversity%20offsetting%20green%20paper.pdf.

Defra (Department for Environment, Food and Rural Affairs). (2016). *Defra's payments for ecosystem services pilot projects 2012–15: Review of key findings*. Department for Environment, Food and Rural Affairs. London, UK. Available at: https://assets.publishing.service.gov.uk/government/uploads/

system/uploads/attachment_data/file/578005/pes-pilot-review-key-findings-2016.pdf.

Demeritt, D. (1998). Science, social constructivism and nature. In B. Braun & N. Castree (Eds.), *Remaking reality: Nature at the millennium.* London and New York: Routledge.

Dempsey, J., & Robertson, M. M. (2012). Ecosystem services tensions, impurities, and points of engagement within neoliberalism. *Progress in Human Geography, 36,* 758–779.

Dempsey, J., & Suarez, D. C. (2016). Arrested development? The promises and paradoxes of "selling nature to save it". *Annals of the American Association of Geographers, 106*(3), 653–671.

Department for Communities and Local Government. (2013). *Appeal by Bellway Homes (North East) Ltd.* Land at Whitehouse Farm, West Moor, Newcastle-upon-Tyne. Application ref: 11/02337/FUL. APP/W4515/A/12/2175554. Available at: https://acp.planninginspectorate.gov.uk/ViewCase.aspx?Caseid=2175554&CoID=0.

Department for Transport. (2017a). *Revised draft airports national policy statement: New runway capacity and infrastructure at airports in the South East of England.* Presented to Parliament pursuant to Section 9(2) of the Planning Act 2008. London: Department for Transport.

Department for Transport. (2017b). *Appraisal of sustainability: Draft airports national policy statement.* Consultation draft. Exeter: Department for Transport and WSP, Parsons Brinckerhoff.

Department of Defense and EPA (Environmental Protection Agency). (2008). *Compensatory Mitigation for Losses of Aquatic Resources; Final Rule.* U.S. Army Corps of Engineers, DoD and Environmental Protection Agency.

Department of Employment. (1986). *Building business, not barriers.* London: HMSO.

Department of the Environment. (1985). *Lifting the burden.* London: HMSO.

Department of the Environment. (1995). *The Thames gateway planning framework (RPG9A).* London: HMSO.

Dick, P. (1969). *Ubik.* US: Doubleday.

Dickinson, E. (2011). Displaced in nature: The cultural production of (non-) place in place-based forest conservation pedagogy. *Environmental Communication: A Journal of Nature and Culture, 5*(3), 300–319.

Dickinson, S., & Berner, P. O. (2010). Ambatovy project: Mining in a challenging biodiversity setting in Madagascar. In S. M. Goodman & V. Mass

(Eds.), *Biodiversity, exploration, and conservation of the natural habitats associated with the Ambatovy project*. Malagasy Nature 3 (pp. 2–13). Antananarivo: Association Vahatra.

Dourojeanni, M., Barandarián, A., & Dourojeanni, D. (2010). Amazonía peruana en 2021: Explotación de recursos naturales e infraestructura ¿Qué está pasando? ¿Qué es lo que significa para el futuro? *Bois et Forets des Tropiques, 305,* 77–82.

Dressler, W., & Roth, R. (2011). The good, the bad, and the contradictory: Neoliberal conservation governance in rural Southeast Asia. *World Development, 39,* 851–862.

Duménil, G., & Lévy, D. (2001). Costs and benefits of neoliberalism: A class analysis. *Review of International Political Economy, 8,* 578–607.

Duménil, G., & Lévy, D. (2004). *Capital resurgent: Roots of the neoliberal revolution.* Cambridge, MA and London, UK: Harvard University Press.

Duménil, G., & Lévy, D. (2013). *The crisis of neoliberalism.* Cambridge, MA and London, UK: Harvard University Press.

Dyer, S., Grant, J., Lesack, T., et al. (2008). *Catching up: Conservation and biodiversity offsets in Alberta's boreal forests.* Ottawa: Canadian Boreal Initiative.

EC (European Commission). (2011). *Our life insurance, our natural capital: An EU biodiversity strategy to 2020.* Communication from the Commission to the European Parliament, the Council, the Economic and Social Committee and the Committee of the Regions, COM (2011) 244 Final, Brussels.

EFTEC and IEEP. (2010). *The use of market-based instruments for biodiversity protection—The case of Habitat Banking.* Technical Report for European Commission DG Environment, EFTEC and IEEP, London.

Ekers, M., & Loftus, A. (2013). Revitalizing the production of nature thesis: A Gramscian turn? *Progress in Human Geography, 37,* 234–252.

Elnaschie, S. (2013). *Rio Tinto: Compensation manipulation in Southeast Madagascar.* Available at: http://www.ejolt.org/2013/11/rio-tinto-compensation-manipulation-in-southeast-madagascar/.

Engels, F. (1844). *The condition of the working-class in England.* Moscow: Progress Publishers.

Engels, F. (1872). The housing question. Part III, "Supplement on proudhon and the housing question". In K. Marx & E. Engels, *Selected works* (Vol. 2). International Publishers.

Engels, F. (1940). *Dialectics of nature.* New York: International Publishers.

Engels, F. (2017 [1878]). *Anti-dühring.* São Paulo: Boitempo Editorial.

England Biodiversity Group. (2011). *ThinkBIG: How and why landscape-scale conservation benefits wildlife, people, and the wider economy*. Available at: http://publications.naturalengland.org.uk/publication/30047.

Environment Bank. (2012a). *Conservation credit requirement scoping analysis of Whitehouse Farm*. Killingworth: The Environment Bank Ltd.

Environment Bank. (2012b). *Offset search: Offset for Whitehouse Farm*. Killingworth: The Environment Bank Ltd.

Environment Bank. (2012c). Biodiversity offsetting to compensate for nightingale habitat loss at Lodge Hill, Kent. Ripon: The Environment Bank Ltd.

Environment Bank. (2014a). *Biodiversity offsetting information sheet 4 February 2014, guidance for developers*.

Environment Bank. (2014b). *Biodiversity offsetting scheme, Whitehouse Farm*. Killingworth: The Environment Bank Ltd.

Escalas, A., Ayllón, R., Escobedo Grandez, J., et al. (2017). *Estudios de caso para el desarrollo de la compensación ambiental en el Perú: Madre de Dios*. Lima: CSF.

Escobar, A. (1999). After nature: Steps to an antiessentialist political ecology. *Current Anthropology, 40,* 1–30.

European Environment Agency (EEA). (2017). *About urban environment*. Available at: https://www.eea.europa.eu/themes/sustainability-transitions/urban-environment.

European Investment Bank. (2017). *Environmental and social data sheet TAP—Trans Adriatic Pipeline*. Available at: http://www.eib.org/attachments/registers/66312481.pdf.

Fabiani, J. L. (1985). Sciences des écosystèmes et protection de la nature. In A. Cadoret (Ed.), *protection de la nature. Histoire et idéologie* (pp. 75–93). Paris: L' Harmattan.

Fairhead, J., Leach, M., & Scoones, I. (2012). Green Grabbing: A new appropriation of nature? *The Journal of Peasant Studies, 39,* 237–261.

Featherstone, D. (2005). Towards the relational construction of militant particularisms: Or why the geographies of past struggles matter for resistance to neoliberal globalisation. *Antipode, 37,* 250–271.

Featherstone, D. (2013). "Gramsci in action": Space, politics, and the making of solidarities. In M. Ekers, G. Hart, S. Kipfer, et al. (Eds.), *Gramsci: Space, nature, politics* (pp. 65–82). Chichester: Wiley.

Featherstone, D., Ince, A., Mackinnon, D., Strauss, K., & Cumbers, A. (2012). Progressive localism and the construction of political alternatives. *Transactions of the Institute of British Geographers, 37,* 177–182.

Felli, R. (2014). On climate rent. *Historical Materialism, 22,* 251–280.

Felli, R. (2015). Environment, not planning: The neoliberal depoliticisation of environmental policy by means of emissions trading. *Environmental Politics, 24,* 641–660.

Ferguson, N. (2014). From coal pits to tar sands: Examining labour migration between the Athabasca oil sands and an Atlantic Canadian region. *Just Labour, 17–18,* 106–118.

Fine, B. (2005). From actor-network theory to political economy. *Capitalism Nature Socialism, 16,* 91–108.

Fine, B., & Lapavitsas, C. (2000). Markets and money in social theory: What role for economics? *Economy and Society, 29,* 357–382.

Fine, B., & Milonakis, D. (2009). *From economics imperialism to freakonomics: The shifting boundaries between economics and other social sciences.* London: Routledge.

Fish and Wildlife Service. (2003). *Federal guidance for the establishment, use, and operation of conservation banks.* Washington, DC: United States Department of the Interior.

Fisken, T. D. (2012). *The turn to the political: Post-Marxism and Marx's critique of politics* (Doctoral dissertation). UC Berkeley.

Fletcher, R. (2012). Using the master's tools? Neoliberal conservation and the evasion of inequality. *Development and Change, 43,* 295–317.

Fletcher, R. (2014). Orchestrating consent: Post-politics and intensification of Nature TM Inc. at the 2012 World Conservation Congress. *Conservation & Society, 12,* 329–342.

Foster, J. B. (1999). Marx's theory of metabolic rift: Classical foundations for environmental sociology. *American Journal of Sociology, 105,* 366–405.

Foster, J. B. (2000). *Marx's ecology: Materialism and nature.* New York: Monthly Review Press.

Foster, J. B. (2002). *Ecology against capitalism.* New York: Monthly Review Press.

Foster, J. B. (2012). *Rio + 20: Here we go again, Amandla!* Available at: http://www.amandlapublishers.co.za/special-features/the-green-economy/1317-rio–20-here-we-go-again–by-john-bellamy-foster.

Foster, J. B., Clark, B., & York, R. (2009). The midas effect: A critique of climate change economics. *Development and Change, 40,* 1085–1097.

Foster, J. B., Clark, B., & York, R. (2011). *The ecological rift: Capitalism's war on the earth.* New York: NYU Press.

Friedman, T. L. (2005). *The world is flat: A brief history of the twenty-first century.*

Gamarra, M. J. C., Lassoie, J. P., & Milder, J. (2018). Accounting for no net loss: A critical assessment of biodiversity offsetting metrics and methods. *Journal of Environmental Management, 220,* 36–43.

Gandy, M. (2003). *Concrete and clay: Reworking nature in New York City.* Cambridge: MIT Press.

Gardner, T. A., Von Hase, A., Brownlie, S., Ekstrom, J. M. M., Pilgrim, J. D., Savy, C. E., et al. (2013). Biodiversity offsets and the challenge of achieving no net loss. *Conservation Biology, 27,* 1254–1264.

Gerritsen, E. J. (2009). *The global infrastructure boom of 2009–2015: Strategic economic consequences for America, China and the global economy* (CRGP Working Paper 48). Stanford, CA: Collaboratory for Research on Global Projects (CRGP).

Ghimire, K., & Pimbert, M. P. (1997). *Social change and conservation.* London: Earthscan.

Gibbons, P., & Lindenmayer, D. B. (2007). Offsets for land clearing: No net loss or the tail wagging the dog? *Ecological Management & Restoration, 8,* 26–31.

Gibbons, P., Macintosh, A., Constable, A. L., & Hayashi, K. (2018). Outcomes from 10 years of biodiversity offsetting. *Global Change Biology, 24,* 643–654.

Gómez-Baggethun, E., De Groot, R., Lomas, P. L., et al. (2010). The history of ecosystem services in economic theory and practice: From early notions to markets and payment schemes. *Ecological Economics, 69,* 1209–1218.

Gonçalves, B., Marques, A., Soares, A. M. V. D. M., & Pereira, H. M. (2015). Biodiversity offsets: From current challenges to harmonized metrics. *Current Opinion in Environmental Sustainability, 14,* 61–67.

Gordon, A., Bull, J. W., Wilcox, C., & Maron, M. (2015). Perverse incentives risk undermining biodiversity offset policies. *Journal of Applied Ecology, 52,* 532–537.

Gramsci, A. (1971). *Selections from the prison notebooks* (Q. Hoare & G. Nowell-Smith, Eds. & Trans.). New York: International Publishers.

Gramsci, A. (1992). *Prison notebooks* (J. A. Buttigieg, Trans.) New York, NY: Columbia University Press.

Gray, M., & Barford, A. (2018). The depths of the cuts: The uneven geography of local government austerity. *Cambridge Journal of Regions, Economy and Society, 11,* 541–563.

Greco, E. (2014). Land, property and rent. In A. Fredriksen, S. Bracking, E. Greco, J. Igoe, R. Morgan, & S. Sullivan (Eds.), *A conceptual map of the study of value* (Working Paper No. 2, pp. 41–44). Leverhulme Centre for the Study of Value, University of Manchester, Manchester, UK.

Greco, E. (2015). *Value or rent? Leverhulme centre for the study of value* (Working Paper No. 12). University of Manchester.

Greco, E., & Apostolopoulou, E. (2019). Value, rent and nature: The centrality of class. *Dialogues in Human Geography, 10*(1).

Griffiths, V. F., Bull, J. W., Baker, J., et al. (2018). No net loss for people and biodiversity. *Conservation Biology*. https://doi.org/10.1111/cobi.13184.

Grimm, N. B., Faeth, S. H., Golubiewski, N. E., et al. (2008). Global change and the ecology of cities. *Science, 319,* 756–760.

Gunderson, R., Stuart, D., & Petersen, B. (2017). Ideological obstacles to effective climate policy: The greening of markets, technology, and growth. *Capital and Class, 42,* 133–160.

Hackett, R. (2015). Offsetting dispossession? Terrestrial conservation offsets and First Nation treaty rights in Alberta, Canada. *Geoforum, 60,* 62–71.

Hackett, R. (2016). 'Shell games', displacement and the reordering of boreal landscapes in Alberta, Canada. *Area, 48,* 153–160.

Hadjimichalis, C. (2011). Uneven geographical development and socio-spatial justice and solidarity: European regions after the 2009 financial crisis. *European Urban and Regional Studies, 18,* 254–274.

Hahs, A. K., McDonnell, M. J., McCarthy, M. A., et al. (2009). A global synthesis of plant extinction rates in urban areas. *Ecology Letters, 12,* 1165–1173.

Hallstrom, L., Stonechild, J., & Reist, W. (2015). Alberta. In S. Markey, S. Breen, A. Lauzon, et al. (Eds.), *State of rural Canada report* (pp. 17–24). CRRF: Canada.

Hannis, M., & Sullivan, S. (2012). *Offsetting nature? Habitat banking and biodiversity offsets in the English Land Use Planning System.* Green House, UK.

Haraway, D. J. (2016). *Staying with the trouble: Making kin in the Chthulucene.* Durham: Duke University Press.

Harvey, D. (1985). *The Urbanization of Capital*. Baltimore: John Hopkins University Press.

Harvey, D. (1989). From managerialism to entrepreneurialism: The transformation in urban governance in late capitalism. *Geografiska Annaler: Series B, 71,* 3–17.

Harvey, D. (1996). *Nature, justice and the geography of difference*. Oxford: Blackwell.

Harvey, D. (2001). Cosmopolitanism and the banality of geographical evils. In J. Comaroff & J. L. Comaroff (Eds.), *Millenial capitalism and the culture of neoliberalism* (pp. 271–309). Durham and London: Duke University Press.

Harvey, D. (2005). *A brief history of neoliberalism.* Oxford: Oxford University Press.

Harvey, D. (2006). *The limits to capital.* London and New York: Verso.

Harvey, D. (2007). Neoliberalism as creative destruction. *The Annals of the American Academy of Political and Social Science, 610,* 21–44.

Harvey, D. (2008). The right to the city. *New Left Review, 53,* 23–40.

Harvey, D. (2010a). *A companion to Marx's capital.* London and New York: Verso.

Harvey, D. (2010b). *The enigma of capital: And the crises of capitalism.* London: Profile Books.

Harvey, D. (2011). Roepke lecture in economic geography—Crises, geographic disruptions and the uneven development of political responses. *Economic Geography, 87,* 1–22.

Harvey, D. (2012). *Rebel cities: From the right to the city to the urban revolution.* London and New York: Verso.

Harvey, D. (2014). *Seventeen contradictions and the end of capitalism.* London, UK: Profile Books.

Harvey, D. (2016). Neoliberalism is a political project. *Jacobin.* Available at: https://www.jacobinmag.com/2016/07/david-harvey-neoliberalism-capitalism-labor-crisis-resistance/.

Heathrow Airport Ltd. (2018). *Heathrow 2.0.* Carbon Neutral Growth Roadmap. Available at: https://your.heathrow.com/wp-content/uploads/2018/12/Heathrow-Carbon-neutral-growth-online.pdf.

Heathrow Expansion Consultation. (2018). *Our approach to the natural environment.* Heathrow Airport Limited. Available at: https://www.heathrowconsultation.com/wp-content/uploads/2018/01/6763-Expansion-Natural-Environment-v5-KL-002.pdf.

Herrera, L. M. G., Smith, N., & Vera, M. Á. M. (2007). Gentrification, displacement, and tourism in Santa Cruz de Tenerife. *Urban Geography, 28,* 276–298.

Heynen, N. (2014). Urban political ecology I: The urban century. *Progress in Human Geography, 38,* 598–604.

Heynen, N., Kaika, M., & Swyngedouw, E. (Eds.). (2006). *In the nature of cities: Urban political ecology and the politics of urban metabolism.* London and New York: Routledge.

Heynen, N., McCarthy, J., Prudham, S., & Robbins, R. (Eds.). (2007). *Neoliberal environments: False promises and unnatural consequences.* London and New York: Routledge.

Heynen, N., & Robbins, P. (2005). The neoliberalization of nature: Governance, privatization, enclosure and valuation. *Capitalism Nature Socialism, 16*, 5–8.

HM Government. (2010). *The coalition: Our programme for government*. London: Cabinet Office.

HM Government. (2011). *The natural choice: Securing the value of nature*. Available at: https://assets.publishing.service.gov.uk/government/uploads/system/uploads/attachment_data/file/228842/8082.pdf.

HM Government. (2013). *Biodiversity offsetting* (p. 1126). Defra: Impact Assessment.

HM Government. (2016). *Natural Capital Committee: Terms of reference*. Available at: https://assets.publishing.service.gov.uk/government/uploads/system/uploads/attachment_data/file/517123/ncc-terms-of-reference.pdf.

HM Government. (2017). *Cutting red tape: Review of house building*. Available at: https://assets.publishing.service.gov.uk/government/uploads/system/uploads/attachment_data/file/591039/house-building-cutting-red-tape-review.pdf.

HM Treasury. (2006). *Long-term opportunities and challenges for the UK: Analysis for the 2007 comprehensive spending review*. London, UK.

HM Treasury. (2013). *Infrastructure plan 2013*. London, UK.

Hobbs, R. J., Hallett, L. M., Ehrlich, P. R., & Mooney, H. A. (2011). Intervention ecology: Applying ecological science in the twenty-first century. *Bioscience, 61*, 442–450.

Hodge, I. D., & Adams, W. M. (2012). Neoliberalisation, rural land trusts and institutional blending. *Geoforum, 43*, 472–482.

Holmes, G., & Cavanagh, C. J. (2016). A review of the social impacts of neoliberal conservation: Formations, inequalities, contestations. *Geoforum, 75*, 199–209.

Hough, P., & Robertson, M. (2009). Mitigation under Section 404 of the Clean Water Act: Where it comes from, what it means. *Wetlands Ecology and Management, 17*, 15–33.

HS2. (2013). *London-West Midlands environmental statement* (Vol. 5). Technical Appendices. Scope and Methodology Report Addendum (Ct-001-000/2). Available at: https://www.gov.uk/government/uploads/system/uploads/attachment_data/file/260153/Vol5_Scope_and_methodology_report_addendum_CT-001-000.2.pdf.

Huber, M. (2009). Energizing historical materialism: Fossil fuels, space and the capitalist mode of production. *Geoforum, 40*, 105–115.

Huber, M. (2017). Value, nature and labor: A defense of Marx. *Capitalism Nature Socialism, 28,* 39–52.

Hutton, J., Adams, W. M., & Murombedzi, J. C. (2005). Back to the barriers? Changing narratives in biodiversity conservation. *Forum for development studies, 32,* 341–370.

Ibañez, D., & Katsikis, N. (2014). Editorial. In D. Ibañez & N. Katsikis (Eds.), *New geographies 06: Grounding metabolism* (pp. 2–9). Cambridge: Harvard University Press.

ICMM (International Council on Mining and Metals). (2005). *Biodiversity offsets—A briefing paper for the mining industry.* London, UK. Available at: https://www.ecosystemmarketplace.com/wp-content/uploads/archive/documents/Doc_417.pdf.

ICMM (International Council on Mining and Metals) and IUCN (International Union for Conservation of Nature). (2012). *Independent report on biodiversity offsets.* Prepared by the Biodiversity Consultancy. Available at: www.icmm.com/biodiversity-offsets.

IEEP. (2014). *Study on specific design elements of biodiversity offsets: Biodiversity metrics and mechanisms for securing long term conservation benefits.* DG ENVIRONMENT: ENV.B.2/ETU/2013/0060r. Final Report. A Report Submitted by ICF Consulting Services in association with IEEP and Associated Experts. London, UK.

IEEP. (2016). *Study on specific design elements of biodiversity offsets: Biodiversity metrics and mechanisms for securing long term conservation benefits.* DG ENVIRONMENT: ENV.B.2/ETU/2013/0060r.

Igoe, J., & Brockington, D. (2007). Neoliberal conservation: A brief introduction. *Conservation and society, 5,* 432–449.

Igoe, J., Neves, K., & Brockington, D. (2010). A spectacular eco-tour around the historic bloc: Theorising the convergence of biodiversity conservation and capitalist expansion. *Antipode, 42,* 486–512.

International Monetary Fund (IMF). (2011). *Regional economic outlook, Western Hemisphere*. Washington, DC: IMF.

IUCN, The Biodiversity Consultancy, & IGF. (2018). *Understanding government biodiversity offset policies in the mining sector*. Available at: https://www.iucn.org/sites/dev/files/content/documents/understanding_government_biodiversity_offset_policies_in_the_mining_sector_november_2017.pdf.

Ives, C. D., & Bekessy, S. A. (2015). The ethics of offsetting nature. *Frontiers in Ecology and the Environment, 13,* 568–573.

Jacob, C., Vaissiere, A. C., Bas, A., et al. (2016). Investigating the inclusion of ecosystem services in biodiversity offsetting. *Ecosystem Services, 21,* 92–102.

Jähnig, S. C., Lorenz, A. W., Hering, D., Antonis, C., Sundermann, A., Jedicke, E., et al. (2011). River restoration success: A question of perception. *Ecological Applications, 21,* 2007–2015.

Jameson, F. (1991). *Postmodernism, or, the cultural logic of late capitalism.* Durham, NC: Duke University Press.

Jasanoff, S. (2011). Constitutional moments in governing science and technology. *Science and engineering ethics, 17,* 621–638.

Jonas, A. E. G., Gibbs, D., & While, A. (2011). The new urban politics as a politics of carbon control. *Urban Studies, 48,* 2537–2554.

Jones, P. (2009). Saving the planet or selling off the atmosphere? Emissions trading, capital accumulation and the carbon rent. *Marxist Interventions, 1,* 9–22.

Kaika, M. (2005). *City of flows.* New York and London: Routledge.

Kaika, M., & Swyngedouw, E. (2014, May 14). Radical urban political-ecological imaginaries: Planetary urbanization and politicizing nature. *Derive Journal.*

Kallis, G., & Swyngedouw, E. (2018). Do bees produce value? A conversation between an ecological economist and a Marxist geographer. *Capitalism, Nature, Socialism, 29,* 36–50.

Katz, C. (1998). Whose nature, whose culture?: Private productions of space and the 'preservation' of nature. In B. Braun & N. Castree (Eds.), *Remaking reality: Nature at the millennium* (pp. 45–62). London and New York: Routledge.

Katz, C. (2001). On the grounds of globalization: A topography for feminist political engagement. *Signs: Journal of Women in Culture and Society, 2,* 1213–1234.

Katz, C. (2004). *Growing up global: Economic restructuring and children's everyday lives.* Minneapolis: University of Minnesota Press.

Katz, C. (2011). Accumulation, excess, childhood: Toward a countertopography of risk and waste. *Documents d'anàlisi geogràfica, 57,* 47–60.

Katz, C., & Kirby, A. (1991). In the nature of things: The environment and everyday life. *Transactions of the Institute of British Geographers, 16,* 259–271.

Keil, R. (2003). Progress report: Urban political ecology. *Urban Geography, 24,* 723–738.

Keller, R., & Turek, M. (1998). *American Indians and National Parks.* Tucson: University of Arizona Press.

Kelly, A. B. (2011). Conservation practice as primitive accumulation. *Journal of Peasant Studies, 38,* 683–701.

Kemf, E. (1993). *Indigenous peoples and protected areas: The law of mother Earth*. London: Earthscan.

Klein, N. (2014). *This changes everything: Capitalism vs. the climate*. New York: Simon and Schuster.

Knox-Hayes, J. (2010). Constructing carbon market spacetime: Climate change and the onset of neo-modernity. *Annals of the Association of American Geographers, 100*, 953–962.

Kruczynski, W. L. (1990). Mitigation and the sector 404 program: A perspective. In J. A. Kusler & M. E. Kentula (Eds.), *Wetland creation and restoration: The status of the science* (pp. 549–554). Washington, DC: Island Press.

Laitila, J., Moilanen, A., & Pouzols, F. M. (2014). A method for calculating minimum biodiversity offset multipliers accounting for time discounting, additionality and permanence. *MEE, 5*, 1247–1254.

Lansing, D. M. (2012). Performing carbon's materiality: The production of carbon offsets and the framing of exchange. *Environment and Planning A, 44*, 204–220.

Lapavitsas, C., Kaltenbrunner, A., Labrinidis, G., Lindo, D., Meadway, J., Michell, J., et al. (2010, September). *The Eurozone between austerity and default*. Research on Money and Finance (RMF) Occasional Report. Available at: http://www.researchonmoneyandfinance.org/.

Lapavitsas, C., Kaltenbrunner, A., Lindo, D., Meadway, J., Michell, J., Painceira, J. P., et al. (2011, November). *Breaking up? A route out of the Eurozone crisis*. Research on Money and Finance (RMF) Occasional Report 3. Available at: http://www.researchonmoneyandfinance.org/.

Lapeyre, R., Froger, G., & Hrabanski, M. (2015). Biodiversity offsets as market-based instruments for ecosystem services? From discourses to practices. *Ecosystem services, 15*, 125–133.

Latour, B. (1987). *Science in action: How to follow scientists and engineers through society*. Cambridge: Harvard University Press.

Latour, B. (2005). *Reassembling the social: An introduction to actor-network theory*. Oxford: Oxford University Press.

Lawton, J. H., Brotherton, P. N. M., Brown, V. K., Elphick, C., Fitter, A. H., Forshaw, J., et al. (2010). *Making space for nature: A review of England's wildlife sites and ecological network*. Report to Defra.

Layzer, J. A. (2012). *Open for business: Conservatives' opposition to environmental regulation*. Cambridge: MIT Press.

Lees, L. (2012). The geography of gentrification: Thinking through comparative urbanism. *Progress in Human Geography, 36*, 155–171.

Lefebvre, H. (1968). *Le Droit à la Ville*. Paris: Anthropos.

Lefebvre, H. (1970). *La révolution urbaine* (Vol. 216). Paris: Gallimard.
Lefebvre, H. (1976). Reflections on the politics of space. *Antipode, 8,* 30–37 (M. J. Enders, Trans.).
Lefebvre, H. (1991). *The production of space*. Oxford: Blackwell.
Lefebvre, H. (1996). *Writings on cities* (Vol. 63[2]). Oxford: Blackwell.
Lefebvre, H. (2003 [1990]). From the social pact to the contract of citizenship. In S. Elden, E. Lebas, & E. Kofman (Eds.), *Henri Lefebvre: Key writings* (pp. 238–254). New York: Continuum.
Leitner, H., Peck, J., & Sheppard, E. S. (2007). *Contesting neoliberalism: Urban frontiers*. New York and London: Guilford Press.
Levins, R., & Lewontin, R. (1985). *The dialectical biologist*. Cambridge, MA: Harvard University Press.
Lockhart, A. (2015). Developing an offsetting programme: Tensions, dilemmas and difficulties in biodiversity market-making in England. *Environmental Conservation, 42,* 335–344.
Loftus, A. (2012). *Everyday environmentalism: Creating an urban political ecology*. Minneapolis: University of Minnesota Press.
Loftus, A. (2014). Against a speculative leftism. In J. Wilson & E. Swyngedouw (Eds.), *The post-political and its discontents: Spaces of depoliticization, spectres of radical politics* (pp. 229–243). Edinburgh, UK: Edinburgh University Press.
Lohmann, L. (2005). Marketing and making carbon dumps: Commodification, calculation and counterfactuals in climate change mitigation. *Science as Culture, 14,* 203–235.
Lohmann, L. (2010). Neoliberalism and the calculable world: The rise of carbon trading. In K. Birch & V. Mykhnenko (Eds.), *The rise and fall of neoliberalism* (pp. 77–93). Zed Books: London and New York.
Lohmann, L. (2012). Financialization, commodification and carbon: The contradictions of neoliberal climate policy. *Socialist Register, 48,* 85–107.
Lohmann, L., Hällström, N., Österbergh, R., & Nordberg, O. (2006). *Carbon trading: A critical conversation on climate change, privatisation and power* (p. 362). Uppsala: Dag Hammarskjöld Foundation.
Lynch-Stewart, P., Niece, P., Rubec, C., & Kessel-Taylor, I. (1996). *The federal policy on wetland conservation: Implementation guide for federal land managers*. Ottawa: Environment Canada.
MacDonald, C. (2008). *Green Inc.: An environmental insider reveals how a good cause has gone bad*. New York: Lyons Press.

MacDonald, K. I., & Corson, C. (2012). 'TEEB begins now': A virtual moment in the production of natural capital. *Development and Change, 43,* 159–184.

MacKenzie, D. (2009). Making things the same: Gases, emission rights and the politics of carbon markets. *Accounting, Organizations and Society, 34,* 440–455.

Maclennan, D., & O'Sullivan, A. (2013). Localism, devolution and housing policies. *Housing Studies, 28,* 599–615.

Madanipour, A. (2009). Newcastle upon Tyne. In J. Punter (Ed.), *Urban design and the british urban renaissance* (pp. 132–147). Abingdon: Routledge.

Madanipour, A. (Ed.). (2010). *Whose public space international case studies in urban design and development*. London, UK: Routledge.

Madsen, B., Carroll, N., Kandy, D., & Bennett, G. (2011). *State of biodiversity markets report: Offset and compensation programs worldwide*. Washington, DC: Forest Trends.

Madsen, B., Carroll, N., & Moore Brands, K. (2010). *State of biodiversity markets report: Offset and compensation programs worldwide*. Washington, DC: Forest Trends.

Maestre-Andrés, S., Calvet-Mir, L., & Apostolopoulou, E. (2018). Unravelling stakeholder participation under conditions of neoliberal biodiversity governance in Catalonia, Spain. *Environment and Planning C: Politics and Space*. https://doi.org/10.1177/2399654417753624.

Magdoff, F. (2002). Capitalism's twin crises: Economic and environmental. *Monthly Review*, 54. Available at: https://www.monthlyreview.org/0902magdoff.htm.

Malm, A. (2016). *Fossil capital: The rise of steam power and the roots of global warming*. Brooklyn: Verso Books.

Malm, A., & Hornborg, A. (2014). The geology of mankind? A critique of the Anthropocene narrative. *The Anthropocene Review, 1*(1), 62–69.

Mandle, L., Tallis, H., Sotomayor, L., et al. (2015). Who loses? Tracking ecosystem service redistribution from road development and mitigation in the Peruvian Amazon. *Frontiers in Ecology and the Environment, 13,* 309–315.

Maron, M., Hobb, R. J., Moilanen, A., Matthews, J. W., Christie, K., Gardner, T. A., et al. (2012). Faustian bargains? Restoration realities in the context of biodiversity offset policies. *Biological Conservation, 155,* 141–148.

Maron, M., Ives, C. D., Kujala, H., Bull, J. W., et al. (2016). Taming a wicked problem: Resolving controversies in biodiversity offsetting. *BioScience, 66,* 489–498.

Marsden, T., & Sonnino, R. (2008). Rural development and the regional state: Denying multifunctional agriculture in the UK. *Journal of Rural Studies, 24,* 422–431.

Martinez-Alier, J. (1995). Political ecology, distributional conflicts and economic incommensurability. *New Left Review, 211,* 70–88.

Marx, K. (1859). *A contribution to the critique of political economy*. Moscow: Progress Publishers.

Marx, K. (1887). *Capital: A critique of political economy* (Vol. I). Moscow: Progress Publishers. Available at: https://www.marxists.org/archive/marx/works/1867-c1/.

Marx, K. (1894). *Capital: A critique of political economy* (Vol. III). New York: International Publishers. Available at: https://www.marxists.org/archive/marx/works/1894-c3/.

Marx, K. (1899). *Value, price and profit*. London: Swan, Sonnenschein & Co.

Marx, K. (1910). *Capital: A critique of political economy* (Vol. II). Chicago: Charles H. Kerr & Co. and London: Swan Sonnenschein & Co.

Marx, K. (1964). *The economic and philosophic manuscripts of 1844*. New York: International Publishers.

Marx, K. (1973 [1857]). *Grundrisse*. London: Penguin.

Marx, K. (1990). *Capital: A critique of political economy* (Vol. I, Ben Fowkes, Trans.). London: Penguin Classics.

Marx, K., & Engels, F. (1970). *The German ideology* (Vol. 1). New York: International Publishers.

Marx, K., & Engels, F. (2002). *The communist manifesto*. London: Penguin.

Massey, D. (1984). *Spatial divisions of labour: Social structures and the geographies of production*. Basingstoke: Macmillan.

Massey, D. (2005). *For space*. London: Sage.

Mathiesen, K. (2015). *How and where did UK lose city-sized area of green space in just six years?* Available at: https://www.theguardian.com/environment/2015/jul/02/how-where-did-uk-lose-green-space-bigger-than-a-city-six-years.

May, J., Hobbs, R. J., & Valentine, L. E. (2017). Are offsets effective? An evaluation of recent environmental offsets in Western Australia. *Biological Conservation, 206,* 249–257.

McAfee, K. (1999). Selling nature to save it? Biodiversity and green developmentalism. *Environment and Planning D: Society and Space, 17,* 133–154.

McAfee, K. (2012). The contradictory logic of global ecosystem services markets. *Development and Change, 43,* 105–131.

McAfee, K. (2016). Green economy and carbon markets for conservation and development: A critical view. *International Environmental Agreements: Politics, Law and Economics, 16,* 333–353.

McBrien, J. (2016). Accumulating extinction. In J. W. Moore (Ed.), *Anthropocene or capitalocene?* (pp. 116–137). Oakland: PM Press.

McCarthy, J. (2006). Neoliberalism and the politics of alternatives: Community forestry in British Columbia and the United States. *Annals of the Association of American Geographers, 96,* 84–104.

McCarthy, J. (2012). The financial crisis and environmental governance "after" neoliberalism. *Tijdschrift voor economische en sociale geografie, 103,* 180–195.

McCarthy, J., & Prudham, S. (2004). Neoliberal nature and the nature of neoliberalism. *Geoforum, 35,* 275–283.

McCarthy, M. A., Parris, K. M., van der Ree, R., et al. (2004). The habitat hectares approach to vegetation assessment: An evaluation and suggestions for improvement. *Ecological Management & Restoration, 5,* 24–27.

McDonough, E. (2015). *Global flows, local conflicts and the challenge of urban governance: Managing the city-airport interface in London, UK*. BIVEC/GIBET Transport Research Days 2015: Current Issues in Transportation Research.

Mead, D. L. (2008). History and theory: The origin and evolution of conservation banking. In R. Bayon, J. Fox, & N. Carroll (Eds.), *Conservation and biodiversity banking: A guide to setting up and running biodiversity credit trading systems* (pp. 9–31). London, UK and Sterling, VA: Earthscan.

Mercer and Inter-American Development Bank (IDB). (2017). *Crossing the bridge to sustainable infrastructure investing*. Inter-American Development Bank. Available at: https://publications.iadb.org/en/publication/12717/crossing-bridge-sustainable-infrastructure-investing-exploring-ways-make-it.

Merrifield, A. (2013). The urban question under planetary urbanization. *International Journal of Urban and Regional Research, 37,* 909–922.

Mitchell, D., Attoh, K., & Staeheli, L. (2015). Whose city? What politics? Contentious and non-contentious spaces on Colorado's Front Range. *Urban Studies, 52,* 2633–2648.

Moilanen, A., & Kotiaho, J. S. (2018). *Planning biodiversity offsets.* Nordic Council of Ministers.

Moore, J. W. (2017). The Capitalocene, Part I: On the nature and origins of our ecological crisis. *The Journal of Peasant Studies, 44,* 594–630.

Moreno-Mateos, D., Maris, V., Béchet, A., et al. (2015). The true loss caused by biodiversity offsets. *Biological Conservation, 192,* 552–559.

Moreno-Mateos, D., Power, M. E., Comín, F. A., & Yockteng, R. (2012). Structural and functional loss in restored wetland ecosystems. *PLoS Biology, 10*(1), e1001247.

Muradian, R., Arsel, M., Pellegrini, L., et al. (2013). Payments for ecosystem services and the fatal attraction of win-win solutions. *Conservation letters, 6,* 274–279.

Muradian, R., & Rival, L. (2012). Between markets and hierarchies: The challenge of governing ecosystem services. *Ecosystem Services, 1,* 93–100.

Muûls, M., et al. (2016). *Evaluating the EU emissions trading system: Take it or leave it? An assessment of the data after ten years* (Grantham Institute Briefing Paper 21). Imperial College London.

Nash, R. (1989). *The rights of nature: A history of environmental ethics.* Wisconsin: University of Wisconsin Press.

Nayak, A. (2003). Last of the 'real geordies'? White masculinities and the subcultural response to deindustrialisation. *Environment and Planning D: Society and Space, 21,* 7–25.

Neves, K., & Igoe, J. (2012). Uneven development and accumulation by dispossession in nature conservation: Comparing recent trends in the Azores and Tanzania. *Tijdschrift voor Economische en Sociale Geografie, 103,* 164–179.

Newell, R., Pizer, W., & Raimi, D. (2013). Carbon markets 15 years after Kyoto: Lessons learned, new challenges. *The Journal of Economic Perspectives, 27,* 123–146.

NPPF. (2012). *National planning policy framework.* London: Department for Communities and Local Government. Available at: https://www.gov.uk/government/uploads/system/uploads/attachment_data/file/6077/2116950.pdf.

NTC (North Tyneside Council). (2015a). *Indices of multiple deprivation.* Available at: https://my.northtyneside.gov.uk/sites/default/files/web-page-related-files/IMD%202015%20Headline%20Analysis%20and%20Mapping%20-%20Data%20Files.pdf.

NTC (North Tyneside Council). (2015b). *Green space strategy.* Available at: http://my.northtyneside.gov.uk/sites/default/files/web-page-related-files/Green%20Space%20Strategy%202015%20%28Sub%20Header%29.pdf.

NTC (North Tyneside Council). (2017). *North Tyneside local plan.* Available at: https://my.northtyneside.gov.uk/sites/default/files/web-page-related-files/North%20Tyneside%20Local%20Plan%202017-2032.pdf. Accessed 7 November 2019.

O'Connor, J. (1998). *Natural causes.* New York: Guilford.

Ojeda, D. (2012). Green pretexts: Ecotourism, neoliberal conservation and land grabbing in Tayrona National Natural Park, Colombia. *Journal of Peasant Studies, 39,* 357–375.

Ordner, J. (2018). Petro-politics and local natural resource protection: Grassroots opposition to the keystone XL pipeline in Nebraska. In E. Apostolopoulou & J. Cortes-Vazquez (Eds.), *The right to nature: Environmental justice, social movements and neoliberal natures* (pp. 29–42). Abingdon: Routledge-Earthscan.

Organisation for Economic Co-operation and Development (OECD). (2018). *Cost-benefit analysis and the environment: Further developments and policy use.* Paris: OECD Publishing. Available at: https://doi.org/10.1787/9789264085169-en.

Oxford Economics. (2013). *The economic impact of the trans-Adriatic pipeline on Albania: A report for TAP AG.*

Padoch, C., Brondizio, E., Costa, S., et al. (2008). Urban forest and rural cities: Multi-sited households, consumption patterns, and forest resources in Amazonia. *Ecology and Society, 13*(2), 2.

Palmer, M. A., & Filoso, S. (2009). Restoration of ecosystem services for environmental markets. *Science, 325*(5940), 575.

Paloniemi, R., Apostolopoulou, E., Cent, J., Bormpoudakis, D., Scott, A., Grodzińska-Jurczak, M., et al. (2015). Public participation and environmental justice in biodiversity governance in Finland, Greece, Poland and the UK. *Environmental Policy and Governance, 25,* 330–342.

Paloniemi, R., Apostolopoulou, E., Primmer, E., Grodzinska-Jurczak, M., Henle, K., Ring, I., et al. (2012). Biodiversity conservation across scales: Lessons from a science–policy dialogue. *Nature Conservation, 2,* 7–19.

Parkers, D., Newell, G., & Cheal, D. (2003). Assessing the quality of native vegetation: The 'habitat hectares' approach. *Ecological Management & Restoration, 4,* 29–38.

Pawliczek, J., & Sullivan, S. (2011). Conservation and concealment in SpeciesBanking.com, US: An analysis of neoliberal performance in the species offsetting industry. *Environmental Conservation, 38,* 435–444.

Peck, J. (2012). Austerity urbanism: American cities under extreme economy. *City, 16,* 626–655.

Peck, J., Theodore, N., & Brenner, N. (2012). Neoliberalism resurgent? Market rule after the great recession. *The South Atlantic Quarterly, 111,* 2.

Peck, J., & Tickell, A. (2007). Conceptualizing neoliberalism, thinking Thatcherism. *Contesting Neoliberalism: Urban Frontiers, 26,* 50.

Peet, R. (2013). Comparative policy analysis: Introduction. *Human Geography, 6,* 1–10.

Peet, R., & Watts, M. (2000). *Liberation ecologies: Environment, development, social movements*. London: Routledge.

Perreault, T. (2013). Dispossession by accumulation? Mining, water and the nature of enclosure on the Bolivian Altiplano. *Antipode, 45,* 1050–1069.

Pierce, J., Martin, D., & Murphy, J. (2011). Relational place making: The networked politics of place. *Transactions of the Institute of British Geographers, 36,* 54–70.

Pilgrim, J. D., Brownlie, S., Ekstrom, J. M. M., Gardner, T. A., von Hase, A., ten Kate, K., et al. (2013). A process for assessing offsetability of biodiversity impacts. *Conservation Letters, 6,* 376–384.

Poulton, D. W. (2014). *Biodiversity offsets, a primer for Canada*. Sustainable Prosperity and Institute of the Environment, Ottawa.

Price, N. (2012). *Proof of evidence of nick price on behalf of West Moor Residents' association Rule 6 party* (Planning Inspectorate Reference: APP/W4515/A/12/2175554).

Prudham, S. (2009). Pimping climate change: Richard Branson, global warming, and the performance of green capitalism. *Environment and planning A, 41,* 1594–1613.

Purcell, M. (2014). Possible worlds: Henri Lefebvre and the right to the city. *Journal of Urban Affairs, 36,* 141–154.

Purcell, M., & Tyman, S. K. (2015). Cultivating food as a right to the city. *Local Environment, 20,* 1132–1147.

Purcell, T. F., Loftus, A., & March, H. (2019). Value–rent–finance. *Progress in Human Geography*. https://doi.org/10.1177/0309132519838064.

QMM (QIT Madagascar Minerals). (2009, March). *A mine at the rescue of the unique biodiversity of the littoral zone of Fort-Dauphin*. QIT Madagascar Minerals SA Press kit. Available at: http://www.riotintomadagascar.com/english/media.asp.

Quétier, F., & Lavorel, S. (2011). Assessing ecological equivalence in biodiversity offset schemes: Key issues and solutions. *Biological Conservation, 144,* 2991–2999.

Quétier, F., Regnery, B., & Levrel, H. (2014). No net loss of biodiversity or paper offsets?—A critical review of the French no net loss policy. *Environmental Science & Policy, 38,* 120–131.

Re:Common. (2019). *Turning forests into hotels: The true cost of biodiversity offsetting in Uganda*. Available at: https://www.recommon.org/eng/turning-forests-into-hotels-the-true-cost-of-biodiversity-offsetting-in-uganda/.

Robbins, P. (2012). *Lawn people: How grasses, weeds, and chemicals make us who we are*. Philadelphia: Temple University Press.

Robertson, M. (2000). No net loss: Wetland restoration and the incomplete capitalization of nature. *Antipode, 32,* 463–493.

Robertson, M. (2004). The neoliberalization of ecosystem services: Wetland mitigation banking and problems in environmental governance. *Geoforum, 35,* 361–373.

Robertson, M. (2006). The nature that capital can see: Science, state, and market in the commodification of ecosystem services. *Environment and Planning D: Society and Space, 24,* 367–387.

Robertson, M. (2012). Measurement and alienation: Making a world of ecosystem services. *Transactions of the Institute of British Geographers, 37,* 386–401.

Rotherham, I. D. (2015). *The rise and fall of countryside management: A historical account*. Earthscan: Routledge.

Rubec, C. D., & Hanson, A. R. (2009). Wetland mitigation and compensation: Canadian experience. *Wetlands Ecology and Management, 17,* 3–14.

Ruddick, S., Peake, L., Tanyildiz, G. S., et al. (2017). Planetary urbanization: An urban theory for our time? *Environment and Planning D: Society and Space, 36,* 387–404.

Ruhl, J. B., & Salzman, J. E. (2006). *The effects of wetland mitigation banking on people* (FSU College of Law, Public Law Research Paper No. 179). Available at SSRN: https://ssrn.com/abstract=878331.

Sayer, D. (1987). *The violence of abstraction: The analytic foundations of historical materialism.* Basil: Blackwell.

Schmid, C., Karaman, O., Hanakata, N. C., et al. (2018). Towards a new vocabulary of urbanisation processes: A comparative approach. *Urban Studies, 55,* 19–52.

Schoukens, H., & Cliquet, A. (2016). Biodiversity offsetting and restoration under the European Union Habitats Directive: Balancing between no net loss and deathbed conservation? *Ecology and Society, 21*(4), 10.

Seagle, C. (2012). Inverting the impacts: Mining, conservation and sustainability claims near the Rio Tinto/QMM ilmenite mine in Southeast Madagascar. *Journal of Peasant Studies, 39,* 447–477.

Sears, P. (1964). Ecology—A subversive subject. *Bioscience, 1,* 11–13.

Seto, K. C., Fragkias, M., Güneralp, B., et al. (2011). A meta-analysis of global urban land expansion. *PLoS One, 6*(8), e23777.

Seto, K. C., Güneralp, B., & Hutyra, L. R. (2012). Global forecasts of urban expansion to 2030 and direct impacts on biodiversity and carbon pools. *Proceedings of the National Academy of Sciences, 109,* 16083–16088.

Sharma, S. (2009). Baring life and lifestyle in the non-place. *Cultural Studies, 23,* 129–148.

Smith, N. (1979). Toward a theory of gentrification a back to the city movement by capital, not people. *Journal of the American Planning Association, 45,* 538–548.

Smith, N. (1998). Nature at the millenium: Production and re-enchantment. In B. Braun & N. Castree (Eds.), *Remaking reality: Nature at the millennium* (pp. 269–282). London and New York: Routledge.

Smith, N. (2006a). Foreword. In N. Heynen, M. Kaika, & E. Swyngedouw (Eds.), *In the nature of cities: Urban political ecology and the politics of urban metabolism.* London and New York: Routledge.

Smith, N. (2006b). Nature as accumulation strategy. In L. Panitch & C. Leys (Eds.), *Socialist register 2007: Coming to terms with nature* (pp. 16–36). London: Merlin.

Smith, N. (2008). Comment: Neo-liberalism: Dominant but dead. *Focaal, 2008,* 155–157.

Smith, N. (2010). *Uneven development* (3rd ed.). New York: Verso.

Smith, N. (2005). Foreword. In N. Heynen, M. Kaika, & E. Swyngedouw (Eds.), *In the nature of cities: Urban political ecology and the politics of urban metabolism* (pp. xi–xv). Abingdon: Routledge.

Smith, N., & Katz, C. (1993). Grounding metaphor: Towards a spatialized politics. In M. Keith & S. Pile (Eds.), *Place and the politics of identity* (pp. 66–81). London: Routledge.

Soja, E. W. (2010). *Seeking spatial justice.* Minneapolis: University of Minnesota Press.

Soja, E. W. (1980). The socio-spatial dialectic. *Annals of the Association of American Geographers, 70,* 207–225.

Soja, E. W., & Kanai, J. (2006). The urbanization of the world. In N. Brenner (Ed.), *Implosions/explosions: Towards a study of planetary urbanization* (pp. 142–159). Berlin: Jovis.

Solly, R. (2016). *A struggle between a Tethered Donkey and a Tiger*. London Mining Network. Available at: http://londonminingnetwork.org/2016/08/cerrejon-coal-a-struggle-between-a-tethered-donkey-and-a-tiger/.

Sparke, M. (2000). "Chunnel visions": Unpacking the anticipatory geographies of an Anglo-European Borderland. *Journal of Borderlands Studies, 15,* 187–219.

Spash, C. L. (2010). The brave new world of carbon trading. *New Political Economy, 15,* 169–195.

Spash, C. L. (2015). Bulldozing biodiversity: The economics of offsets and trading-in nature. *Biological Conservation, 192,* 541–551.

Stockdale, A. (2010). The diverse geographies of rural gentrification in Scotland. *Journal of Rural Studies, 26,* 31–40.

Stokstad, E. (2008). New rules on saving wetlands push the limits of the science. *Science, 320*(5873), 162–163.

Stuart, D., Gunderson, R., & Petersen, B. (2019). Climate change and the Polanyian counter-movement: Carbon markets or degrowth? *New political economy, 24,* 89–102.

Sullivan, S. (2006). Elephant in the room? Problematising 'new'(neoliberal) biodiversity conservation. *Forum for Development Studies, 33,* 105–135.

Sullivan, S. (2013). After the green rush? Biodiversity offsets, uranium power, and the 'calculus of casualties' in greening growth. *Human Geography, 6,* 80–101.

Sullivan, S., & Hannis, M. (2015). Nets and frames, losses and gains: Value struggles in engagements with biodiversity offsetting policy in England. *Ecosystem Services, 15,* 162–173.

Sweezy, P. M. (2004). Capitalism and the environment. *Monthly Review, 56,* 86–93.

Swyngedouw, E. (1996). The city as hybrid: On nature, society, and cyborg urbanization. *Capitalism Nature Socialism, 7,* 65–80.

Swyngedouw, E. (2000). Authoritarian governance, power, and the politics of rescaling. *Environment and Planning D: Society and Space, 18,* 63–76.

Swyngedouw, E. (2004). Globalisation or 'Glocalisation'? Networks, territories and rescaling. *Cambridge Review of International Affairs, 17,* 25–48.

Swyngedouw, E. (2005). Governance innovation and the citizen: The Janus face of governance-beyond-the-state. *Urban Studies, 42,* 1991–2006.

Swyngedouw, E. (2010). Impossible sustainability and the post-political condition. In M. Cerreta, M. Concillio, & G. Monno (Eds.), *Making strategies in spatial planning* (pp. 185–205). The Netherlands: Springer.

Swyngedouw, E. (2012). Rent and landed property. In B. Fine & A. Saad-Filho (Eds.), *The Elgar companion to Marxist economics* (pp. 310–315). Northhampton: Edward Elgar.

Swyngedouw, E. (2015). Politicizing urban political ecologies. In T. Perreault, G. Bridge, & J. McCarthy (Eds.), *The Routledge handbook of political ecology* (p. 609). London and New York: Routledge.

Swyngedouw, E., & Ernstson, H. (2018). Interrupting the anthropo-obscene: Immuno-biopolitics and depoliticizing ontologies in the anthropocene. *Theory, Culture & Society, 35,* 3–30.

Swyngedouw, E., & Heynen, N. C. (2003). Urban political ecology, justice and the politics of scale. *Antipode, 35,* 898–918.

Takacs, D. (1996). *The Idea of biodiversity: Philosophies of paradise*. Baltimore: Johns Hopkins University Press.

TAP (Trans Adriatic Pipeline). (2013a). *ESIA Albania*. Available at: https://www.tap-ag.com/resource-library/reference-documents/esia-documents/esia-albania-in-english.

TAP (Trans Adriatic Pipeline). (2013b). *ESIA Greece*. Available at: https://www.tap-ag.com/resource-library/reference-documents/esia-documents/esia-greece-in-english.

TAP (Trans Adriatic Pipeline). (2013c). *ESIA Italy*. Available at: https://www.tap-ag.com/resource-library/reference-documents/esia-documents/esia-italy-in-english.

TAP (Trans Adriatic Pipeline). (2015). *Environmental, socioeconomic, and cultural heritage overview*. Available at: https://www.tap-ag.com/resource-library/reference-documents/project-finance-disclosure/environmental-social-and-cultural-heritage-overview.

TAP (Trans Adriatic Pipeline). (2016). *Biodiversity offset strategy*. CAL00-C5577-640-Y-TRY-0001. RSK.

TEEB. (2010). *The economics of ecosystems and biodiversity: Mainstreaming the economics of nature: A synthesis of the approach, conclusions and recommendations of TEEB*. Geneva: TEEB.

Teherán, S. (2014). Impactos socioterritoriales de la explotación minera en los departamentos del Cesar y La Guajira. In G. Rudas & D. Hawkins (Eds.), *La minería de carbón a gran escala en Colombia: Impactos económicos, sociales, laborales, ambientales y territoriales* (pp. 44–67). Friedrich-Ebert-Stiftung: Bogotá.

Temper, L., Demaria, F., Scheidel, A., et al. (2018). The Global Environmental Justice Atlas (EJAtlas): Ecological distribution conflicts as forces for sustainability. *Sustainability Science, 13,* 573–584.

ten Kate, K., Bishop, J., & Bayon, R. (2004). *Biodiversity offsets: Views, experience, and the business case*. Gland, Switzerland and London, UK: IUCN and Insight Investment.

The Biodiversity Consultancy. (2013). *Government policies on biodiversity offsets*. Cambridge: The Biodiversity Consultancy Ltd.

Thoft-Christensen, P. (2012). Infrastructures and life-cycle cost-benefit analysis. *Structure and Infrastructure Engineering, 8,* 507–516.

Thomas, P. (2006). Modernity as 'passive revolution': Gramsci and the fundamental concepts of historical materialism. *Journal of the Canadian Historical Association, 17,* 61–78.

Thomas, P. (2009). *The Gramscian moment: Philosophy, hegemony and Marxism.* Leiden: Brill.

TransCanada Pipelines Limited. (2018). *Applications for the white spruce pipeline project Fort McKay area.* Alberta Energy Regulator. Available at: https://www.aer.ca/documents/decisions/2018/2018ABAER001.pdf.

Treweek, J. with contributions from: Ten Kate, K., Butcher, B., Venn, O., Garland, L., Wells, M., Moran, D., Thompson, S. (2009). *Scoping study for the design and use of biodiversity offsets in an English Context*. Final Report to Defra (Contract NE 0801). Available at: https://www.cbd.int/financial/offsets/unitedkingdom-scoping.pdf.

Trimedia. (2009). *Lodge Hill stage two public consultation—Workshops: Feedback analysis.* Available at: http://www.medway.gov.uk/pdf/Statement_of_Community_Involvement_Final_Evidence_.pdf.

Tucker, G. M., Allen, B., Conway, M., Dickie, I., Hart, K., Rayment, M., et al. (2014). *Policy options for an EU no net loss initiative.* Report to the European Commission (with Annexes), Institute for European Environmental Policy (IEEP), London.

UK NEA. (2011). *The UK national ecosystem assessment: Technical report*. Cambridge: UNEP-WCMC.

UNEP/CBD/COP. (2012). *Report of the eleventh meeting of the conference of the parties to the Convention on Biological Diversity*. Available at: http://www.cbd.int/doc/?meeting=cop-11.

UNFCCC. (2015). *Adoption of the Paris agreement*. Geneva: United Nations Office.

United Nations (UN). (2018). *World urbanization prospects* (The 2018 Revision). New York: UN.

Vergara, K., Figallo, M., & Glave, M. (2014). Infraestructura en la Amazonía peruana: Una propuesta para proyectar cambios en la cobertura boscosa en la carretera Pucallpa-Cruzeiro do Sul. In R. Barrantes & M. Glave (Eds.), *Amazonía peruana y desarrollo económico* (pp. 161–208). Lima: IEP and GRADE.

Vicente, A., Martin, N., Slee, D., et al. (2011). Carbón para el mundo, retroceso para La Guajira. In *PBI Colombia, Boletín informativo no 18* (pp. 32–36). Colombia: PBI.

Villarroya, A., Barros, A. C., & Kiesecker, J. (2014). Policy development for environmental licensing and biodiversity offsets in Latin America. *PLoS One, 9,* e107144.

Virah-Sawmy, M., Ebeling, J., & Taplin, R. (2014). Mining and biodiversity offsets: A transparent and science-based approach to measure "no-net-loss". *Journal of Environmental Management, 143,* 61–70.

Virilio, P. (1997). *Open sky* (J. Rose, Trans.). New York: Verso.

Vlachou, A. (2005a). Debating sustainable development. *Rethinking Marxism, 17,* 627.

Vlachou, A. (2005b). Environmental regulation: A value-theoretic and class-based analysis. *Cambridge Journal of Economics, 29,* 577–599.

Wachsmuth, D. (2014). City as ideology: Reconciling the explosion of the city form with the tenacity of the city concept. *Environment and Planning D: Society and Space, 32,* 75–90.

Walker, S., Brower, A. L., Stephens, R. T. T., & Lee, W. G. (2009). Why bartering biodiversity fails. *Conservation Letters, 2,* 149–157.

Wanner, T. (2015). The new 'passive revolution' of the green economy and growth discourse: Maintaining the 'sustainable development' of neoliberal capitalism. *New Political Economy, 20*(1), 21–41.

Weinhold, B. (2011). Alberta's oil sands: Hard evidence, missing data, new promises. *Environmental Health Perspectives, 119,* A126–A131.

Wheeler, D. P., & Strock, J. M. (1995). *Official policy on conservation banks.* Issued Jointly by the Resources Agency and California Environmental Protection Agency.

White, B., Borras, S. M., Jr., Hall, R., Scoones, I., & Wolford, W. (2013). The new enclosures: Critical perspectives on corporate land deals. In *The new enclosures: Critical perspectives on corporate land deals* (pp. 13–42). London and New York: Routledge.

Wilkinson, J. (2008). In-lieu fee mitigation: Coming into compliance with the new Compensatory Mitigation Rule. *Wetlands Ecology and Management, 17,* 53–70.

Wilson, E. O., & Peter, F. M. (1988). *Biodiversity.* Washington, DC: National Academy Press.

Wilson, J., & Swyngedouw, E. (Eds.). (2014). *The post-political and its discontents: Spaces of depoliticization, spectres of radical politics.* Edinburgh, UK: Edinburgh University Press.

Winter, A. (2018). Illegal camping in "stolen native land". In E. Apostolopoulou & J. Cortes-Vazquez (Eds.), *The right to nature: Environmental justice, social movements and neoliberal natures* (pp. 263–276). Abingdon: Routledge-Earthscan.

Womble, P., & Doyle, M. (2012). The geography of trading ecosystem services: A case study of wetland and stream compensatory mitigation markets. *Harvard Environmental Law Review, 36*(1), 229–296.

World Bank. (2009). *Global Economic Prospects*. Washington, DC: World Bank.

WRM & Re:Common. (2016). *Rio Tinto in Madagascar: A mining destroying the unique biodiversity of the littoral zone of Fort Dauphin*. WRM and Re:Common.

Index

A

abstract labor 116, 128
abstract nature 325
abstract value 325
access to green space 98, 207
accountability 56, 190, 268, 270, 298
accounting 10, 11, 16, 35, 39, 84, 255, 268, 269, 275, 322, 329
accumulation by dispossession 38
activist 8, 20, 45, 225, 229, 231, 237, 247, 248, 340
actor-network theory 128
additionality 16, 81, 82, 89, 126
Africa 60, 80, 83, 141, 143
ahistorical 94, 326
air pollution marketable permit systems 47
Akyem Mining Project 143
Albania 146
Alberta 145, 146
alienation 39, 341
Amazon 143, 144, 319, 324
amenity 78, 203, 207, 242, 322
ancient woodland 81, 185, 249, 251, 275, 278, 308
annihilation of space with time 244
Anthropocene 8, 333, 339, 341
Anthropo-obScene 333
anti-capitalistic 339
apocalyptic 7, 121, 206
appropriation 40, 114, 116, 117, 119, 120, 129, 130, 323
arable 235, 237, 254, 292
area ratios 79, 83
artifactual 93
Asia 60
asocial 11, 96, 100

E. Apostolopoulou, *Nature Swapped and Nature Lost*,
https://doi.org/10.1007/978-3-030-46788-3

assets 5, 6, 10, 17, 30, 49, 58, 131, 132, 167, 170, 178, 197, 330, 336
asteroid mining 330
austerity 14, 35, 38, 195, 218, 233, 234, 256, 331, 335, 338
austerity localism 229, 231
austerity politics 18, 172, 230, 240
Australia 50, 61, 77, 80, 81, 83, 126, 319, 320
authoritarian/authoritarianism 44, 45, 335
authoritative managerial class 34
authority 39, 43, 47, 56, 227, 270, 280–282
autocratic 44

B

bailout packages 37
banking industry 44
Belgium 61
Belt and Road Initiative (BRI) 320
benchmark 79, 84
best practice 51, 56, 187
Big Society 171–173
BINGOs 8, 15, 330, 331
BioBanking 46, 50, 58, 77, 126, 130, 197
biodiversity 7–9, 11, 12, 14–16, 39, 40, 46, 52
biodiversity accounting 86, 87
The Biodiversity Consultancy (TBC) 53, 62, 64
biodiversity credits 77, 174
biodiversity governance 33–35, 56, 171, 176, 253
biodiversity loss 15, 17, 19, 34, 58, 78, 84, 87–90, 96, 97, 124, 136, 167, 186, 202, 203, 206, 249, 255, 258, 302, 329, 336, 337
biodiversity offsets 9, 46, 49–51, 57, 58, 60–62, 77, 129, 133, 136, 140–142, 174, 176, 178, 180, 223, 320
biodiversity units 77, 80, 91, 94, 128, 179, 180, 186, 207, 223, 239, 251, 252, 274, 325, 338
birds 238, 242, 243, 292, 294, 303, 308, 310
Birmingham airport 306
bourgeoisie 120, 321, 327, 340
bourgeois political 120, 332, 340
brownfield 181, 236, 276, 300, 307
built environment 135
bureaucratic 189, 223
business 3, 11–13, 32, 42, 48, 86, 121, 132, 165, 168–171, 176, 178, 201, 218–220, 223, 224, 229, 232, 236, 247, 252, 268, 282, 313, 319, 334
Business and Biodiversity Offsets Program (BBOP) 46, 51–57, 77–81, 83–86, 89–92, 96, 126, 143, 169, 177, 187, 254

C

calculation 16, 59, 77, 80, 81, 84, 85, 87, 90, 92, 93, 96, 98–100, 128, 184–187, 191, 200, 202, 235, 238, 239, 250, 251, 254, 278, 296, 303, 314, 332
calculative devices 128
Canada 50, 55, 58, 61, 145, 146, 319

capital accumulation 11, 17, 29–31, 38, 40, 43, 94, 101, 113, 114, 123, 129–131, 133, 189, 190, 199, 234, 245, 322, 324, 328, 330, 340
capital investment 31, 217
capitalism 4, 6, 8, 11, 12, 15, 20, 30, 31, 33, 37, 38, 40–43, 45, 90, 93, 94, 97, 98, 100, 101, 111–114, 116, 117, 119–125, 128, 129, 135, 231, 244, 256, 301, 325–330, 332–335, 338, 339, 341
capitalism's environmental contradictions 39, 94
capitalist class 29, 30, 34, 124, 329
capitalist commodity production 43, 94, 329
capitalist natures 31
capitalists 6, 8, 114, 120, 122, 123, 128, 131, 255, 329, 338
Capitalocene 7
capital valorization 131
carbon 4, 7, 9, 10, 12, 13, 16, 40, 82, 89, 92, 124, 126–128, 131, 132, 147, 168, 325, 330, 333, 338
carbon offsets/carbon offsetting 13, 14, 16, 35, 46, 51, 89, 100, 117, 124, 133, 201, 283, 320
Cerrejón mine 144
China 319, 320
Chthulucene 7
city borders 319
civil society 32, 34, 43, 44, 51, 146, 171, 172
class analysis 205, 327
class contradictions 30
class politics 332
class power 30, 38
class struggle 40, 122, 234, 340
clean water 337
climate change 8, 12, 34, 90, 124, 139, 240, 329, 330, 336
climate emergency 336
climate strike 45, 329, 336
CO_2 336
coal export 236
Coalition Government 168, 172, 174, 176, 218, 230
coercion 41, 43, 44, 240
Colombia 144
colonialism 327
commodification 33, 64, 324
commodities 17, 94, 113–115, 117, 118, 120, 124–126, 338
commodity forms 117, 125, 338
commodity production 31, 114, 126, 127
common sense 43, 284, 313, 322, 329
communities 4, 82, 98, 140, 141, 144, 146, 147, 170, 204, 230, 231, 233, 322, 324, 329, 332
comparability 98
compensation 48, 51, 57, 58, 60–64, 87–89, 99, 100, 126, 132, 146, 174, 179, 184, 186, 187, 189, 193, 195, 196, 201, 203, 205, 221–224, 226–228, 233, 237–239, 242, 243, 249, 251, 253–255, 268, 271, 272, 275, 276, 279, 284–287, 289, 292–294, 297, 303, 304, 308, 309, 320, 321, 323, 324, 327, 328, 332

compensation mechanisms 46, 60, 61, 64, 87, 136, 320, 321, 323, 324, 331
compensatory mitigation 46, 48, 49, 60, 61, 64
competition 4, 31, 40, 101, 122, 171, 199, 340
competitive 31, 32, 170, 171
compliance offsets 63
concrete abstraction 43, 329
condition of production 118, 130
condition scoring method 180
connectivity 77, 93, 175, 250
consensus 12, 16, 44, 91, 171, 240, 331, 336
consent 41, 43, 44, 240, 335, 337
conservation areas 95, 123
conservation austerity 187
conservation banking 6, 46, 49, 51, 58–60, 94, 96, 132, 193–195, 198, 203, 279, 282, 300, 330, 333, 334
conservation brokers 44, 56, 165, 166, 187, 188, 190, 191, 193, 238, 255, 267
conservation–development conflicts 91, 189
conservation NGOs 34, 139, 141, 145, 165, 178, 181, 189, 194, 277, 301, 331
conservation organizations 34, 170, 194, 242, 299
conservation partnerships 33, 171
Conservative 133, 172, 174, 220, 225, 230, 276, 304, 327
constructionism 93
the construction of equivalence 16, 77, 86
construction rush 319
consultants 5, 35, 56, 60, 92, 169, 184, 187, 189, 191, 192, 194, 195, 204, 226, 227, 233, 238, 250, 251, 253, 284, 296, 303, 304
consumption 113, 135, 146, 236, 337, 338
Convention on Biological Diversity 18, 39
corporate interests 34, 328
corporate social responsibility 139, 329
cost–benefit analysis 200, 201
counterfactual scenario 81, 82, 89, 126
counter-revolutionary project 30
countertopography 324
countryside 5, 135, 136, 233, 236, 240, 242, 310, 311, 315, 324, 339
Coventry 176, 229, 302, 306, 307, 309, 313, 314
Coventry Warwickshire Gateway 165, 166
credits 6, 7, 11–14, 48–50, 58, 59, 85, 87, 92, 98–100, 124–126, 128, 130–132, 174, 175, 177, 180, 184, 185, 188, 191, 193, 195–200, 207, 210, 235, 237, 239, 269, 280–283, 301, 308, 320, 322, 325, 328, 330, 333
credit trading 46
cultural importance of place 98
cultural values 57, 77, 78, 84, 96, 270, 271, 294, 295
culture 94, 111, 147, 200, 202, 203
currencies 77, 84, 86
cuts 35, 172, 229, 247

D

de-alienate 342
debt economy 30
debt trap 38
decentralization 230
decision-making 46, 56, 87, 92, 169, 190, 192, 218, 230, 240, 274, 329
deforestation 82, 143, 144, 337
Defra 77, 80, 81, 126, 165, 167, 168, 170–180, 184, 185, 193, 198, 199, 205, 206, 210, 218–220, 224, 249, 250, 284, 288, 302, 304
Defra Green Paper 177, 179, 193, 224, 304, 331
Defra metric 81, 176, 179, 180, 183, 185, 186, 191, 202, 203, 205–207, 224, 227, 237, 238, 243, 250, 253, 269, 284, 286, 291
deindustrialization 236, 241
depoliticization 91, 189, 235, 240
deregulation 30–32, 38, 44, 47, 220, 222, 230, 236, 247
development site 17, 58, 64, 77, 79, 97, 130, 179, 185, 186, 194, 204, 224, 226, 236, 242, 243, 288, 290–294, 322, 326
devolution 35, 230
dialectical 42, 43, 93, 111, 112, 121, 332, 338, 341
dialectics 40, 41, 44, 112, 240, 322
dialectics of space 137
differentiation 95, 100, 120
disempower 30, 45, 231, 240
displacement 112, 123, 137, 140, 145, 236, 324
dispossession 137, 145, 324
distinctiveness score 80, 250
domination 41, 43–45, 338
donor agencies 34
double grabbing 64, 141, 322
dualism 15, 100
dystopian 7, 11, 20, 92, 255, 318, 335, 336, 338, 339, 341

E

Earth Summit 10, 39
eco-friendly 40, 139
ecological compensation 59, 322
ecological equivalence 14, 17, 57, 59, 77, 78, 84, 86, 87, 94, 207
ecological frontier 330
ecological gains 14, 77, 126
ecological losses 7, 14, 19, 46, 77, 80, 84, 86, 97, 99, 138, 185, 199, 202, 204, 225, 235, 237, 240, 250, 320, 322, 325, 328, 331–333
ecological modernization 122
ecological network 170, 175, 178, 250
ecologies of fear 338
eco-modernist 339
econometric 329
economic coercion 240, 338
economic crisis 18, 37, 38, 40, 41, 64, 88, 122, 217, 228, 234, 269, 339
economic development 9, 14, 18, 36, 40, 46–48, 64, 82, 87, 88, 91, 96–98, 134, 146, 165, 186, 218, 222–225, 228, 244, 254, 256, 305, 320, 322, 326
economic geographies 244, 322

economic growth 9, 14, 42, 98, 168, 172, 190, 204, 233, 253–256, 276, 309, 321, 322, 324, 328, 331, 335, 339, 340
economic recession 101, 218, 227, 239, 283
The Economics of Ecosystems and Biodiversity (TEEB) 10, 11, 46
economic valuation 11, 35, 39, 42, 43, 46, 115, 117, 125, 130, 169, 197, 320, 325, 329
ecosocialists 111, 117, 340
eco-spatial 328
ecosystem 7, 18, 39, 40, 51, 57, 77–80, 82, 84, 86, 89, 92, 93, 95, 98, 99, 122–124, 136, 144, 146, 147, 167, 199, 200, 207, 209, 310, 320, 324, 328, 334, 335, 337
ecosystem degradation 39, 42, 90, 96
Ecosystem Marketplace 51
Ecosystem Markets Task Force 169
ecosystem services 9–12, 16, 18, 35, 39, 46, 58, 77, 92, 131, 132, 141, 144, 167–169, 176–178, 205–210, 268, 288, 300, 301, 328
ecotourism 33, 39, 46
eco-utopia 5
efficiency 189, 193, 201, 221, 329
elites 30, 44
emancipation 100
emissions 12, 13, 147, 325, 337
emissions trading 100, 127, 240
emplacement 137
enclosure 33, 39, 135
endangered species 5, 6, 322
energy 5, 6, 12, 13, 45, 62, 117, 123, 135, 136, 139, 178, 223, 337
energy sector 60, 62, 219, 321
England 15, 19, 59, 80, 81, 83, 85, 88, 92, 126, 130, 131, 165–168, 170, 171, 175–178, 186, 187, 192, 194, 199, 201, 202, 204–206, 217, 218, 233–236, 240–242, 244–246, 254–257, 267, 278, 292, 321, 323, 331–334
entrepreneurial 29, 34, 171, 244
Environment Agency 210, 220
environmental backlash 64
environmental catastrophe 5, 121
environmental crisis 339
Environmental Impact Assessment (EIA) 61, 62, 140, 190, 223, 249
environmentalism 15, 46, 87, 166, 168, 192, 254, 327–329, 331
environmentalists 8, 18, 40, 41, 48, 165, 255, 268, 326
environmental justice 14, 138, 141, 189
Environmental Justice Atlas 45, 340
environmental legislation 40, 124, 190, 201, 218, 256
environmental markets 10, 49, 62
Environmental Markets Exchange (EME) 130, 176
environmental NGOs 14, 34, 165, 268, 270, 271, 277, 321
environmental politics 31, 43, 48, 240, 332, 338, 340
environmental protection 9–11, 14, 33, 40, 46, 48, 223, 249, 254, 320, 326, 328

Environmental Protection Agency (EPA) 49, 132, 201
environmental regulation 40, 47–49, 122, 224
Environment Bank 52, 165, 166, 169, 174, 176, 177, 187, 190, 193, 225, 227, 237, 238, 242, 255, 279, 281, 300
epistemology 189
equivalence 14, 16, 77, 84, 85, 94, 99, 125, 128, 186, 199, 202, 288, 325, 332
equivalent natures 92, 99, 199, 334
Essex pilot 166, 227
ethical turn 91
EU Biodiversity Strategy 11, 58
EU Emissions Trading Scheme 13
Europe 13, 18, 29, 59–61, 146, 197, 234, 270, 287, 289, 319
European Union (EU) 13, 34, 37–39, 57, 58, 61, 83, 223, 234, 270, 290, 319
exchange 7, 14, 19, 31, 32, 84, 92, 93, 95, 98, 113, 115, 120, 123–125, 135, 136, 174, 193, 196–200, 239, 322, 330, 338
exchangeability 89, 96, 182
exchange rules 16, 91
exchange value 43, 94, 115–117, 120, 125, 126, 128, 131, 325, 329, 338
exclusion 35, 44, 56, 130, 231, 240, 247
executive housing 235, 236
expertise 92, 220, 231, 239
expert knowledge 44, 205, 328
experts 44, 91, 92, 169, 174, 187, 189, 190, 220, 233
exploitation 10, 38, 40, 42, 97, 114, 120, 123, 127, 128, 187, 229, 324, 335, 340
externalities 46, 87, 192, 329
external nature 89, 94, 121, 326
Extinction Rebellion 41
extractive industry 223
extractivism 138

F

farmland 136, 143, 210, 233, 235, 237, 273, 281, 290, 292, 307, 311
fast-policy regimes 240
fetishized 222, 239
fetishized nature 132, 334
fictitious capital 131, 132
finance 13, 37, 51, 62, 219
2008 financial crash 9, 14, 18, 35, 37, 39, 44, 64, 100, 166, 217, 240, 256, 319, 321, 328, 335
financial crisis 37
financialization 325
fiscal austerity 171
Fish and Wildlife Service 47, 48, 54, 126
flanking mechanisms 32
flat world 200
flood 142, 168, 177, 178, 210, 278
food 12, 139, 144, 310, 335
food security 136, 140, 337
foreclose 189, 233, 255
forest 51, 55, 85, 89, 91, 140, 141, 143, 145, 233, 322
Forest Trends 51, 55, 62, 92
fossil fuel 13, 41
France 61, 62, 78, 79, 200

free appropriation of nature 116, 117
free gifts of nature 117
funding 14, 35, 145, 167, 170, 171, 175, 177, 183, 187, 194, 226, 331
fungibility 89
futurism 333, 340

G

gateway 229, 302, 305, 308, 311–313
generalized urbanization 134
gentrification 324
geoengineering 116
geographical differentiation 100
geographical space 95
Germany 50, 61, 62, 78, 79, 83, 319
Ghana 54, 55, 143
Global Inventory of Biodiversity Offset Policies (GIBOP) 62
Global South 18, 90, 127, 138, 145, 209, 325, 341
governance 12, 16, 30, 31, 33–35, 38, 44, 56, 170, 190, 326
grassland 194, 254, 258, 273, 275, 280, 288, 291, 292, 296
grassroots 248, 251
Greece 9, 40, 44, 146, 147, 313
green belt 233, 240, 302
green capitalism 9, 40
green economy 9, 10, 38, 168, 170
greenfield 233, 236, 240, 300
green grabbing 39, 40, 97, 139
green growth 9, 170
green spaces 11, 15, 92, 94, 97, 203, 226, 233, 236, 322, 324, 330, 335, 339
greenwash 41, 138, 145, 329

H

habitat area 79
habitat banking 58, 92, 194, 195, 280, 281, 300
habitat condition score 80
habitat descriptors 78
habitat extent 80
habitat hectare 79, 80, 83, 85, 177, 185
habitat loss 82, 83, 284, 288, 337
habitat restoration 127, 145, 175
Habitats and Birds Directives 61, 147, 218, 224
habitat structure 57, 78
habitat variables 79
Handmaid's Tale 6
Hands off Thaxted 165
Heathrow airport 147
hedgerow 237, 238, 310, 313
hegemony 9, 14, 15, 30, 41, 43–45, 88, 240, 332
heritage 203
heterogeneity 325
hierarchy/hierarchies 43, 59, 85, 172, 230, 269
hinterland 135, 319
historical-geographical 16, 18, 101
historical-materialist 111
historic bloc 42, 43
homogeneity 98, 325
housebuilding 217
housing crisis 217, 335
housing industry 44, 232, 233, 284

housing market 217
HS2 high-speed rail link 81, 165, 166, 217, 234, 246–249, 251–253, 275, 276
Hybrid Bill 246–248, 252

I

idealism 93
ideology 8, 34, 42, 43, 93, 94, 117, 172, 327, 329
Impact Mitigation Regulation (IMR) 50
imperialism 4, 327
income distribution 30
indigenous 127, 144, 146, 319, 324, 327
individuation 93
industrialized 12, 307
industrial park 307
industries 8, 15, 31, 33, 34, 44, 56, 59, 60, 64, 89, 139, 142, 165, 199, 233, 254, 321, 326, 327, 330, 331
infrastructure 6, 8, 44, 57, 61, 123, 134, 135, 138, 143–145, 147, 166, 170, 183, 191, 200, 203, 205, 207, 217, 219, 223, 244, 246, 253, 276, 319, 320, 322, 324, 331, 337, 339
injustice 17, 45, 100, 138, 204, 325, 332, 336
in kind 80
innovative 7, 19, 58, 168, 175, 223, 254, 326
in perpetuity 256–258, 272
integral state 43
intellectual terrorism 189
interchangeability 98
interest-bearing capital 132
Intergovernmental Science-Policy Platform on Biodiversity and Ecosystem Services (IPBES) 34, 337
internalization 87, 329
International Monetary Fund (IMF) 10, 38, 44, 134
International Union for Conservation of Nature (IUCN) 62, 82, 141, 321
interviewees 165, 168, 172, 175, 180–184, 186–188, 190–192, 198, 202, 206–210, 225, 226, 233, 243, 248, 251, 252, 256, 257, 268–279, 281, 282, 284–288, 290–292, 294–296, 298–302, 304, 306–314
intra-capital competition 122, 245
intrinsic 180, 207, 250, 341
investment 12, 33, 38, 51, 134, 135, 139, 144, 169, 170, 174, 175, 183, 201, 219, 232, 239, 319, 320, 327, 334, 336
irreplaceability 96
Italy 61, 146, 147

K

Kalagala biodiversity offset 141, 142
Kent 241, 242, 272
Kyoto Protocol 10, 12, 13

L

labor 30, 31, 111–120, 125–130, 132, 141, 144, 145, 334, 338, 341, 342
Labor government 167, 172, 173

labor movement 31
labor power 11, 29, 112, 113, 115, 117, 122, 129
labour theory of value 17, 117, 125, 128, 323
Lambeth 165
land 5, 6, 17, 33, 35, 38–40, 45, 49, 60–62, 77, 92, 99, 101, 112, 116, 123, 126, 130–133, 135, 136, 139–144, 146, 169, 188, 193–195, 197–199, 203, 205, 210, 217, 221, 223, 224, 226, 228, 232, 233, 236–238, 243, 245, 247, 250, 257, 272, 273, 275, 281, 283–285, 288, 290–294, 296, 300, 304, 309, 311, 313, 319, 321–323, 325, 327, 330, 331, 333, 334, 337, 340
land acquisition 330
land extraction 324
land grabbing/land grab 39, 40, 44, 64, 135, 138–140, 248, 322
landowners 5, 6, 49, 87, 130, 131, 146, 170, 174, 178, 187, 188, 190, 193, 195–197, 199, 226, 227, 257, 281, 330
landownership 131, 198
land price 192, 199, 323
landscape 5, 36, 51, 77, 78, 83, 95, 97, 99, 119, 146, 170, 175, 177, 178, 181, 194, 202, 203, 209, 218, 233, 250, 300, 309, 312, 314, 315, 323
Land Securities 242–244
land speculation 131, 198, 279
land swaps 79
land use 64, 98, 100, 130–132, 135, 143, 146, 196, 198, 199, 205, 219, 225, 247, 279, 294, 322, 323, 332, 337
land use change 64, 98, 100, 134, 143, 196, 205, 225, 322, 323, 332, 337
LaTerre Bank 48
Lawton Report 168
leadership 43
legitimacy 37, 190, 335
liberalization 38
libertarian 172, 201
license to trash 139, 222, 228, 268, 270, 329
like-for-like 77, 80, 81
limits to growth 329
livelihoods 17, 44, 45, 78, 98, 137, 140, 142, 143, 147, 148, 205, 208, 209, 319, 320, 322, 323, 337, 339
local authorities 165, 170, 174, 178, 188, 226, 232, 247, 252, 295, 331
local communities 56, 90–92, 140, 146, 147, 165, 166, 189, 195, 204, 226, 229, 232, 238, 239, 248, 270, 294, 295, 305, 309, 314, 332, 333, 339
localist 171, 231
locality 146, 171, 271
Local Nature Partnerships (LNPs) 170, 171, 177, 178
Local Planning Authority (LPAs) 176, 220, 232
local plans 218, 229, 230, 232, 233, 239, 244, 257, 295, 304, 305, 311
local traditions 98
Lodge Hill 165, 166, 199, 241–245, 255, 272, 296–299

London 7, 31, 81, 187, 198, 241, 245–247, 249, 277, 324
loss of rights 35, 40, 204

M

51m 165, 247, 252
Madagascar 83, 85, 89, 90, 127, 139, 140
Mad Max 6
managerialism 328
marginalize 240, 332
market 9, 10, 13, 14, 18, 29–34, 39, 40, 42, 43, 46, 48–50, 55, 58–61, 64, 87, 88, 113, 115, 123, 125, 127, 128, 130–132, 166, 169, 173, 175, 176, 182, 187, 192–194, 196–199, 221, 236, 239, 283, 319, 329, 341
market-based 7, 9, 10, 39, 47, 49, 51, 62, 92, 124, 136, 172, 174, 175, 221, 323, 326
market-based environmentalism 41–44, 328, 332
market-based instruments 9, 34, 35, 39, 46, 117, 133, 225, 255, 330
market failure 192
market-friendly reregulation 30, 32, 47, 221
marketization 32, 35, 230
market logic 12, 171, 197, 240, 331
market proxies 32, 34
Marxist 18, 99, 111, 120, 125, 340
Marxist political economy 125
materialist theory of ideology 332
materiality 99
means of production 112–116, 118, 119, 122, 123
measurability 329
Medway 241–245, 298
megaprojects 217, 253
metabolism 18, 111, 112, 137, 148
metrics 16, 59, 78–80, 83, 84, 89, 91, 92, 174, 181, 182, 205, 225, 237, 243, 258, 274, 275, 286, 296, 300, 308, 314, 328
Millennium Ecosystem Assessment (MEA) 167
minerals 60, 62, 169, 178, 321, 337
mining 8, 45, 57, 60, 62, 89, 139–141, 143, 145, 236, 301, 321
mitigation 13, 46–50, 58–61, 63, 79, 83, 87, 126, 139, 174, 186, 197, 221, 224, 251, 269, 270, 286, 287, 289–291, 309, 323
mitigation hierarchy 50, 59, 224, 227, 243, 269, 286, 289, 290, 303, 321
mode of production 42, 43, 112, 113, 119–122
monetary 47, 98, 116, 167, 196, 206, 209, 301, 337, 338
multipliers 16, 79–81, 185, 206, 238, 250, 275, 283–285

N

national parks 95, 112, 326, 327
National Planning Policy Framework (NPPF) 218, 219, 227, 230, 239, 256, 286, 289
National Planning Practice Guidance (NPPG) 232
natural capital 10–12, 16, 18, 19, 35, 39, 46, 167–170, 186,

295, 296, 301, 320, 322, 326, 329, 330, 334, 336
Natural Capital Coalition 11
Natural Capital Committee 169
Natural England 165, 167, 170, 171, 179, 180, 196, 220, 222, 243
Natural Environment White Paper 168
natural resource management 34, 39
natural resources 6, 38, 40, 43, 90, 119, 122, 130, 138, 144, 146, 168, 209, 330
nature conservation 5, 8, 9, 11, 15, 18–20, 32, 33, 42, 94, 122, 123, 223, 294, 326, 341
Nature Improvement Areas (NIAs) 170, 171, 177, 178, 223
nature–society relationships 10, 11, 20, 31, 33, 39, 41, 111, 114, 136, 326, 328, 335, 336, 339–341
neoclassical 12, 125, 197
neoclassical economics 12, 125, 197
neo-fascism 335
neoliberal biodiversity governance 33, 171, 176
neoliberal conservation 8, 16, 18, 20, 31–36, 46, 47, 101, 165, 172, 187, 239, 324
neoliberal hegemony 45, 240, 337
neoliberal ideology 29, 38
neoliberalism 8, 10, 29–32, 35, 37, 38, 40, 43–45, 137, 245, 329, 335, 337
neoliberalization 30, 31, 33, 35, 36, 38
neoliberalization of nature 20, 31, 32, 35, 38, 41, 64, 245, 320, 332
neoliberal nature 36, 244, 245, 337
neoliberal reform 30, 31
neoliberal restructuring 168, 173, 217
neoliberal rhetoric 47, 336
neoliberal spatial governance 230
neo-Malthusian 4, 338
net gain 7, 14, 56, 57, 61, 64, 88, 166, 189, 223, 238, 243, 249, 288
Newcastle 231, 235, 237, 240
New South Wales Biodiversity Banking 50
nightingale 199, 241–243, 272
No Net Loss (NNL) 7, 14, 16, 19, 48, 50, 57, 61, 77, 83, 84, 94, 96, 145, 180, 189, 208, 239, 243, 249, 251, 253, 268–270, 288, 291, 297, 298, 322, 330
non-metaphysical 332
non-place 95, 96, 333, 335
non-state actors 34, 123
North America 58, 60, 146
North Tyneside 166, 199, 235, 236, 239, 241
numerical scores 58, 59, 77, 84, 87, 89, 97

O

offsetability 16, 86
offset metrics 16, 77–81, 83–86, 88, 90, 93, 94, 128, 179, 183, 185, 208, 224, 238, 330
offset site 58–60, 64, 77, 79, 80, 93, 94, 96–98, 125, 140, 145,

175, 182, 185, 187, 193, 195, 198, 199, 203, 204, 224, 226, 237–239, 243, 257, 291–294, 322, 323, 326, 334, 338
off-site mitigation 237, 290, 291
off-the-shelf 7, 320
oil and gas 8, 41, 57, 145, 329
Olympic Games 9, 241, 313
on-site 59, 290, 291, 302
ontological 93, 97
open space 203, 204, 235, 247, 331
operational landscapes 136
opportunistic 19, 36, 251
oppositional politics 245, 331, 332
out of kind 80

P

Paris Climate Agreement 13
performative economics 128
philanthropy 172, 173
philosophy of praxis 340
pilots 83, 131, 165–167, 172, 175, 176, 178, 181, 182, 193, 195, 199, 222, 227, 228, 237, 250, 254, 257, 279, 299
pipelines 45, 145, 146, 319
place 6, 14, 17, 19, 35–37, 40, 45, 47, 48, 85, 92, 95–101, 112, 113, 115, 119–121, 131, 133, 136, 137, 140, 142, 148, 195, 199, 200, 202–204, 207, 209, 210, 223, 226, 229, 230, 235, 239, 240, 244, 245, 252, 255, 256, 272, 289, 295, 299, 301, 303, 305, 308, 311–313, 319–324, 328, 332, 334, 335, 341
placeless 100
place-making 137
planetary urbanization 135, 136, 138, 330, 339
planning 4, 38, 44, 47, 59, 87, 88, 166, 169, 176, 178, 181–184, 188, 190, 192, 194, 195, 202–205, 217, 218, 220–228, 230, 231, 238, 242, 243, 245, 253, 255–257, 268–270, 273, 274, 277–279, 281, 282, 287, 290, 291, 298–300, 302–304, 306, 312–314, 321, 322, 332, 334, 335
planning conditions 220
planning permission 64, 130, 219, 225–227, 229, 232, 234, 235, 237, 239, 247, 253, 271, 280, 297, 312, 331
policy expertise 187
the political 8, 12, 14, 17, 19, 20, 29, 30, 33–36, 41, 42, 47, 85, 87–90, 92, 120, 123, 127, 128, 172, 190, 201, 204, 217, 225, 240, 243, 267, 268, 276, 288, 289, 324, 331, 332, 335, 338–341
political ecology 112, 136, 138, 324
political power 45
politics 12, 14, 16, 18, 42, 89, 90, 182, 189, 221, 240, 247, 332, 340, 341
politics of measurement 16
the polluter pays 201, 320
polluters 15, 321, 331
population density 323
populist 205
positionality 267
positivism 99
post-apocalyptic 339

postcolonial 324
post-crisis 38, 234, 246, 321
post-political 171, 339
post-productive countryside 236
power 34, 35, 37, 44, 56, 85, 112, 137, 138, 172, 205, 209, 228, 230, 233, 246, 247, 252, 319, 328
power relationships 35
pragmatism 4, 8, 56, 257, 338
precautionary principle 288
price 10, 13, 29, 32, 45, 59, 94, 120, 123–126, 128–130, 132, 133, 195–199, 201, 210, 236, 238, 242, 280–283, 308, 320, 322, 323, 325, 326, 334, 338
private land 204, 247, 272, 280, 294, 334
private sector 11–14, 19, 32, 34, 47, 60, 62, 139, 165, 169–171, 187, 190, 219, 283
privatization 11, 30–33, 35, 38, 47, 91, 132, 187, 324, 335
pro-development 175, 222, 229, 255, 321, 329, 331
production 5, 9, 14, 16, 17, 36, 45, 60, 90, 95, 99–101, 113–120, 122, 124, 125, 127, 129–131, 133, 135–137, 197, 199, 205, 236, 245, 256, 267, 281, 323, 325, 328, 329, 332, 338–341
production of space 20, 98, 100, 245, 332
profit 5, 30, 33, 40, 87, 113, 119–121, 124, 130–132, 141, 198, 232, 279, 282, 287, 330
pro-market localism 230
property rights 29, 31, 32, 341
protected areas 11, 33, 34, 50, 123, 139, 141, 144, 320, 322, 323, 327, 330, 341
protectionism 326
protection offsets 81
protest 45, 237, 334
protest movements 335
public access 64, 96, 123, 204, 238, 243, 272, 273, 280, 294, 333, 334
public assets 40
public consultation 92
public debate 92, 186, 189, 231, 233, 238, 239, 243, 332
public inquiry 237, 244, 304, 305, 312
public opposition 244, 247, 249, 255
public–private partnerships 33–35
public property 38
public right 201, 294
public sector 32, 172, 188, 283

Q

quantitative balance 97, 186

R

radical alternatives 245, 341
radical praxis 20, 340
reactionary 19, 20, 245, 326, 327
real and formal subsumption of nature 117, 321
real estate business 35
recreation 145, 146, 185, 203, 207, 226
red tape 217, 218, 220, 222

reductionism 15, 87, 89, 93, 99, 200, 204–206, 328
reformist 245, 329
reforms 30, 122, 176, 230, 232, 329
regeneration 241, 244, 324, 337
regulation 30, 32, 46, 47, 49, 64, 87, 122, 124, 198, 202, 208, 217–221, 224, 229, 233, 255, 334
relations of production 42, 43, 112, 117, 129
relocation 143, 224, 323, 327, 332
remaking 135, 148, 205, 321, 328
rent extraction 64, 324, 325
rentiers 130, 131
rent seeking 330
reordering of the landscape 17, 134, 137, 138, 235, 322
reproducible 98, 200, 340
rescaling 33–35, 38, 171, 176
resistance 16, 18, 45, 166, 201, 248, 327, 340
resource grabbing 40
restoration ecology 82, 84, 322
restoration offsets 82
retrenchment 38
revanchist 335, 336, 341
rhetoric 15, 19, 36, 90, 99, 140, 173, 218, 221, 225, 228, 231, 242, 247, 253–257, 276, 336
rights of way 247, 294
right to nature 137, 234, 339–341
right to the city 234, 339
Rio+20 summit 39
Rio Tinto mine 83, 85, 89
roll back 30–32, 38, 40, 47
roll-out 30
rural 20, 35, 134–136, 146, 209, 227, 236, 244, 302, 319, 321, 323–325, 332, 334
rural community 203
rural gentrification 236

S

S160 agreements 226
Save Gosforth Wildlife Campaign 165
scale(s) 33–35, 100, 120, 136, 137, 168, 170, 171, 176–180, 250, 280, 285, 290, 300, 321, 328
scarcity 124
sci-fi 4, 6, 340
second nature 120
sections of capital 12, 40, 100, 134, 142, 199, 205, 233, 255, 322, 324, 330, 334
self-regulation 33
service economy 236
simplistic 89, 181, 182, 197
simulation 4
social aspects 205, 209, 210, 238, 240, 295, 327, 332
social control 101, 341
social-ecological configurations 205, 328
social-ecological metabolism 134
social enterprise 172
social-environmental movements 15, 45, 190, 341
social impacts 36, 64, 166, 202, 204, 210, 240, 313
social inequality 30, 138, 235, 325
social license to operate 57, 64
socially necessary labor time 115, 116, 125

social majority 205, 236, 335, 341
social movements 20, 40, 336, 337, 341
social provision 35, 47
social relations 42, 98, 127, 128, 131, 136, 137, 341
social reproduction 47, 322, 335, 339
social space 98
social ties 98
socio-ecological conflicts 131, 240
socionatures 14, 16, 19, 20, 99, 119, 133, 136–138, 143, 200, 205, 321, 324, 325, 328, 339, 341
socio-spatial dialectic 137
solidarity 45, 172, 245, 341
southern Europe 37
space 4, 11, 14, 19, 20, 31, 33, 35, 43, 86, 91, 95, 97–100, 120, 124, 128, 135, 137, 184, 199, 204, 217, 236, 267, 320, 322, 328, 330, 332, 341
SPAs 292
spatial 20, 45, 89, 95, 97, 98, 135–138, 141, 183, 186, 204, 230, 244, 250, 283, 320, 322, 332
spatialized abstractions 325
spatial organization of capitalism 101
spatial redistribution 323
special economic zone (SEZs) 229, 334
special nature zones 335
species 5, 6, 39, 49, 51, 58, 77, 79, 80, 83, 87, 90, 96, 100, 119, 124, 126, 132, 144, 146, 147, 174, 179, 181, 183, 186, 194, 196, 199, 200, 203, 205, 207, 208, 221, 229, 237, 242, 250, 271, 272, 283–286, 288, 290–293, 302, 328, 334, 337
species banking 5, 6, 9, 47, 49
species composition 57, 78
species extinction 337
speculation 13, 101, 131, 132, 199, 280
SSSI 243, 244, 275, 289, 290, 294, 299
state/quasi-state 3, 29–32, 34, 35, 37, 43, 44, 47, 48, 51, 57, 60, 64, 83, 88, 91, 118, 122–124, 133, 171, 173, 174, 177, 187, 189, 190, 196, 198, 204, 221, 230, 231, 249, 255, 290, 299, 305, 306, 335, 337, 341
state-market-civil society relations 31
stock of resources 94, 326
STOP HS2 165, 247, 248, 252
structure 6, 29, 34, 42, 82, 85, 113, 128
subsistence livelihoods 141, 142
superstructures 42
surplus capital 31
surplus-labor 113, 114, 118
surplus value 95, 113–115, 117–120, 124, 127, 129, 130
surrogates 78–80, 86
sustainability 8, 9, 39, 122, 124, 133, 225, 278, 329, 337
sustainable development discourse 9, 42
Switzerland 61, 62, 83
synthetic biology 116
System for Environmental and Economic Accounts (SEEA) 10

T

tar sands 145
taxation 31, 334
technocracy 186
technocrats 15, 19, 44, 94, 96, 99, 189, 190, 205, 225, 226, 330, 333
techno-managerial 240, 338, 339
territorial injustice 230
territorialization 245
Thames Gateway project 241, 244
Thameslink project 166
Thaxted 186, 298
theory of hegemony 332
theory of rent 17, 101, 125, 129, 130, 322, 323
There Is No Alternative (TINA) 33, 88, 190, 256, 336
Town and Country Planning Act 226
tradeable rights 48
trade-offs 207
traffic 147, 203, 306
trainline 165, 234, 246–249, 253
Trans-Adriatic-Pipeline (TAP) 146, 147
translocation 291
transparency 64, 185, 190, 251
transportation sector 60, 62, 321
transport geographies 241, 244, 245
Troika 44
tyranny of pragmatism 254, 338

U

Uganda 54, 141–143, 210
UK 4, 17, 18, 29, 31, 38, 41, 53, 56, 59, 61, 62, 77, 81, 92, 147, 165–172, 174–176, 181, 182, 188, 192, 195, 198, 208, 209, 219, 222, 223, 225, 228, 233, 234, 242, 245, 246, 253, 268, 272, 278, 279, 286, 287, 289, 291, 320
UK Biodiversity action plan 174, 179
UK National Ecosystem Assessment (NEA) 18, 167, 168
unaccountable 44
uncertainty 80, 185, 238, 277, 280
undemocratic 44
unemployment 236, 241
unequal distribution 45, 121
uneven development 38, 45, 99, 100, 138
unevenness 16, 17, 95, 99, 100, 120, 131, 202, 325
un-green grabbing 39–41, 97, 139, 322
uniqueness 98
units of measurement 77, 79
Uplands Ecosystem Service Pilots 167
urban 11, 19, 20, 97, 134–136, 143, 181, 203, 235–237, 241, 245, 248, 321–324, 332, 334, 335, 337
urban development 14, 17, 18, 97, 137, 138, 144, 147, 165, 166, 224, 233, 234, 240, 246, 253, 309, 322, 324, 328, 331, 335, 339, 340
urban expansion 134, 136, 143, 223, 319, 322, 339
urban fabric 143, 319
urbanization 17, 18, 64, 134–138, 143, 145–147, 166, 223, 234,

245, 319, 321, 324, 325, 332, 337, 340
urban land cover 319
urban political ecology 134, 138, 324
urban redevelopment 241, 324
urban-rural 319
urban studies 138, 324
USA 58, 83, 289
US Clean Water Act 47, 49
US Endangered Species Act 49
use value 43, 78, 94, 113–117, 119, 120, 125, 126, 129, 130, 197, 325, 329, 338, 341

V

valuation 11, 16, 47, 83, 116, 128, 196, 206, 208, 209, 255, 337
value 9–11, 16, 39, 43, 56, 61, 79, 81–85, 87, 91, 94, 96, 97, 111, 113, 115–119, 124–133, 137, 168, 169, 176, 179, 180, 193, 197, 198, 200, 201, 203, 205–208, 219, 223, 225, 238, 250, 254, 268, 271, 273, 276, 292, 295, 300, 301, 308, 313, 320, 322, 323, 325, 329, 334, 341
value of labor 129
value of labor power 129
variegation of neoliberalization 35, 36
Victoria Native Vegetation Offsets 50
violence 325
virtual natures 5, 339
voluntary 13, 34, 46, 50, 139, 145, 166, 171, 176, 182, 198, 253
volunteering 172
vulnerability 96

W

Wales 58, 83, 126, 130, 171, 192, 197
Warwickshire, Coventry and Solihull pilot 166
wasteland 6, 20
wealth 11, 30, 115–118, 120, 122, 201, 319
Wealth Accounting and the Valuation of Ecosystem services (WAVES) 10
welfare state 30, 124, 173
West Moor Residents Association 165
wetland 7, 47–50, 58, 79, 82, 197, 233, 323, 333
wetland banking 49, 50, 58, 100
Whitehouse Farm 166, 198, 235–240, 243, 244
wildlife habitat 175
win-win 94, 172, 192, 193, 228, 254–256, 297, 328, 336
working class 45, 335, 340
World Bank 10, 13, 134, 139, 142, 201, 334

Y

Yellowstone 112, 327

The manufacturer's authorised representative in the EU is Springer Nature Customer Service Centre GmbH, Europaplatz 3, 69115 Heidelberg, Germany. If you have any concerns regarding our products, please contact ProductSafety@springernature.com

Printed and bound by CPI Group (UK) Ltd, Croydon, CR0 4YY

05/10/2026

02228125-0014